Collaborating to Meet Language Challenges in Indigenous Mathematics Classrooms

Mathematics Education Library

VOLUME 52

For further volumes:
http://www.springer.com/series/6276

Tamsin Meaney · Tony Trinick ·
Uenuku Fairhall

Collaborating to Meet Language Challenges in Indigenous Mathematics Classrooms

Springer

Dr. Tamsin Meaney
Charles Sturt University
School of Education
Wagga Wagga New South Wales
Australia
tamsin.meaney@mah.se

Tony Trinick
University of Auckland
Faculty of Education
PO Box 92109
1020 Auckland
Fisher Bldg.
New Zealand
t.trinick@auckland.ac.nz

Uenuku Fairhall
Te Kura Kaupapa Māori O Te
Koutu
PO Box 379
Rotorua
New Zealand
uen_fai@koutu.school.nz

ISBN 978-94-007-1993-4 e-ISBN 978-94-007-1994-1
DOI 10.1007/978-94-007-1994-1
Springer Dordrecht Heidelberg London New York

Library of Congress Control Number: 2011935852

Printed on acid-free paper

Springer is part of Springer Science+Business Media (www.springer.com)

He mihi aroha

He whakahokinga mihi te pukapuka nei ki ngā tāngata katoa i puta atu ai. He mihi ki a koutou me tō koutou kaha i ora Māori mai ai ngā mahi tätaitai. Hei aha? Hei whītiki mō ā tätou tamariki, mō ä tātou mokopuna.

Nō reira, tēnā ra koutou o Te Koutu! Kaiako mai, mätua mai, tamariki mai, hoa mai – tenā koutou katoa!

He mihi hoki ki a koe, e Piri, i piri tahi ai matou tokotoru.

Kei wareware hoki ō mātou hoa rangatira. Tēnā koutou i manawanui mai nā!

Tenā hoki koe, e Te Reo! Te taonga e manahau ai te ngäkau. Auē, taukuri ē!

Acknowledgements

This book is dedicated to the teachers, students, and wider community of Te Kura Kaupapa Māori o te Koutu, both past and present. Without their enthusiasm and commitment, there would have been no school, no Māori-medium mathematics register research, and no projects to report upon in this book. We take inspiration from them and celebrate their achievements.

For us, as the authors, our connection with Te Koutu has been valuable not just work-wise but also for the friendships that we have forged over the years. It has involved us in much discussion (heated at times), much laughter, huge amounts of food, and a few bottles of red wine. When we began working together, we did not realise that a decade later we would still be talking and working with the teachers at Te Koutu. The longevity of the work being done in this school makes it a remarkable project to write about but even more fascinating to be part of.

Over the years, our understanding of what was occurring at Te Koutu has become more insightful from the many conversations that we have had with colleagues and students both at Te Koutu and in the teacher education programmes that we have taught. Thank you for your insights.

We also acknowledge the role that Bill Barton, presently professor at the University of Auckland, has played in introducing us to each other in 1997 and the initial academic influence he had on our research, which is reported in this book.

Another substantial influence on our research has been the funding we received from the Ministry of Education's Teaching and Learning Research Initiative for the work that we carried out between 2005 and 2007. This funding was provided specifically for research projects based around issues highlighted by teachers and schools and which could be done in collaboration with university researchers.

There have been many, many people who have contributed to the collection, transcription, and translation of the data. Without you the book would have been the poorer. We know that many of you have also been inspired by the school, and we hope that seeing the data presented in a book makes those routine tasks worthwhile.

Finally to our families, both immediate and extended, we would like to express our thanks. They have put up with our colonisation of each other's free time, and sometimes they have had to endure conversations of which they may have been able to make no sense. We do not expect you to read this book but thank you for your patience whilst it has been created. Without your support, it would not have been started, let alone completed.

"Nāu anō au!"

Contents

Chapter 1
Introduction

Ko tā te rangatira kai he kōrero.
Language is the food of chiefs.

For many students around the world, learning mathematics is not a simple activity. The complexity is increased when the language in which it is being taught is not the students' first language. When the language of instruction is an Indigenous[1] language that has had to develop very quickly a mathematics register with specialised vocabulary and grammatical expressions for discussing Western mathematics, the challenges students face are even greater and can seem almost insurmountable. Yet at Te Kura Kaupapa o Te Koutu (Te Koutu for short), a Māori-immersion school in Aotearoa[2]/New Zealand, students are learning mathematics through *te reo Māori*, the Māori language, and achieving at high levels. Given that Māori students in mainstream education traditionally do not perform as well as their peers (Meaney, McMurchy-Pilkington, & Trinick, 2008), we argue that fluency in *te reo Māori* has enabled the students at Te Koutu to become mathematical chiefs.

This is a book of hope as it shows how collaboration can contribute to overcoming significant challenges. Around the world, school communities struggle with how to ensure that all students achieve their mathematical potential. In many cases, the focus for these schools is not on language issues, but on student achievement, on teacher content knowledge, mathematics curriculum and so on. Irrespective of the issue, there is always a need for many people to work together at different levels of the educational system for important changes to happen. By describing the process of how collaboration, between groups of people at different levels of the education

---

[1] When *Indigenous* refers to a specific group of people or languages, we have used a capital to indicate that they have the same status to other groups of languages such as European. When *indigenous* refers generically to something or someone who originated in a specific area, then it has not been capitalised.

[2] We have used the established tradition in Māori research publications of italicising Māori words which are not proper nouns. Proper nouns are not italicised. On the first occasion, a translation is provided. A glossary of Māori terms as well as acronyms is provided.

T. Meaney et al., *Collaborating to Meet Language Challenges in Indigenous Mathematics Classrooms*, Mathematics Education Library 52,
DOI 10.1007/978-94-007-1994-1_1, © Springer Science+Business Media B.V. 2012

system, who all helped fuel shifts in teaching practice, we show how educational achievement can be enhanced.

Wero and the Story of Māui

Experiencing and meeting challenges has resonance with aspects of Māori culture as it does with all cultures. Here, we describe two connections that we used in conceptualising the purposes of this book. These are the *wero* and the story of the great hero Māui.

As part of the welcoming ceremony into a village, the *pōwhiri*, visitors are greeted by a mock challenge, *wero* (Matthews, 2004). Up to three strong and fearsome warriors would greet the visitors.

> The modern *wero* is the abbreviated descendant of a whole series of war-like evolutions that were once performed whenever strangers met. In peace and war strangers were greeted with the same ritual forms, because an unknown group might always be planning treachery, and a display of strength could dissuade them. Early observers of these encounters remarked that it was almost impossible to distinguish peaceful overtures from warlike ones, and just to be sure groups who were meeting for the first time went armed and in full strength. (Salmond, 1975, p. 132)

Although Te Koutu faced a different set of challenges than those described above, the idea that mathematics came fully armed to integrate with traditional Māori cultural values had resonance for us as we proceed through the book. Until there is a negotiation of intent, as symbolised by the *wero* (Irwin, 1992), there will be no real integration of knowledge systems. Indeed, it could be said that the part played by mathematics in Māori students' underachievement was of duplicitous intent which arose because of the lack of formal negotiation. We hope to make explicit some of the negotiation about the role of mathematics in education systems designed for Māori children by focussing on the use of language.

In the introductions of each of the book's four parts, we have woven the legend of Māui, a Polynesian hero prominent in Māori legends. Many stories of Māui and his exploits appear with variations throughout the Pacific. However, the underlying morals and role model remain much the same. Māui was the youngest of the five sons of his mother Taranga. His mother believed Māui was still-born and in her grief wrapped him in a bundle of hair (*tikitiki*) and cast him upon the ocean. He floated in the hair knot to a beach where he was found by his ancestor Tamanui-ki-te-rangi, who raised him as his own. When old enough Māui began to question his origins and eventually set out to find his family and carve an identity for himself. Over the course of his life he faced many challenges, but overcame them using his personal attributes, sometimes with, and sometimes without, the support of his community. Māui was known for challenging established practices and protocols and for not accepting the current state of affairs as being the only possible outcome. His way of acting was not viewed positively by many of the establishment, and huge disagreements and fights were an accepted component when new ideas were introduced.

It is traditional in Māori culture to provide an analogy or metaphor to promote an understanding of some of the nuances within themes that are difficult to raise in other ways. Therefore, in keeping with this tradition, we have chosen to use *wero* and the story of Māui to enrich our discussion of the challenges described in this book.

We outline some of the language challenges in learning mathematics that were faced by Te Koutu and how these were met by teachers, researchers and community members working together. Predominantly, we examine issues around the development, teaching and use of the mathematics register. Halliday (1978) defined a "register" as follows:

> a set of meanings that is appropriate to a particular function of language, together with the words and structures which express these meanings. We can refer to a 'mathematics register', in the sense of the meanings that belong to the language of mathematics (the mathematical use of natural language, that is: not mathematics itself), and that a language must express if it is being used for mathematical purposes. (p. 195)

The mathematics register is the vocabulary and grammatical expressions in *te reo Māori* that enable discussion and learning of mathematical ideas. One of the functions of education is to support students to move from everyday conversational language to the more technical language of specific subjects (Herbel-Eisenmann, 2002; Schleppegrell, 2007), and learning mathematics at a *kura kaupapa Māori* is no exception to this. A focus on language is reinforced by the school's commitment, since its inception, to the revival of *te reo Māori*. The consequence of this commitment has meant that the school (including teachers, parents, students and the wider community) has had to consider how to overcome the challenges from using *te reo Māori* for the teaching and learning of mathematics. This book contains 11 chapters that document the process by which some of these challenges have been met and are being overcome.

The School and the Data

The setting is a Māori-immersion school. This form of schooling is discussed more fully in Chapter 3. However, it is useful at this point to provide a few details about the specific school at the centre of the study: Te Kura Kaupapa Māori o Te Koutu. Te Koutu is located in a regional city in New Zealand with most parents working in urban jobs. There are classes for children from 5 to 18 years of age in year levels from 0 to 13. At Te Koutu, with the exception of the specialised English language classes, all classes are taught in *te reo Māori*. Spanish is taught as a foreign language from the first year of school, and children receive English lessons from Year 7 (around 11 years old). Students only use it, or Spanish, whilst at *kura* (school) and are encouraged to use it when communicating with their friends outside of school hours.

Although the parents made active decisions to place their children in a Māori-immersion situation, they themselves may not be fluent in *te reo Māori* – the reasons for this are discussed in Chapter 3. As well, *te reo Māori* is a minority language in New Zealand, and so many of the children's experiences outside school would be conducted in English. These include activities that use their mathematical understandings such as going to fast-food restaurants and playing computer games. Although English may be the first language for almost all students, the students do have some fluency in *te reo Māori* when they start school. Most children gain this from attending a Māori-immersion preschool, *kōhanga reo,* and from Māori-speaking parents or grandparents at home. The children at this school participate in some traditional Māori cultural experiences both inside and outside the *kura.* However, many of their experiences outside of school are not very different from their non-Māori peers.

Some of the younger teachers had their own schooling and/or their teacher education in *te reo Māori*, but for most teachers in *kura kaupapa Māori,* such as Te Koutu, all their education has been in the medium of English. The majority of teachers teaching mathematics at Te Koutu have had to learn the mathematics register in *te reo Māori,* whilst they were teaching it to their classes (this is discussed in more detail in Chapter 9). The bilingual existence of the students, their families and their teachers has resulted in the students being in a linguistically complex situation.

The data comprise interviews and questionnaires from more than 12 teachers and over 200 students, aged 5 to 18, with photos and videos from mathematics classrooms, as well as a collection of more than 2000 writing samples. It also includes interviews with family members and meeting notes.

The different data sets come from a variety of research projects about mathematics education that have occurred in the school since 1998. In 1998–1999, we were interested in considering how a mathematics curriculum could be developed that was more culturally responsive. This project involved having joint meetings between teachers and parents, and although no curriculum eventuated, the discussions provided rich information about teaching mathematics in a Māori-immersion school (Meaney, 2001). From 2005 to 2006, we documented the strategies that teachers used to support students gaining the mathematics register (Meaney, Fairhall, & Trinick, 2007). In 2007, we concentrated on how the teachers increased the quantity and quality of students' writing in mathematics (Meaney, Trinick, & Fairhall, 2009b). Since then, we have continued to collect data from parents, teachers, and students, as well as from policy documents and research reports in order to make sense of how the emphasis on revitalising the Māori language has permeated the teaching of mathematics and the impact of this on students' learning.

The book provides descriptions of different kinds of challenges based on the input from participants, such as teachers, either indirectly from us observing their teaching practice or directly from having discussions with them on some of the issues. The interpretation given in the following chapters will not be an accurate reflection of the views of all participants. There is always a differential in power between those whose comments we interpret and ourselves as authors. Nonetheless, we also have been participants in the research and so cannot be considered only as external commentators. For example, Uenuku Fairhall, the principal of Te Koutu, was involved in the development of the mathematics register in the 1980s and 1990s with Tony Trinick (see Barton, Fairhall, & Trinick, 1998). As university researchers, Tony Trinick and Tamsin Meaney have ongoing interests in this area of research and have spent considerable time working with the staff at the *kura*. Some of the teachers had been students of Tony whilst doing their teacher education through Auckland College of Education and later the University of Auckland. Tamsin has been involved with the school since 1998.

The teachers, students and parents have come and gone over the years of the project. Consequently, each participant is not given a pseudonym but rather an initial that identifies his or her role (T for teacher, P for parent and S for student) and a number. For teachers we have provided the Year level that they taught when the extract was recorded. This is a long-running research project, making anonymity difficult to maintain in published material. Pseudonyms used consistently would not contribute to maintaining this anonymity. Using numbers concentrates the reader on what is being said rather than who has said it. The projects belong to everyone who has participated, and it was important all participants felt comfortable that what has been included is a appropriate portrayal of the range of views.

From synthesising the material from the different projects, it is clear that the collaboration between participants to meet the different challenges has had long-term implications for working with Indigenous students and schools. Collaboration at this school has resulted in positive achievement in mathematics for Indigenous students, a process which is problematised in Chapter 4. Generally Māori children do not perform as well as their non-Māori peers in the New Zealand education system, so the achievement of students at Te Koutu is to be celebrated (Meaney, McMurchy-Pilkington, et al., 2008). Therefore, our purposes for writing this book are different to those of the original research projects. Consequently, we have reanalysed data using a range of additional theoretical lenses to show how we worked through the different challenges.

Using Case Studies

In reanalysing the data to illustrate how challenges were met and overcome, it was important to consider how to retain the complexity of the context but not be overwhelmed by it so that our understanding about the different kinds of challenges did not become restricted. We have chosen to do this by using a series of case studies. Case studies as one kind of ethnography place a primacy on the situated meaning and contextualised experience as the basis for explaining and understanding social behaviour including language (Woods, 1992; Nunan, 1992; Brewer, 2000).

Case studies present complete descriptions of a phenomenon within its context (Yin, 2003). They are described as being "bounded" (Merriam, 1988, p. 29) in that a specific situation is isolated for an intense study. This boundedness contributes to a rich understanding of the case, but by its nature can simplify some of the complexity by considering some variables to be outside of its boundaries, and therefore not part of what is to be researched. For example, it is often assumed that the relevant community of interest has been identified, but this may not be the case (Creswell, 1998).

Community, formal organisation, informal group, and individual-level perceptions may all play a causal role in the subject under study, and the importance of these may vary by time, place, and issue. There is a possibility that an ethnographic focus may overestimate the role of community culture and underestimate the causal role of individual, psychological or sub-community (or for that matter, extra-community) forces.

To overcome a simplification of the complexity in the situation, we have provided a series of case studies grouped particularly around kinds of challenges. The book has four parts that are based on political, mathematical, community and pedagogical challenges. In each part, two or three chapters identify different challenges, and these are presented as case studies. These sets of challenges allow us to show how teachers, parents, students and other community members operated within a formal education system to support students to learn mathematics in a way that connected them to the Māori language and culture. The grouping of case studies allows an in-depth consideration, whilst the combination of the different groups of challenges shows their interrelatedness.

The Complexity of Learning Mathematics in an Indigenous Language

The challenges that Te Koutu faced and are still facing in teaching mathematics through an Indigenous language are multifaceted and complex. Therefore, it was important that our theoretical framework reflected this complexity. Consequently, we considered the process of meeting challenges to be an *activity* in the sense of Leont'ev (1978) or *practice* in the sense of Schatzchi (2005). Theories around these ideas provide insights into how the challenges arise from the interactions between individuals and the wider socio-historical-political context.

In recent years, research into mathematics education has expanded from simply being seen as something that goes on inside a student's head or as something that only occurs in a classroom (Valero, 2009). Valero (2002) suggested that a focus on the mathematics learner or the mathematics teacher as the entry into understanding mathematics education is reductive because learner and teacher are no longer seen as human beings with whole lives. Looking at what the teacher or the learner does in the mathematics classroom can result in what occurs outside the classroom being ignored, even though it also may affect the learning process. In describing the use of narratives in research on teachers' lives, Goodson (1999) warned that there was a need to take the context of those lives into the narratives in order for the meaning of the narratives to make sense. Valero (2009) also warned of the risks of ignoring the context when researching the mathematics interaction. For example, Yup'ik teachers discussed how their Western teacher education resulted in them teaching in English so that they were perceived as being "real" teachers even though they and their students were fluent in the Yup'ik language (Lipka, Mohatt, & Ciulistet Group, 1998). The value given to their teacher education, and its portrayal of "normal" teachers, restricted in the classroom the Yup'ik teachers' beliefs about the usefulness of traditional ways of interacting. The destructive influence of their teacher education would not be recognised if the focus remained on the children's cognitive grasp of mathematics within the classroom setting. Considerations of this kind will affect the learning interactions that occur in a classroom. Valero (2009) illustrates some of the complexity that contributes to classroom practices in Fig. 1.1.

In the diagram, Valero concentrated on the network of social practices suggesting that each one affects the teaching dyad of teacher, learner and mathematics. There are an increasing number of examples that consider the complexity of influences on classroom interactions. Lange (2008) showed how public discourse influenced the perceptions of a teacher and the children she taught about what was normal family interactions around homework; and this then affected expectations and interpretations of what occurred. However, diagrams such as Fig. 1.1 seem to view mathematics education interactions as being a-historical, or at least there is no acknowledgement of how the history of each component may be interconnected.

In regard to understanding how some groups of students, such as Māori, underachieve in mathematics, there has been an acceptance of the impact that the socio-political situation of the society has had on mathematics learning. However, it is difficult to show how the relationship between society and classroom interactions contributes to inequitable outcomes for some groups of students. Through his theory of pedagogic device, Bernstein (1990, 2000) attempted to explain the

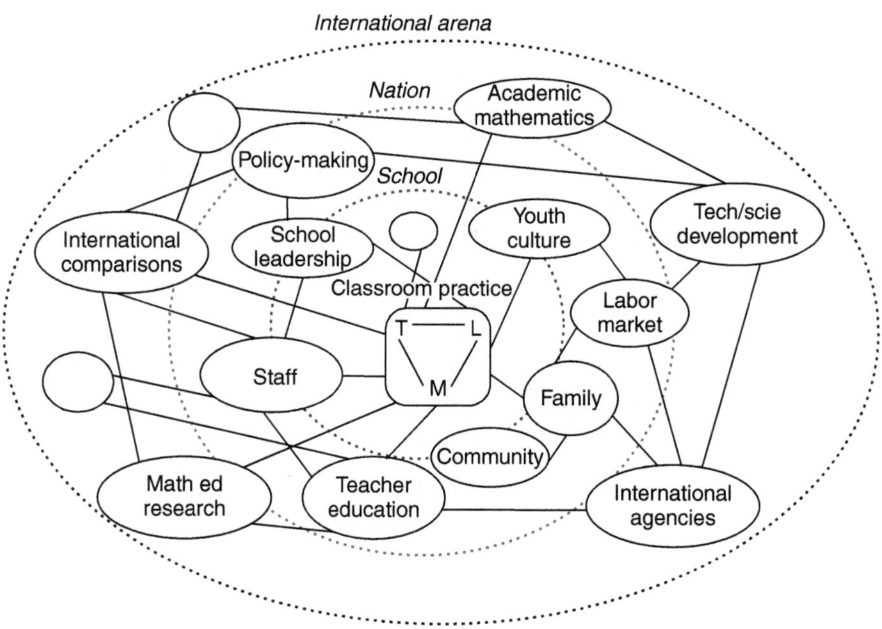

Fig. 1.1 A representation of the "network of the social practices of mathematics education" – T is for teacher, L is for learner and M is for mathematics (from Valero, 2009)

educational rules or principles that underpin knowledge transformation within education systems by describing a pedagogic device. He stated, "the device continuously regulates the ideal universe of potential meanings in such a way as to restrict or enhance their realisations" (2000, p. 27). However, by regulating consciousness, or the ways that people become aware of how they value some knowledge more than others, the rules of the device can produce a gap between abstract meanings and immediate contexts, allowing for the unthinkable to become thinkable. The historical development of ideas is restricted to the time between the acceptance of specific policies and the implementation of the actual teaching–learning interaction.

In contrast to the set of static rules that make up the pedagogic device (Bernstein, 2000), theories that deal with practices as social entities consider how the knower and the process of coming to know are intricately linked. It is the social practices, through which people learn and concurrently change the way that they interact with the world, that enable alternative courses of action to be seen as possibilities (Radford & Empey, 2007). The historical influence on these practices is built into the analysis because the simultaneous development of the knower and the process of knowing is done within the context of societal views on what knowledge is and how learning of it is expected to occur. However, at the moment there are "limited explanations as to how practices emerge and evolve over time" (Antonacopoulou, 2008, p. 115). Radford and Empey's (2007) account of the development and influence of new mathematics ideas during the Middle Ages and the Renaissance in both the West and the Islamic World would be one example of this.

Individual adoption of certain practices is rarely what is analysed in practice theory, rather it is the way that the practices adapt and change as they are instituted within specific situation that connects knowers to the process of coming to know. Although some commentators on practice theory see the practitioner as the driver for change through the practice and its purpose (Antonacopoulou, 2008), many others problematise the relationship between the individual and the collective. For example, Gergen (1985) suggested, "knowledge is not something that people possess in their heads; rather, it is something that people do together". In a more detailed description of a practice, Reckwitz (2002) stated the following:

> Moreover, the practice as a 'nexus of doings and sayings' (Schatzchi) is not only understandable to the agent or the agents who carry it out, it is likewise understandable to potential observers (at least within the same culture). A practice is thus a routinized way in which bodies are moved, objects are handled, subjects are treated, things are described and the world understood. To say that practices are 'social practices' then is a tautology; a practice is social, as it is a 'type' of behaving and understanding that appears at different locales and at different points of time and is carried out by different body/minds. (p. 250)

In developing an understanding of how challenges around using *te reo Māori* for teaching mathematics are met and overcome, it is useful to consider this process as a practice. This is because it enables the practitioners to be connected to the practice and its purposes through the social context. Practices are social entities and their purposes are thus extra-individual. Reckwitz (2002) stated, "practice theory encourages us to regard the ethical problem as the question of creating and taking care of social routines, not as a question of the just, but of the 'good' life as it is expressed in

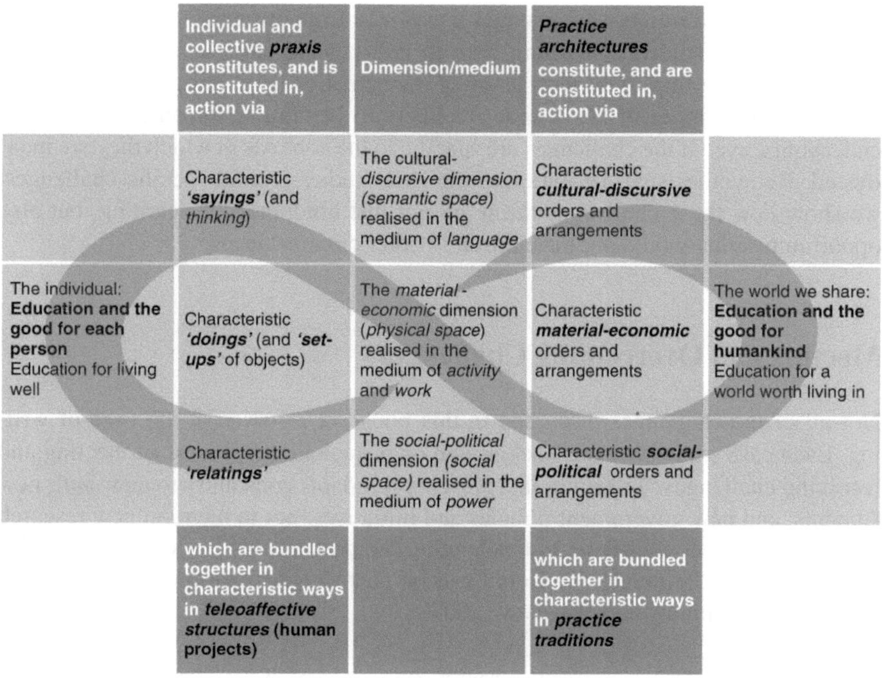

Fig. 1.2 The individual and collective purposes of education constituted in praxis and practice architectures (from Kemmis, 2009)

certain body/understanding/things complexes" (p. 259). Practice theories are more aligned to considerations of how meeting and overcoming challenges can lead to a better life than to Bernstein's concentration on how social injustice is embedded within the classroom practices through the implementation of the pedagogical device.

Figure 1.2 illustrates one way of perceiving how individual and extra-individual components contribute to practices (Kemmis, 2009). Kemmis and Grootenboer (2008) described three extra-individual structures and processes – culturally-discursive, material-economic and social-political – that "shape dispositions and actions, both in the educator's general response to a particular situation or setting, and in relation to their particular responses at particular moments" (p. 50). These processes were described as "practice architectures". The practices of saying, doings and relatings, which mediate the shaping of individuals and structures that make up practice architectures, often are not separate entities but are bundled together. Understanding how different factors combine to facilitate or constrain people's meeting of challenges involves considering how individuals interact via these extra-individual dimensions of language, work and power.

The complexity of the situation and the necessity of thinking across a number of layers determined our choice to utilise a series of case studies. All educational endeavours face a variety of challenges that evolve from different sources and which merge to affect students' learning both in and outside classrooms. In this book, we look at challenges to do with learning mathematics in an Indigenous language. However, the sources of challenges are likely to be similar for other educational endeavours, even if the challenges are specific to the contexts in which they are manifested. By synthesising the data into the case studies around specific challenges, we show how these challenges throw up not just hindrances to learning, but also opportunities that would not have arisen in other circumstances.

Meeting and Overcoming Challenges

Not all of the challenges discussed in this book are resolved at the time of writing. Each case study provides background on a stage in the process of meeting and resolving challenges. Te Koutu, like all schools, adapts constantly to new staff, new students, and new government policies and initiatives, not to mention new research findings. These all contribute to challenging the possibilities for how mathematics could or should be taught. Te Koutu can be considered a learning community as described by Anthony and Walshaw (2007):

> The classroom learning community is neither static nor linear. We can more usefully think of it as nested within an evolving systems network. This system might be described as an ecology in which the activities of the teacher and the students—as well as those of the centre/school and the home/community—are mutually constituted through the course of interactions. Thus, "teaching and learning coexist in a web of economic, social, and cultural differences". (Hamilton & McWilliam, 2001, pp. 17, 21)

In discussing the continuation of professional growth by teachers several years after a professional development project, Franke, Carpenter, Levi and Fennema (2001) noted the importance of collaboration in fostering a learning community. They felt that teachers needed "time to develop relationships with others that they can talk with in ways that meet their needs and push their thinking" (p. 685). In research on successful home–school partnerships in New Zealand, Bull, Brooking and Campbell (2008) reiterated the need for time, but also commitment. They also referred to the importance of the way that power was shared between the different participants.

Te Koutu has provided spaces for community members to talk about the challenges within different situations. These spaces support the development of pathways for overcoming different sets of challenges. This does not mean that the discussions are harmonious and always reach consensus. Far from it, debate and disagreement illustrate the sharing of power but also enable new ideas to arise. Meeting and overcoming challenges is messy, and the case studies illustrate this messiness and the necessity of it.

We do not discuss all the challenges that were met by the school, the wider community and society around the teaching of mathematics in an Indigenous language.

Choices had to be made about what to include, and so we have selected material that we felt best exemplified the process of meeting and overcoming challenges.

There is little research on how challenges that impact on education can be met and resolved. For example, in Lipka et al. (1998), there was much discussion about the sorts of challenges that were faced but little information about how challenges were met and overcome. This was because the process itself was not the object of the research. In this book, it is the description of the challenges and the specifics of how they are being overcome that are the foci.

We propose that meeting challenges requires going through a number of stages. These stages include the following: recognising the challenge, resisting it, identifying unintentional opportunities or small activity spaces provided by the challenges that allow different kinds of agency to be nurtured, and finally a reorganisation of the system so that what is considered "normal" is redefined. In the following chapters, we describe different challenges and the process that we, as a collaborative research group, went through to meet them. As Smith (2003b) suggested in critiquing linear models of development:

> Māori experience tends to suggest that these elements may occur in any order and indeed may all occur simultaneously. It is important to note as well ... the idea of simultaneous engagement with more than one element. (pp. 16–17)

Resistance is not considered a negative aspect of meeting challenges. Using the ideas of Foucault, McMurchy-Pilkington (2008) stated, "resistance arises out of the exercise of power, and the notion of resistance links to people's ability as human agents to act in social situations" (p. 617). Certainly, the different chapters show clearly that not all the teachers or members of the community connected with Te Koutu were going through the same steps at the same time on the same challenges. Instead, different school members saw the challenges in different lights and reacted to them accordingly. In looking at a dynamic organism such as a school, there will always be differences, and it is how these differences play out in the learning experiences offered to the students that is of interest.

Overview of the Chapters

Part I of the book is on political challenges. Although all challenges are political as they are to do with how power is distributed and enacted, to some degree they are more obvious when looking at how Indigenous groups interact with mainstream society through the formal education system. The three chapters in this part deal with specific political challenges to mathematics education in *kura kaupapa Māori*. Chapter 2 looks at the development of a mathematics register as a consequence of the push by Māori to simultaneously revitalise their language and improve educational outcomes for their children. Chapter 3 describes the history of Te Kura Kaupapa Māori o Te Koutu and the challenges that the parents and teachers faced in setting it up and continuing to run it. In Chapter 4, we explore the provision of bilingual examinations for students at the end of high school. This is a continuing challenge that is being met at a political level, but where listening to students' views could provide activity spaces for moving forward.

Part II concerns mathematical challenges faced by the school. In Chapter 5, we explore how the features of *te reo Māori*, both traditional ones such as logical connectives and newly coined vocabulary that has a transparent meaning, can support students learning mathematics. Writing in mathematics is explored in Chapter 6. It examines students' and teachers' perceptions of the contribution of writing to students' understanding of mathematics. Chapter 7 focuses on the development of students' understanding about probability using *te reo Māori*. Probability was identified by the teachers as an area of mathematics that students learning in *te reo Māori* struggled with.

Part III is about meeting community challenges. Chapter 8 examines how the use of the mathematics register has begun to permeate Māori radio and television. It also discusses how students who have now left Te Koutu found learning mathematics in *te reo Māori* affected their subsequent learning or use of mathematics as adults. Chapter 9 looks at the impact of the teachers' own fluency in *te reo Māori* on discussing mathematics. Few teachers who teach in *te reo Māori* had their own mathematics learning experiences in this language. The lack of teachers with strong *te reo Māori* language skills has long been identified as a significant community challenge in providing a strong education in Māori-medium schooling.

Part IV is on how the teachers taught mathematics through the second language of most students. In Chapter 10 we discuss in detail the model for mathematics register acquisition (MRA) that was used in understanding how the teachers supported students' learning of the mathematics register. In Chapter 11 we describe the Māori approaches to teaching that support the simultaneous learning of language and mathematics. It critically examines the concept of Māori pedagogical practices. Chapter 12 outlines the reasons behind teachers making changes to their teaching practices as a consequence of being involved in the research projects.

The final chapter of the book looks specifically at how the collaboration operated and identified its main features. The chapter ends by considering how the roles of different participants, teachers, parents, students, researchers and other community members worked in conjunction with the stages in meeting the challenges discussed in the first chapter of the book.

Part I
Meeting Political Challenges

Māui

Māui was drawn on by the singing and chanting, the slapping and stamping, all carried clearly beyond the village palisades through the porous night. Squirming easily between the posts of the palisade, he found everyone congregated in the meeting house. He silently slipped inside, keeping to the shadows cast against the walls by the glistening bodies of the performers and their spellbound audience facing one another across the fire pit.

Māui inched closer as four strapping warriors, obviously brothers, stood to do the *haka*. Their war dance was powerful, their voices strong and clear. The menacing facial grimaces and deliberate hand actions delighted the people before them, no one as much as their proud mother Taranga who sat in the midst of the tribe with her only daughter. Around her Māui quickly noted other faces, expressing awe, admiration, desire, and even envy, but he was more intent on watching the mother. His heart quickened with expectation.

The *haka* ended to loud cheers until Taranga stood up and walked past the fire pit to her sons. At the same time, but unnoticed, Māui stood up in the shadows and eased forward to stand at the far end of the line of brothers. The tribe watched silently as Taranga moved from son to son, counting them off and proudly reciting their names, all called Māui, but with different epithets.

"*Ka tahi*, that's one – Māui-mua. *Ka rua*, that's two – Māui-pae. *Ka toru*, that's three – Māui-roto. *Ka whā*, that's four – Māui-taha. Ah, *ka rima* – and that's five", she said indulgently to the wild-looking youth that had placed himself alongside her much taller, and much handsomer, sons. "Who are you? And where do you come from?" She looked back at the tribe, inviting their laughter.

Her smile vanished when the young urchin said, "*Nāu anō au*, I'm yours. My name is Māui-pōtiki, Māui-the-youngest".

Taranga decided to humour the boy, though her voice now had an edge, clearly heard by the rest of the tribe who were silently intent on the drama playing out before them. She continued, "But I have only four sons. *Ka tahi, ka rua, ka toru, ka whā*. See, there are only four".

"I'm sorry", said the boy, though he sounded more angry than sorry, "But I thought I *was* your son. As you heard, my name is Māui-pōtiki. I was born prematurely to an older woman beside the sea. She thought me dead and so cut off her...".

Taranga grabbed his arm and quickly pulled him outside. The tribe was disappointed with the interruption. Her sons stood where they were, bewildered, not knowing what to do. Their mother had given them no indication to follow her and the scruffy stranger out into the dark, and they were used to obeying.

This part examines the political challenges implicit in the establishment of Māori-medium education. These challenges were twofold. Māori were challenged as they learnt how to set up and run Māori-medium schools and classes, whilst the system itself was challenged by having to negotiate with these schools and their beliefs about what should be valued in providing an education to Māori children. During the 1970s, faced with significant language shift to English and rapid decline in the Māori-speaking population, there were a number of initiatives to revitalise *te reo Māori* (Reedy, 2000). One of these was the development of schooling in the medium of Māori. The birth of Māori-medium education generally and *kura kaupapa Māori* (Māori-medium schools, usually set up by parents) specifically, like that of Māui-pōtiki's birth, was not without its challenges. One of the major challenges was how to finance and sustain these schools, which being outside of the Ministry of Education's education system meant that they received no state funding. Much time and effort have gone into their establishment, so that by 2010 they have become an accepted part of the New Zealand/Aotearoa education system. However, the challenge to the status quo by the establishment of these schools left many, including the New Zealand Ministry of Education, unsure of how to proceed. Others such as educational researchers seemed to be caught almost unaware of the changes which were being initiated and were left on the sidelines watching the drama unfold.

We explore three interrelated challenges that reveal the interactions between the wider society and Māori in relationship to the three situations: establishment of Māori-medium education, the mathematics register for use in Māori-medium education, and the impact of bilingual exams on Māori students' achievement at the end of high school. In all three situations, Māori people, whether teachers, parents, students, or wider community members, have been politicised by the experience. Māui in approaching and confronting his mother about his origins can also be seen as politicised in that he no longer was willing to accept the status quo of being an "unrecognised" son. When a minority group such as Māori start to question why a current situation is as it is, then they begin to exert their power in a positive way. This can lead to changes to the situation. However, for this to happen there is a need to move beyond identifying problems and apportioning blame. Apple (1992) stated the following:

> Thinking critically is not necessarily a natural occurrence. It doesn't automatically arise simply because one is told to look for problems. Rather, such an awareness is built through concentrated efforts at a relational understanding of how gender, class, and race power actually work in our daily practices and in the institutionalised structures we now inhabit. (p. 418)

Often apportioning blame leads to simplistic categorisation of participants within an educational system as "good" or "bad". For example, Freire (1996) divided people into oppressors and the oppressed and was emphatic that oppressors were unable to participate in the reflection and action, which makes up praxis. Darder (1991) distinguished between those from dominant and subordinate cultures. These distinctions, although helpful in highlighting differences, are not adequate for understanding the complexity of the situation. Knijnik (2000) in discussing the work of Popkewitz stated that dividing the world into oppressors and oppressed "cover[s] up the actions and practices of individuals through which power also operates" (p. 5).

Instead, political challenges should be considered as manifestations of power distribution and use. To try to understand what enabled or constrained the implementation and operation of different aspects of Māori-medium education, we take up Foucault's ideas about the fluctuating nature of power and knowledge. Michel Foucault believed that power "needs to be considered as a productive network which runs through the whole social body, much more than a negative instance whose function is repression" (Gordon, 1980, p. 119). One of his central ideas was that it was the fluctuation of power between people, as knowledge was discussed and accepted, which affected what happened in different situations.

We have combined these ideas with Kemmis and Grootenboer's (2008) description of praxis architectures to show how these influences can be "unpacked" and the opportunities for change recognised. As is illustrated in Figure 1.1, Kemmis (2009) perceived education as something that should contribute to the good of each individual engaged in it as well as to the good for humankind as a whole. This good is achieved through the interaction of individuals and collectives through society-constructed practices.

> [O]rganisations, institutions and settings, and the people in them, create *practice architectures* which prefigure practices, enabling and constraining particular kinds of sayings, doings and relatings among people within them, and in relation to others outside them. The way these practice architectures are constructed shapes practice in its cultural-discursive, social-political and material-economic dimensions, giving substance and form to what is and can be actually said and done by, with and for whom. (Kemmis & Grootenboer, 2008, pp. 57–58)

If education is considered as a practice, then an individual's performance of this practice is constrained by the characteristics that have been historically embedded within education. However, options exist for alternative enactments of education that could contribute to the practice itself being reconfigured. Thus, it is within the enactment of the practice that both the individual and the practice can be affected. This recognition of a two-way change process is beneficial because an analysis using practice architecture provides indications of how the good of a practice could be increased without simplifying the actual reality of its enactment.

Although both ideas of power and praxis architecture are relevant to the other kinds of challenges discussed in this book, to some degree they are more obvious when looking at how Indigenous groups interact with mainstream society in the case studies highlighted in this part. Mathematics education, and how it is constituted in schooling, is controlled by and at the same time controls the knowledge that is valued by societies and minority communities. These are complex interactions with different players taking on sometimes contradictory roles as they juggle to determine the most appropriate knowledges and practices for a specific period of time.

For example, although the instigation of Māori-medium education and the development of a Māori mathematics register was a major initiative, little research based in schools was undertaken at the time. Instead, the research focussed more on the policy level. Part of the reason for this could have been a sense that the new initiatives would be short-lived and so were not worth documenting. The other reason was that most educational researchers at this time were not Māori, or did not speak Māori, and so had no easy pathways for working with those communities who were establishing the new Māori-medium education system. The development of Māori educational research is commensurate with the development of Māori-medium education. Māori were no longer prepared to allow others to have the sole rights to comment on issues that they felt were intrinsically theirs.

> Kura Kaupapa Māori, (Māori language immersion schools) in particular are somewhat wary about research taking place on their sites and of researchers in general. This stems from past experiences where much research has been conducted in these settings with little direct or indirect benefit to the participating schools or immersion education in general. In most instances, research is initiated and conducted by outside parties rather than by the school itself. Allowing external researchers to gather information about the school or its students places them in a vulnerable position. (Rau, 2001, p. 2)

The key players in the development of the Māori mathematics register were the teachers working in Māori-medium situations, in particular those working in secondary schools. They were the ones on the front line with the need to use consistent language to teach abstract mathematical concepts. However, the Māori Language Commission, Te Taura Whiri, also had a role by being the arbitrator of what was appropriate *te reo Māori*. The speakers who made up the commission were perceived as having the power to approve, or not, suggestions for mathematical terms. Yet, the choice of which terms to use remains with individual teachers. As Foucault indicated, power and knowledge were intertwined (Gordon, 1980). The roles that players had in the education system provided them with different choices, and these choices were supported or hindered by the circumstances surrounding them.

The mathematics register in *te reo Māori* and the establishment of Māori-medium schooling system were developed at a certain time, within specific social-political, cultural-discursive, and material-economic orders and arrangements. It was a period when the government, although unsure of how to handle the grassroots development of a new education system, was instigating major educational changes of its own. While language revitilisation efforts were gaining momentum, a neo-liberal transformation began in 1984, with a raft of reforms particularly centred on how state institutions including education were to be structured and managed (Olssen & Mathews, 1997). The Labour Government in 1988 brought in *Tomorrow's Schools* (Lange, 1988), which resulted in major changes to the New Zealand education system. One of the distinctive features of these New Zealand education reforms was the way they were shaped by certain economic and administrative theory, including public choice theory (Boston, Martin, Pallot, & Walsh, 1996). The central tenet of the public choice approach is that human behaviour is dominated by self-interest; therefore, if schools do not meet Māori educational demands, they will "opt out" and choose to go elsewhere. Thus, Māori education became contestable in an open market, with schools and services competing for the provision of services. Ironically, it was this ideology that gave rise to the government's support for government funding of schools like Te Koutu.

The second tenet of *Tomorrow's Schools* was the devolution of control and management of the school to Boards of Trustees, parents, or parent representatives. Previously, New Zealand state primary schools were controlled by a centralised government organisation, the Ministry of Education. The sayings, doings, and relating connected to *Tomorrow's School* clearly influenced the rhetoric used by Te Koutu's parents in Chapter 3. The discussions by parents in 1998 revolved around their wish to be involved in their children's education whilst the comments from 2008 suggest that the parents saw their choice as being one of choosing the best school for their children. However, the acceptance of state funding by *kura* has several stings in the tail. One was the requirement to teach the national curriculum as decreed by the Ministry of Education (Stewart, 2005). This has caused much angst as the schools, teachers, and parents deal with the resulting tension. As Appleby (2002) stated:

> Kura Kaupapa Māori operate a constructivist model of curriculum in which students are 'co-learners' with teachers, and where teacher/student roles are reciprocal. Such a model is collaborative. It is politically transparent and invites critical reflection and an ongoing critique of power relationships. This *narrative pedagogy* is incompatible with a prescribed national curriculum. (p. 114)

Thus the material-economic orders and arrangements that provided the working conditions for schools ultimately contributed to a reduction in *kura kaupapa Māori*'s control over what it taught and how. However, it is unlikely that parents would have agreed to support a school that did not provide their children with qualifications needed to enter the mainstream, adult world of work or further study. What will never be known is whether these outcomes could have been achieved if *kura kaupapa Māori* had been allowed to develop and follow their own curriculum.

The final set of challenges that we examine in this part is to do with the instigation and operation of bilingual examinations, which are sat by students at the end of high school. This case study shows how the right to have the mathematics examinations in *te reo Māori* as well as in English was something that had to be fought for with the New Zealand Qualifications Authority (NZQA). Even once this battle had been won, there were and are issues to do with the quality of the translation and the consequent channelling of the students into using English rather than *te reo Māori*. In this case study, the flow of power and knowledge between different groups is clear. As the controllers of high school examinations for all New Zealand students, NZQA have the ultimate authority to decide which language(s) are to be used. *Wharekura* (Māori-immersion high schools) were able to exert enough pressure to have examinations produced bilingually, because of the obvious inconsistencies of, on the one hand, supporting schools to teach in *te reo Māori,* but, on the other hand, then forcing the children to sit the external examinations in a different language. However, research into the difficulties that students have in sitting the bilingual exams, including the students' own stories, are contributing to a groundswell of concern about the quality of the Māori translations. It is hoped that in the future this will lead to a reappraisal of how bilingual exams are constructed.

Power, therefore, ebbs and flows between NZQA, the translators who produce the examinations, and the students, teachers, and parents who are concerned with the academic qualifications which come from doing them. As the saying, doings, and relatings connected to these examinations change, then the structures such as the social-political order and arrangements also change. With the concession by NZQA to produce bilingual examinations, there was an acceptance by both NZQA and the Māori-medium education system that the Māori-medium education system had the right to challenge NZQA and have their concerns taken seriously. Fortunately, NZQA does not have absolute authority to ignore the concerns of stakeholders. Although there are still battles to be won, it should be easier for changes to be made in the future.

Dealing with and overcoming political challenges has been something that the fledgling Māori-medium education system has had to do since its inception. Like Māui, the people involved in the setting up and maintenance of this education system did not accept the status quo. Instead, different aspects relating to the education of Māori children were questioned. When one set of problems were resolved, others arose. Consequently, there was a constant ebb and flow of power between the participants as certain knowledge took precedence and was viewed as more relevant. With the resolution of each set of problems, the practice architectures in which education was embedded also changed. Hopefully, this will enable easier negotiation and resolutions as Māori continue to work on issues pertinent to their children's education.

Chapter 2
The Development of a Mathematics Register in an Indigenous Language

The introduction of Western mathematics into New Zealand/Aotearoa challenged Māori, in both positive and negative ways. It is most likely that *te reo Māori* traditionally embodied in its semantic structure mathematical concepts, to do with counting, measuring, locating, designing, playing, and explaining. Bishop (1988) argued that these activities were universal and have supported and shaped the development of mathematical concepts in all cultures. The activities and the language required to communicate them were familiar to fluent speakers of Māori up to contemporary times. However, *te reo Māori* grammatical and conceptual structures specifically connected to these cultural activities were not sufficient, by themselves, to express Western mathematical ideas. The situation is similar for other Indigenous languages. Harris (1980), in surveying Aboriginal languages in Australia, found that the languages were adequate for their culture, but these languages had some difficulty expressing particular measurement concepts found in Western mathematics. Since New Zealand/Aotearoa was first colonised, there has been an ongoing debate about whether, and how, *te reo Māori* could be used for learning and using Western mathematical ideas. Much of this debate has been a political one, where issues of power and what knowledge should be valued have led to contrary perspectives being prioritised at different times.

Many of the concerns about using *te reo Māori* in relationship to Western mathematics are relevant for other endangered, Indigenous languages which have had to adapt to incorporate Western mathematical ideas in a relatively short space of time. Broadly defined, an Indigenous language is any language that is "native" to a particular area (Walsh, 2005). Thus, Māori is indigenous to Aotearoa, not withstanding that it originated in the Pacific, whereas English, now used in Aotearoa, is not indigenous. Mandarin could be described as an Indigenous language of China, along with several other less prominent languages. Some Indigenous languages, such as Mandarin, are not endangered languages. Different scholars contest what counts as endangered. For example, Crowley (1998) sees many Indigenous languages in the

This chapter draws heavily on the article: Barton, B., Fairhall, U., & Trinick, T. (1998). Tikanga Reo Tatai: Issues in the development of a Māori mathematics register. *For the learning of mathematics, 18*(1), 3–9.

T. Meaney et al., *Collaborating to Meet Language Challenges in Indigenous Mathematics Classrooms*, Mathematics Education Library 52,
DOI 10.1007/978-94-007-1994-1_2,

Pacific as not being endangered. Conversely, Dixon (1991) sees many of the same languages as endangered!

For many speakers of Indigenous languages like Mandarin, French, English, or Japanese, there is no question about whether mathematics should be taught in their own language. This is not the case for smaller Indigenous groups, particularly those who have been colonised. Colonisers often insisted on their language being the language of instruction citing the Indigenous language as being insufficient for teaching Western domains of knowledge. In his discussion of the teaching of mathematics in Botswana, Berry (1985) suggested that even where a mathematical register was developed in the Indigenous language, there could still be a clash between the different underlying cognitive structures of the mathematics register and the Indigenous language. This could result in children failing to learn mathematics sufficiently to enable them to solve problems (Berry, 1985). Schindler and Davison (1985) argued that it is very difficult and perhaps impossible for native speakers of Navajo to construct an exactly parallel systematic analysis of mathematics concepts to that of English. They noted that Navajo does not have a word for multiply, divide, cosine, sine, etc. (Schindler & Davison, 1985). This argument is also applicable to the English language. The words "cosine" and "sine" were borrowed from other languages when the mathematical ideas were introduced to English speakers. Leap (1982) provided an account of his unsuccessful experience in translating an English language problem into Tewa and suggested that such a translation can result in a very practical mathematics problem becoming "re-written as a topic of senseless speculation" (Leap, 1982, p. 30). This may be an issue of translation rather than the inability of the language itself to express mathematical problems. Literal translations rarely work when dealing with complex subject matter.

In many places, the choice of a language of instruction was connected closely to an assimilatory policy, which tried to ensure that Indigenous students discarded their native languages so they could become like the colonisers (Garrett, 1996). In the 1980s, Māori advocated for schooling in *te reo Māori*, partly because the use of English in schooling had been a means of assimilation (Barton, Fairhall, & Trinick, 1998). On the other hand, the decolonisation of countries can result in the reverse situation being advocated. Setati (1998) provided the example from South Africa where English was promoted as the language for Zulu education for the early years, because the promotion of the Zulu language alone was seen as a continuation of apartheid practices.

The development of mathematics registers takes place in a variety of contexts, and the languages in which they are developed can be affected in different ways. The development process that one Indigenous group pursues and the beliefs and ideologies that underpin the process may be very different to those of another group who seem to face similar issues. Some of these differences are connected to peoples' views on how endangered the language is and the sociocultural contexts of each country. For example, Samoans do not see their language as endangered and so do not necessarily have the same concerns about the "authenticity" of their language that Māori do (Trinick, 2011). Different historical legacies affect the political decisions about what is the most appropriate language of instruction for a particular group of people. Another way of saying this is that the different practice architectures, including those of a socio-political nature, discussed in Chapter 1, prefigure the sorts of practices that are adopted (Kemmis, 2009).

Gibbs and Orton (1994) argued that although mathematical registers can be developed in Indigenous languages, they need to be used by people outside the classroom to be effective. This is true of any register that children are required to learn and use during their schooling. One of the key components of language planning, in Aotearoa and in several other countries, is that language revitalisation must also take place in the wider community, including home and work contexts (Spolsky, 2004). The issue of the use of the mathematics register outside of the school context is discussed in Chapter 8.

The development of a mathematics register for *te reo Māori* has involved two distinct sets of challenges; one externally and the other internally driven. These are related to the initial European colonisation of New Zealand and the resurgence of Māori cultural identity, respectively. The ability to exercise control over the development of the mathematics register corresponded to those times when Māori had or took greater control over their interactions with mathematics. Like the warriors in the mock challenge, *wero*, described in Chapter 1, Māori exhibited its cultural strength to ward off the evil intentions that mathematics could inflict as a pathway of assimilation into European culture. However, there is still some debate over whether they have been successful in stemming the anglicisation of *te reo Māori* through the Trojan horse of the mathematics register (Barton & Fairhall, 1995a; McMurchy-Pilkington & Trinick, 2002). This chapter examines the strategies used to develop the mathematics terminology in *te reo Māori* and the range of issues and tensions associated with this process.

Te Wero Nō Waho – The External Challenge

The introduction of Western mathematical ideas through trade and schooling into *te reo Māori* was an external challenge that drove the expansion of the mathematics register. During this period, Māori retained most of the decision making about which terms were transliterated from English and which terms were modified to include a Western mathematical meaning. Adoption of these terms tended to be immediate and enthusiastic, connected as they were to the new contexts that Māori wished to engage in.

By the time the first missionaries and settlers arrived, Māori had a robust system for educating their children to ensure the survival of their communities. In this education process, some children learnt to use the numeration system, which was generally base 10 (Best, 1907). Numerical calculations were primarily done by counting in ones or skip counting in twos, fives, and tens, sometimes even using vigesimal enumeration (counting in twenties). There were also standardised terms for numbers from 1 to 100 with any number above that being labelled as "a great many", or "a multitude". Upon contact, some of these general terms came to signify an exact quantity, such as *rau* (multitude), and *mano* (indefinitely large number), which came to mean "hundred" and "thousand", respectively.

In the early 1820s, New Zealand/Aotearoa was colonised by European whalers, sealers, and missionaries, predominately from Britain. They were followed by others – including farmers, traders, and land speculators – who established European forms of government and schooling. Still, up until the 1870s, *te reo Māori* remained the dominant language for trade and schooling.

Māori rapidly adopted the mercantile system introduced by the Europeans (Trinick, 2011), possibly because it increased significantly the range and variety of goods that they could access. Trade entailed the use of units of money and measurement, and Māori had to decide how to talk about these. Almost without exception, transliterations of English terms were adopted that involved giving a Māori spelling and pronunciation to the English terms (Barton, Fairhall, & Trinick, 1995). Other languages use what could best be described as "partial transliteration". Adu-Ampona (1975) noted that new Swahili words created for mathematics often incorporated the English phonetic forms. Consequently, mathematics can become an unwitting vehicle for transforming the phonology of a language. However, at this time this was not the case for *te reo Māori*.

Surprisingly, the transliterations were remarkably consistent across the whole of New Zealand/Aotearoa. *Pāuna*, *herengi*, *peni* or pounds, shillings, and penny were some of the monetary terms. Measurement terms ranged from *īnihi* (inch) to *pūhara* (bushel) and *eka* (acre). An exception to the general response of transliteration was the term *mārö* or fathom. In both languages, this measurement was based on the distance between the fingertips when the arms were outstretched. Therefore, there was no need for Māori to adopt a new term.

Schooling for Māori, with the intent to "civilise", was an important feature of colonial policy in New Zealand (Simon, 1998). As with the sphere of trade, schooling engendered many transliterations, including *miriona* (million) and *matipikeihana* (multiplication). Nevertheless, there was more use of traditional Māori terms to signify mathematical activity, including the other basic operations of addition (*huihui*), subtraction (*tango*), and division (*wehe*). The decisions, about which terms to use, appear to have been made by Māori. In 1858, Hënare Taratoa put together a mathematics text with exercises in *te reo Māori*, using the common terms of the day. From this text, it is clear that traditional terms were expanded in meaning so that the Western mathematical concepts were connected to them. This foreshadowed the conscious development of a mathematics register over a hundred and twenty years later.

However, the use of *te reo Māori* in schools changed when the passing of the 1867 Native Schools Act decreed that English was to be the only language used in the education of Māori children. This bill may well have been influenced by the recent cessation of hostilities in the Māori land wars. *Te reo Māori* effectively ceased as the language of instruction for mathematics, and a great opportunity was lost to modernise incrementally the mathematics terminology. It is interesting to speculate how this register might have developed if schooling had remained in *te reo Māori*. Regardless, the transliterated terms for money, weights, and measures continued in vernacular Māori, where, paradoxically, mathematics continued to be known as *mahi whika*, figure work, or as *mahi nama*, number work.

Te Wero Nō Roto – The Internal Challenge

For more than a century, it was generally accepted, by Māori and non-Māori alike, that schooling should, if not could, be only in English. However, the situation changed again with the resurgence of *te reo Māori*. By the 1970s, most Māori had become English-only speakers, leaving *te reo Māori* in a perilous state. Consequently, several different individuals and groups established a range of initiatives to revitalise the language (Reedy, 2000). A key component of this revitalisation was the development of schooling in the medium of Māori (see Chapter 3). Of the many outcomes from this decision, one was the need to develop a more elaborate mathematics register in *te reo Māori*. We see this as an internal challenge as by this time Western mathematics was no longer something new to Māori. Instead the decision to use *te reo Māori* as the language of instruction meant that Māori had set up the conditions whereby a mathematics register was necessary. In some ways, Māori could be considered to be the ones issuing the challenge to the New Zealand government, especially the Ministry of Education and the Māori Language Commission.

Figure 2.1 provides a timeline that outlines the main influences on the elaboration of the mathematics register for this period. There were strong connections between each of the various influences, and it has been quite difficult to separate them out to tell a linear story of the development of the mathematics register in *te reo Māori*. The diagram represents how different influences were operating at the same time that then affected later decisions.

Although Māori had started to develop a register to discuss Western mathematics from first contact, the expansion of this register since the late twentieth century involved deliberate language planning. This had not been the case for the elaboration that had occurred in the nineteenth century. Language planning occurs when there are modifications in vocabulary, grammar, or writing and includes the process of standardisation. Language planning for endangered Indigenous languages includes dealing with the very challenges that endangered the language in the first place. Consequently, Hornberger (1996) argued that Indigenous language policy and planning are not just about language or education, but they remain situated in wider social and political contexts.

Deliberate language planning tends to occur when changes are made in a relatively short period of time. The development of the mathematics register in English took many centuries and involved small incremental changes. Therefore, there was no formal language planning policy (Halliday & Martin, 1993). In New Zealand/Aotearoa, bilingual schools and units emerged in the early 1980s in response to the perilous state of the language highlighted by key studies in the 1970s (Benton, 1991). However, in New Zealand school contexts, bilingual education soon became equated with lower levels of immersion (May & Hill, 2005) and/or tokenistic attempts to revitalise Māori language (McMurchy-Pilkington, 2004). The next

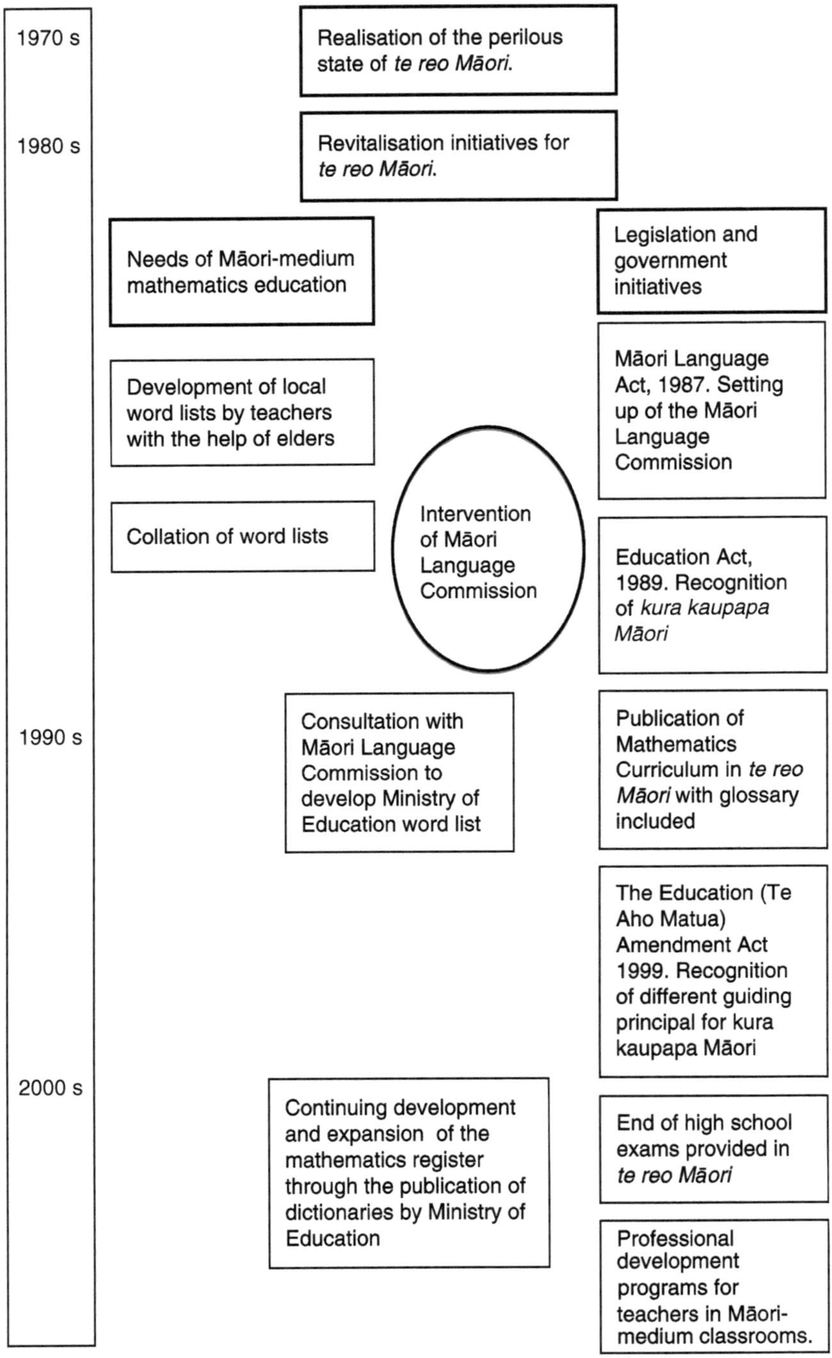

Fig. 2.1 An historical overview of the recent development of the mathematics register

generation of schools developed to address the decline of the Māori language were *kura kaupapa Māori*, such as Te Koutu (see Chapter 3). In the 1980s, *kura kaupapa Māori* were set up as a result of a political movement by Māori parents and the wider community. Since then, Māori mathematics vocabulary has changed from intermittent informal use in primary schools to more formal use in curriculum documents (Barton et al., 1995). Thus, Māori-medium mathematics underwent tremendous modifications with a simultaneous expansion of the specialist mathematics register in a very short space of time. In recent times, this expansion has begun to focus on the standardisation of the Māori mathematics register. The rapidity of these changes has resulted in deliberate language planning through the intervention of the Māori Language Commission.

Te reo Māori is not alone in being the subject of language policy. In several African countries in the 1970s and 1980s, and more recently in South Africa (Schäfer, 2010), attempts were made to expand Indigenous languages so that they could become the language of instruction for teaching mathematics. Clarkson (1991) argued that such attempts did not seem to be driven by research, nor comprehensive theory, but rather were attempts to throw off the colonial yoke. For groups concerned with nationalistic ideals, the use of the colonial language was to be resisted. However, ideology alone does not ensure acceptance of a change in the language of instruction, and in some cases terminology development has met with considerable resistance from the local community (Onyango, 2005). The development of the mathematics register in *te reo Māori* has been one of the few success stories about using Indigenous languages as a language of instruction.

The beginning of this new era of expansion of the mathematics was not, however, part of a formal language planning policy. In many language development situations, the development of new vocabulary is incidental (Spolsky, 2004), and this was the case for mathematical vocabulary in *te reo Māori* in the 1980s. Much of the early creation of *te reo Māori* mathematics terms was done by teachers, working in Māori-immersion situations, with some support from local *kaumātua*, or elders. This resulted in an ad hoc coining of words, using whatever means was at hand (Barton et al., 1998). A profusion of new words was created, many of which were known only in the locality of origin. At this time, the development of the Māori-medium mathematics terminology was a "bottom-up" development based on the belief by the teachers that *te reo Māori* had the resources for discussing mathematics at any level. Many of these teachers were second-language learners of Māori who were caught up in the fervour to save the language. Hornberger (1996) argued that revitalisation can only truly succeed if the community of users have significant involvement in the development. This is in contrast to when decisions are made about the introduction of new terms from what Kaplan (1989) refers to as "top-down" language planning situations. In these situations, the people with power and authority make language-related decisions for groups, often with little or no consultation (Kaplan & Baldauf, 1997).

In their discussion of the creation of new words for the Hawaiian language, Kimura and April (2009) argued that where an Indigenous language needs to be revived, new words are more likely to emerge from a culturally and politicly conscientious group of proactive second-language, Indigenous speakers. This view is partly in response to the difficulty older native speakers have in creating new words without some expertise in the curriculum areas, such as mathematics. The development of the mathematics register for *te reo Māori,* both at this stage and after the intervention of the Māori Language Commission, arose because of the push by teachers. Wherever possible, older native speakers were consulted. Although on the whole this was a successful collaboration, there were tensions, and these are discussed in a later section.

The growth of *kura kaupapa Māori* was affected by three significant legislative decisions: the Māori Language Act 1987, the Education Act 1989, and The Education (Te Aho Matua) Amendment Act 1999 (see Chapter 11 for discussion on Te Aho Matua). The first, the Māori Language Act, declared *te reo Māori* to be an official language of New Zealand. It is interesting to note that New Zealand Sign Language is the only other official language, with English having no official status despite its pre-eminence as the language for communication. This Act was also the direct catalyst for the establishment of the Māori Language Commission, or Te Taura Whiri. The Education Act 1989, which brought in a range of changes connected to *Tomorrow's Schools* (Lange, 1988), legitimised the status of *kura kaupapa Māori* and enabled *kura* to receive state funding. This included and supported the establishment of Te Koutu (see Chapter 3). The Education (Te Aho Matua) Amendment Act 1999 legislated that Te Aho Matua would provide the guiding principles for *kura kaupapa Māori*. These acts all indirectly supported the expansion of the mathematics terms in *te reo Māori*.

The publication of collected mathematics word lists from various regions, such as the one by Barton and Cleave (1989), raised the issue of language change on a national scale. At this point, with huge needs in all subject areas, the Māori Language Commission became involved in the process of enlarging the register (Harlow, 1993) as part of deliberate language planning. The Commission expressed concern that some words in the mathematics word lists were clearly unsuitable and broke their guidelines for good practice (Barton et al., 1995). It decided to try to standardise one list for use in government publications and as a reference point for teachers in an ongoing process of development (Barton et al., 1995). With the combined efforts of the Māori Language Commission and some mathematics teachers, including Tony Trinick and Uenuku Fairhall, the standardisation process was accelerated. The corpus of terms was considerably extended by the development of the Māori-medium curriculum in the 1990s (Trinick, 1994).

The Process of Expanding the Mathematics Register in *Te Reo Māori*

In considering how *te reo Māori* should be expanded to discuss Western mathematics, the Māori Language Commission based its work on international language planning research that suggests that there are five primary areas that account for language health – language usage, status, acquisition, corpus, and awareness. These five areas are "interdependent" (Cooper, 1989). The Commission's goal is to ensure that regeneration efforts are coordinated in each of the five areas that satisfy language health. Consequently, the Commission has supported the development of a wide range of new words especially when school curricula, including mathematics, started to be written in *te reo Māori* (Barton & Fairhall, 1995b). Mathematics is a high status subject in Aotearoa and has been one of the priority learning areas of the Ministry of Education for many years. Consequently, in relation to Māori-medium education, the discipline has received more support and resource in comparison to other learning areas of the school curriculum to develop terminology and the associated register.

As part of language planning, the deliberate expansion of *te reo Māori* to include new terms and grammatical expressions that facilitate the discussion of mathematics is part of corpus planning. Corpus planning focuses on changes by deliberately planning the actual corpus or shape of a language. This may be achieved by creating new words or expressions, modifying old ones, or selecting among alternative forms (Kaplan & Baldauf, 1997). Corpus planning aims to develop the resources of a language so that it becomes an appropriate medium of communication for modern topics and forms of discourse, equipped with the terminology needed for use in administration, education, and so on. Corpus planning is often related to the standardisation of a language, involving the preparation of a normative orthography, grammar, and dictionary for the guidance of writers and speakers in a speech community (Clyne, 1997).

A challenge and constant debate in revitalisation programmes for Indigenous languages concerns the technical principles of corpus planning (Cobarrubias & Fishman, 1983). Should words be borrowed, or should they be created through more natural language development processes such as compounding or descriptive phrases? One of the principles underpinning the Māori Language Commission's development of new vocabulary for Māori-medium education was that new words must be standardised and not loan words from English, and, wherever possible, they also should be short and transparent (Harlow, 1993).

This was a different approach to what had occurred elsewhere. Other than Māori and perhaps Hawaiian, the predominant strategy used in the Pacific, particularly with Polynesian languages, has been to transliterate and/or modify phonologically the existing English language mathematical terms to fit the native vernacular (Begg, 1991). The countries that have followed this approach include Samoa, Tonga, Niue, Rarotonga, and French Polynesia (Begg, 1991). Many languages make use of an international terminology in areas like mathematics. That does not mean that all words are identical, rather they are borrowed from the same source, usually Latin or Greek, and adapted to the sound system of the language concerned. It can be said that the language of modern mathematics is part of the continuum of technical language development that began in Europe in the seventeenth century. Modern mathematics drew heavily on the ancient language stocks in Europe, Asia Minor, and North Africa and represents the cumulative technical language development of diverse peoples over thousands of years (Closs, 1977). The evolution of mathematics is also the evolution of the grammatical resources of the natural languages by which

Western mathematics came to be construed (Halliday, 1993). Consequently, English mathematical notation and terminology have assimilated symbols and terms from many different languages, alphabets, and fonts.

The lexicon committee for Hawaiian used the argument that technical scientific terms are international terms, not specifically English, but mostly developed from Greek or Latin. Subsequently, it was suggested that a great deal of the Hawaiian technical vocabulary consists of "Hawaiianised" borrowings, with the technical terms being adapted to the Hawaiian sound system (Hinton & Hale, 2001).

Nevertheless, Māori did not follow these trends and so did not borrow from English by using transliterations as had been done when first contact was made with Western mathematics. In the nineteenth century, Māori who coined and used the transliterations for weights and measures did not view their language as being endangered. However, by the 1980s the danger facing the language meant that its regeneration was the primary consideration. The change in circumstances for the language influenced the decisions about how the language should be elaborated.

Fishman (2006) noted that corpus planning inevitably becomes an expression of the societal goals, ideologies, and aspirations of the societies and cultures that support it. Language ideology is what people think should be done (Spolsky, 2004). An example is when language policy issues concerning innovation are regularly decided on purist or political rather than pragmatic grounds (Spolsky, 2004). Purism becomes important during a time of language cultivation and modernisation, providing a criterion for the choice of new terms and codification. It often takes the form of removing from the language elements (usually vocabulary) that appear to be foreign or lacking in true authenticity in the linguistic culture in question (Annamalai, 1979). Harlow (1993) highlighted the "puristic" ideologies that have underpinned the development of the science and mathematics terminology for *te reo Māori* as initiated by the Māori Language Commission. Harlow argued that the puristic ideologies were more to do with the status of *te reo Māori* and people's attitudes to its status:

> To preserve the language as a living means of communication entails preserving it in opposition to and distinct from English. If in order to fit Māori for the Māori world, we borrow from English, this looks like a sort of admission of defeat, an admission that in fact Māori is not capable of handling new ideas and topics with its own resources. (Harlow, 1993, p. 129)

Barton et al. (1998) noted that very early in the process of vocabulary development, it was realised that sets of mathematical terms were often structured groups of interrelated words. Thus, a decision was made to purge *te reo Māori* of the nineteenth-century, transliterated term *nama*, for number, and adopt the term *tau* instead. Thus, all words related to number could be prefixed with *tau*. For example, multiple became *taurea*, (*tau* – number, *rea* – multiply) and *taurua* (*tau* – number, *rua* – second) for even numbers. It was important to use this structure so that the vocabulary reflected the subject matter (Barton et al., 1998). Berry (1985) outlined the problems that occurred when the linguistic structures introduced into an Indigenous language did not represent the conceptual structures of mathematics. By connecting related words through a common root, the newly coined words also provided a linguistic clue to their meanings and fulfilled the Māori Language Commission's rule of making new terms transparent in meaning.

The various techniques used for creating new terminology for Māori-medium mathematics is described in Barton et al. (1998) and Harlow (1993). These include the following:

- Derivation by affixation. This is usually a prefix or suffix – that can be attached to a base, stem, or root to form a new word. One of these is the causative prefix *whaka*, which is used extensively in Māori-medium mathematics. *Rūnā* traditionally meant to "pare down" or "reduce". By adding the prefix *whaka*, it becomes *whakarūnā* – to "simplify" – as in "simplify 4/12".
- Gerunding. This occurs when a noun is formed from a verb or vice versa. *Panoni* was a transitive verb meaning to "change", and it has been transformed to a noun to mean "transformation" as in geometric transformation. Traditionally, gerunding was a common practice in *te reo Māori*, but in the creation of the new mathematical terms, this strategy was used to create nouns that were traditionally only used as verbs or vice versa. This strategy, therefore, has implications for how the language as a whole is being changed.
- Calquing. This is when words are created from the translations of the common or original meaning of the mathematical word used in English. For example, the word "chord" is a straight line joining two points on the circumference of a circle and comes from the Greek word (chordê) for a piece of animal gut used as a string. The Māori word for string is *aho*. So the Māori word for a mathematical chord is *aho*.
- Compounding. This is to form new words by combining or putting together old words, for example, *hangarite* as a modern maths term to mean symmetrical. *Hanga* can mean "shape" and *rite* "alike" or "corresponding".
- Resurrection of old words, sometimes with slightly modified meanings. For example *ine* was traditionally a Māori word for "measure". However, it had fallen out of use and was replaced by the transliteration *meha* (measure). *Ine* has now been reinstated through its use in Māori-medium mathematics classrooms. Another example is *tuaka* which had the traditional meaning of "midrib of a leaf" or the "line that runs down the middle joining two mats". In the development of the *te reo Māori* mathematics terms, *tuaka* became the term for an "axis" of a graph.
- Circumlocution. This is where the mathematics term created is an explanation rather than just a single word. Sometimes, it is linguistically too difficult to create a single word to represent a concept. For example *tau e whakareatia ana* (the number that is multiplied) is the Māori mathematics expression for "multiplicand".
- Metaphors. This strategy is closely related to calquing and several of the other strategies described earlier. For example, *kauwhata* is now the word for "graph". A *kauwhata* traditionally was a stage or frame for hanging fish to dry. It now means a frame to "hang" data on. The term for parabola in Māori is *unahi,* whose traditional meaning is fish scale, which is the same shape as a parabola.

> A metaphor suggests a comparison by referring to or describing one thing as something else and was used routinely in everyday and formal Māori oral language. A metaphor paints a picture and is used to state or describe an idea. Metaphors serve as a shorthand description and are also routinely used in mathematics. It was anticipated that students could use the everyday meaning of the terms to support their learning of the mathematical meaning.

The initial involvement of the Māori Language Commission resulted in a set of 600 terms and some grammatical structures (Barton et al., 1995). These were published first as *Nga Kupu Tikanga Pangarau: Mathematics Vocabulary* (Learning Media, 1991), and then an expanded version was included as a glossary at the back of the new Pāngarau (mathematics) curriculum document (Ministry of Education, 1996). A dictionary of mathematical terms was published in 2004 and made available to all schools (Ministry of Education, 2004). There have been several iterations of this mathematics dictionary for use in Māori-medium mathematics education (see Ministry of Education, 2008a, 2008b, 2010). Each iteration has expanded the range of terms and, in some limited cases, resulted in changes. For example, "finite" has been slightly changed from *whai mutunga* to *mutunga*. This is a small grammatical change. Where there has been significant terminology change is with the process terms, such as "define", "identify", "explore", "discuss", and so on. These are cross-curricula terms and initially each curriculum area developed its own terms. When a new curriculum was produced for all subjects, there was a need to ensure that these terms became standardised (Ministry of Education, 2008a, 2008b).

The catalysts for the iterations have been the need to teach mathematics to higher levels, to support the introduction of professional learning initiatives, and to support the introduction of new learning and teaching ideas and theories. For example, many Māori-medium schools including Te Koutu have been involved in the Māori-medium numeracy project: Te Poutama Tau (Trinick & Stevenson, 2007, 2008). The primary aim of the Poutama Tau project is to improve student performance in mathematics through improving the professional capability of teachers. Te Poutama Tau is based upon the number framework developed for New Zealand schools (Ministry of Education, 2006). Discussions about the concepts on the number framework required further development of the mathematics register including the development of a number of mathematical terms. These terms include: "count on" (*tatau ake*), "skip count" (*tatau māwhitiwhiti*), "one-to-one counting" (*tatau pāngatahi*), "inverse operation and place-value partitioning" (*te paheko tau kōaro me te wāwāhi uara tū*), and many more (Ministry of Education, 2010). Using the strategies described earlier, the development of new terminology will need to be continuous if Māori-medium mathematics is to evolve and expand to meet the challenges ahead.

Halliday (1978) argues that the introduction of new vocabulary is not the only aspect of the development of a register. Registers such as those of mathematics also involve new styles of meaning, ways of developing an argument, and of combining existing features into new combinations. Halliday (1978) notes that for the most part, this development of the English mathematics register was not through the creation of entirely new structures but through adapting and elaborating the existing ones. This development has taken place slowly, by more or less natural processes. English has taken at least three or four hundred years to develop its register of mathematics, and it is still developing, for example, the recent addition of stem and leaf graphs to the primary school curriculum for New Zealand schools.

One of the ways that meaning is constructed in mathematics in English is through logical connectives, conjunctions that join ideas together in a logical manner (Meaney, 2007). Examples of these words are "therefore", "because", "so", and "if . . . then". These relationships can be sequential (time), or about reason and purpose, and are heavily embedded in Western mathematics (see Durkin & Shire, 1991 for discussion on the key role of logical connectors in mathematics). A structural issue for many Indigenous languages is the absence of logical connectors. Morris (1975) suggested that in a number of African countries the developing of the teaching language did not seem to cater for logical connectors like "if". In her study of mathematical education and Aboriginal children in Australia, Graham (1988) argued that many languages spoken by children in developing countries lack the mathematics register – both the vocabulary and the logical connectives. In these cases, the introduction of a mathematics register is much harder. Fortunately for *te reo Māori*, there were several logical connectives in traditional *te reo Māori* that could be used in mathematical discussion, and these are discussed in Chapter 5.

The Standardising Process

Since the first compilation of the word lists at the end of the 1980s, standardisation of the mathematics terms has been an aim of the developers of the mathematics register. However, the limited number of teachers who continually discuss mathematical ideas in *te reo Māori* has affected the process of standardisation of the mathematics register terms. These teachers are spread over a large area of New Zealand/Aotearoa and have few opportunities to meet together. As a result, few terms have become solidly integrated into the lexicon with many schools and even individual teachers using local expressions. Words that are used less frequently, such as *whenu* (cosine), are more likely to be standardised, whereas more frequently used words such as multiplication are less likely to be. In the 1980s, *whakarau* was initially coined for multiplication. In recent years, there has been a shift to using *whakarea*, but some teachers have resisted this change. Both the root words *rea* and *rau* mean multiply (in the everyday sense). However, *rau* is also the Māori word for a hundred. Thus *rea* has the advantage that when compounded with *tau* to create *taurea* (multiple), it would not be confused with *taurau*, which means a hundred number rather than multiple. Uenuku Fairhall has been noted as a teacher who had rejected, with good rationales, some of the terms used in Ministry of Education documents (Christensen, Trinick, & Keegan, 2003). A number of older *te reo Māori* speakers are also continuing to use the transliterations that were purged during the standardisation process. These include terms such as *nama* (number), *whika* (figure), *kaute* (count), and so on.

When the first mathematics curriculum written in *te reo Māori* (Ministry of Education, 1996) was released, it included a glossary of terms approved by the Māori Language Commission. However in 2002, Fairhall and Keegan expressed the problems faced by teachers in adapting to this new vocabulary:

> Some of the translations in the Pāngarau Curriculum are not always clear, nor easy to decipher. This is compounded by the use of technical pāngarau language containing a large number of neologisms, many of which have not appeared in other sources. The Pāngarau Curriculum statement advocates some changes in existing pāngarau terminology. As far as it is known the new pāngarau terminology has not yet been adopted by all pāngarau teachers. (p. 2)

The standardisation process takes place amidst conflicting interests prevalent in the social context (Jernudd & Neustupny, 1987; Cooper, 1989). In Aotearoa, there is a tension between revitalising the tribal dialects and the standardisation of terms for all *te reo Māori* users. Some tribal groups have called for the development of mathematics terminology to reflect their own tribal dialect. One of the teachers at Te Koutu expressed it this way:

> That's the thing that everyone needs to look at, is having their own expertise, people from their own *iwi* [tribal grouping] writing the things, not someone else. That's what I'd like, you know, that's what I think is the best thing to do. By going that standardised way, what's going to happen to the *iwi* language? It's going to just end up one plain, immersed, mainstream immersed – *te whakaaro* [afterthought]. (Year 0 Teacher, Meeting November, 2008).

This has implications of course for the development of curriculum resources and the issue of national examinations (see Chapter 4). Although not talking explicitly about mathematical terms, Lozare (1993) argued to the Mohawk standardisation committee that standardisation does not mean the elimination of dialects in favour of a new literary form. Dialects can be preserved in the family and in the community of speakers. However, this is a tension, which is still being resolved in relationship to teaching in Māori-medium education.

Challenges to *Te Reo Māori* from Developing the Mathematics Register

As noted, for the most part, the development of the mathematics register in English was not through the creation of entirely new structures but by adapting and elaborating existing ones. This development has taken place slowly, by a more or less natural process. In contrast, the development of the Māori-medium mathematics register has been part of a deliberate planning process and has occurred at a very rapid pace creating several challenges. Some of these challenges have resulted in losses to traditional ways of discussing mathematical ideas, whilst others have contributed to changes to the semantic structures of the traditional language.

The standardisation process connected to the importation of Western mathematical ideas into an Indigenous language is likely to have an impact on how traditional mathematical ideas are discussed. For example, Lipka (1994) observed that one of the challenges in modernising school mathematics in the Yup'ik communities in southwest Alaska was the traditional use of different words for the same numbers. Similarly, Denny (1986), in discussing the development of a mathematics register for Ojibway (Michigan-USA), noted the context-specific nature of terms. In *te reo Māori*, numerical prefixes are context specific and/or determined by a person's relationship with the objects that are being quantified. For instance, a mountain can be quantified using the same descriptors as those for people only when there is a spiritual relationship between the speaker and the mountain. This relationship is more important than the quantities of the item being discussed. Changing the language, which is used to describe the experience, whether it is a change in natural language, such as English to *te reo Māori*, or a change in register within a language, will influence how that experience is described and therefore what is valued in this experience. Older speakers have expressed their concern about the possible loss or change to traditions and values if the language is changed (Barton et al., 1995). If the language is used to express European cultures and ideas, then perhaps it has lost the traditional Māori values that it was thought to contain.

The development of the mathematics register in an Indigenous language will cause changes in that language as well as the culture. Lipka (1994) noted that the ethnographic records of traditional Yup'ik mathematical activities and contexts differed from contemporary activities. Hurrell (1981) pointed out the difficulty of translating English terms into Sesotho because in that language there is no distinction between nouns and adjectives. Therefore, to develop a modern mathematics register would require language change in Sesotho. For *te reo Māori*, Barton et al. (1998) highlighted that during the early terminology development process, several cultural and linguistic rules were stretched. They provided the example of the development of terms for the mathematics concepts of positive and negative in cases such as numbers. The use of adjectives such as "positive" and "negative" in mathematics is quite distinct from their everyday use. The development of their equivalent terms in *te reo Māori* went through a number of iterations. The original suggestions to use *āe/kao* (yes/no), and the similar *matau/mauī* (right/left), never gained currency. One of the reasons was the distance in meaning from the mathematical idea. Over the years several changes occurred resulting in an entirely new concept: *tau ake/tau iho*. The words *ake/iho* are adverbs that indicate direction (usually up/down) or intensify nouns, adjectives, and pronouns (superior/inferior). In using them as adjectives, a grammatical rule was broken. This linguistic corruption was recognised at the time, but was approved. However, subsequent discussion between the Māori Language Commission and the *te reo Māori* mathematics terminology development team highlighted that this corruption from the mathematics context was being transferred to everyday discourse. In the everyday language of the playground, students were using *ake* in the phrase *kōrero ake* meaning positive talk or positive feeling (Barton et al., 1998). As the terminology development continues, it is quite likely that expansion of the cultural and linguistic rules will continue as well. It may not be possible to limit the changes simply to the mathematics register.

The vast output of new words caused some concern, particularly from the older speakers of Māori in the community (Reedy, 2000). They are often unaware of the rationale that lies behind the creation of new words or the rationale for standardisation. The problem lies in the disparity between their understanding of the traditional use of the word and its new mathematical meaning. Sometimes the jump between the original meaning of the word and its mathematics meaning is too great. For example, words that are changed from verbs to nouns can involve too big a shift in meaning for native speakers. *Panoni* was traditionally the verb (to) "change" in conversational language. In the modernised mathematics register, *panoni* has been changed to a noun to mean a transformation (geometric). As noted by Reedy (2000), Māori frequently turn verbs into nouns, but the semantic shift in this case is great and, therefore, jarring for native speakers.

Sometimes, the development of new terms and expressions did not achieve the clarity and efficiency of meaning that was wanted. This problem is manifested when schooling is organised around year groups. For example, children aged 5–13 are schooled in *kura*. The older children (aged 13–18) are grouped in *wharekura*. Students who transition from *kura* (primary or grade 1–8) to *wharekura* (secondary or year 9–13) find some terms used with a slightly different or expanded meaning. For example, *tau whakawehe* is used to mean the "divisor" in *kura*. However, in *wharekura*, *whakatauwehe* is used to describe factorisation. Traditionally, the meaning of *wehe* is to "pull apart", and this is an appropriate connotation for division. Although factorising can be considered a kind of division (finding the factors), it in fact does not alter the original expression: $2xy - 4y$ is the same as $2y(x - 2)$, even if it looks different. On the other hand, division alters the original amount. The similarity of these terms can be very problematic for students in *wharekura* because the *kura* use provides an inappropriate connotation.

The development of the mathematics register in *te reo Māori* could not achieve both outcomes of appropriately presenting the ideas mathematically as well as keeping the language strong in its traditional forms. All languages change to meet new demands as cultural activities or practices adapt to new situations both culturally and physically. However, these changes tend to be incremental with no formal decision making by agencies such as governments. When large numbers of changes need to be made in very short periods of time, then there is no time for a large number of people to trial out the new terms in order to feel whether they are appropriate. The consequences are that there will be changes which will be regretted by later generations, but perhaps not so much as would have been the case if *te reo Māori* had been allowed to die out.

Meeting Challenges

Developing an Indigenous mathematics register is a challenging proposition when it tries to both support students' learning of mathematics through transparency of meanings and keep the traditional language strong. At various historical times, Māori has responded to this challenge in different ways. In the nineteenth century when the language was the first language of most Māori, transliterations were an acceptable way to enlarge the mathematics register. In the twentieth century, however, the language was in serious decline, and transliterations became unacceptable. This was not the only difference. In the nineteenth century, the development occurred naturally as Māori came in contact with Western mathematics and began to make use of it in trade situations. Schooling in *te reo Māori* also resulted in new terms, but at this time mathematics was arithmetic and did not need the large number of terms required for the different topics taught in the twentieth century. When the first mathematics curriculum was written in *te reo Māori*, the mathematics register had needed to expand in an exponential manner in a very short time. The initiative for this development came from the teachers working in Māori-medium classrooms, who initially had invented their own terms in their individual classrooms. They were helped at first by elders, and then by the Māori Language Commission. Early on, teachers recognised that there was a need to standardise the terms, but this process is the one that has been the hardest to achieve. Although the introduction of new terms and the adjustment of other terms is ongoing, this process has slowed in the last few years.

Legislation also had an impact on this development. In the nineteenth century, the Native Schools Act effectively put a stop to the development of the mathematics register in *te reo Māori*. However, it was the Māori Language Act of 1987 which provided the possibilities for the Māori Language Commission to work with teachers to more rigorously ensure that the new words supported keeping the language strong and making the mathematics clearer.

Once the decision was made by the Māori community to revitalise their language through Māori-medium education, it was essential for the challenge of expanding the mathematics register to be met. In many ways, the norms about whether, or if, *te reo Māori* should be used in teaching mathematics were changed in the twentieth century. Although transliterations were purged from the mathematics register, they provided spaces for opening up opportunities to discuss Western mathematical ideas in *te reo Māori* in the nineteenth century. Developing a mathematics register in *te reo Māori* has been challenging, and the continuing modifications that are occurring concurrently with the standardisation process show that it remains an ongoing challenge, although it is perhaps of a less intense nature than it was in the 1980s. The process has been very much a political one with legislation playing a major role in the legitimisation of *te reo Māori* as a language of instruction. The push for legislation in the 1980s came from Māori as a way of using the governmental process to ensure that their aims for revitalising the Māori language were realised.

Chapter 3
The History of Te Kura Kaupapa Māori o Te Koutu – The Politicisation of a Local Community

E tipu e rea, mō ngā rā o tōu ao;
Ko tō ringa ki ngā rākau a te Pākehā,
Hei ara mō tō tinana,
Ko tō ngakau ngā taonga a ō tipuna Māori
Hei tikitiki mō tō mahunga.
Grow up, oh youth, and fulfil the needs of your generation –
Making use of Pākehā skills for your material well-being,
But cherishing with pride
Your Māori cultural heritage.
Sir Apirana Ngata

Te Kura Kaupapa Māori o Te Koutu was set up in the early 1990s to provide an education in *te reo Māori* for Māori children who had graduated from local *kohanga reo*, literally language nests, or preschools where children were immersed in *te reo Māori*. The story of this school is one of a grassroots initiative in which a community engaged in direct political action, to provide the education that they viewed as important for their children. The need for such action came from a collective realisation that leaving their children in mainstream, English-medium education was likely to contribute to the demise of *te reo Māori*, the loss of connection to the children's Māori culture and academic underachievement. At each step in the establishment of the school, there were challenges. As the composition of the parent group changed and the survival of the school was no longer precarious, some of these challenges have had to be revisited. The story is primarily about the parents, some of whom became teachers or worked in other capacities at the school. Consequently, Uenuku Fairhall's narrative about the setting up of the school is interspersed with extracts from interviews with parents. Within this story, we bring to the fore the notion that mathematics learning was seen not only as a necessary set of knowledge and skills for living in the *Pākehā* (non-Māori) world, but also as a vehicle for reviving the Māori language, traditions, and customs.

Many Indigenous communities perceive a strong link between self-determination and education of their young people. The wish for Māori to use the schooling system for their own purposes was part of this worldwide movement. In the USA in the early 1970s, the Indian Self-Development and Educational Assistance

T. Meaney et al., *Collaborating to Meet Language Challenges in Indigenous Mathematics Classrooms*, Mathematics Education Library 52,
DOI 10.1007/978-94-007-1994-1_3,

Act emphasised this link between education and self-development (Apthorp, D'Amato, & Richardson, 2003). Nonetheless, self-determination was not always focused on academic success. In reviewing a 1989 case study on an American Indian school situated on Navajo land, Apthorp et al. (2003) wrote the following:

> The school's heralded outcomes in the late 1960s were not about academic success, but rather, about improved economic vitality, Navajo pride, and self-governance in the community. School jobs at Rough Rock doubled the local per-capita income. School board members acquired leadership and administrative skills for operating federal programs, which led to new facilities and roads, opening the door to the wider community. (McCarty, 1989, p. 5)

Schooling per se will not necessarily enable children to contribute to their communities achieving self-determination. An Indigenous community "may well recognize that schooling provides the skills necessary to survive in a technological world, but it will also blame the school for alienating students from their home culture, whether deliberately or unintentionally" (Cantoni, 1991, p. 34). Therefore, there is a need to specifically describe how Indigenous culture should be situated within an education system and thus contribute to an Indigenous community's self-determination. The framework for Māori educational advancement advocated by Durie (2003) is based on three fundamental goals or touchstones. These are: to "live as Māori", to "being Māori", and for Māori to actively participate as "citizens of the world". This final goal does not contradict the first one; rather it ensures that Māori children can fulfil the needs of their generation by moving effortlessly between the different spheres of their lives secure in their Māori culture. A strong education is considered a key strategy for easing these transitions.

The use of education to reinforce particular communities' values and beliefs has been an integral aim for education systems throughout the world. However, this role of schooling is perhaps more important when the community is a minority, with significant differences to the mainstream societies who generally established and ran the education system. Dewalt and Troxell (1989) documented how the Old Order Mennonite communities controlled their school curriculum in order to ensure that their children were socialised into the group's "cultural values and ethnic identity" (p. 311). As a consequence of this process, they "maximize[d] the group's economic independence and their resistance to mainstream life-styles and values" (p. 311). Students did arithmetic rather than mathematics and used textbooks published in the 1930s, which were "predominantly mathematical problems with no illustrations and minimal explanation" (p. 315). Children finished their schooling at the eighth grade and were then expected to farm or to keep house. With this schooling, children were unlikely to be able to function in the wider society. Dewalt and Troxell (1989) suggested that "[m]ost ethnolinguistic minority groups attempting to resist mainstream acculturation fail because, in contrast to the case of the Old Order Mennonite . . ., they have not been able to retain economic self-sufficiency, residential independence, and complete control of their own schools" (p. 308). In addition, the Old Order Mennonite communities were successful because their aim was to isolate themselves from the mainstream culture, and schooling was used to reinforce this. Māori have the opposite goal, that of ensuring that their children can operate in the wider society from a solid foundation of their Māori cultural heritage, especially the language.

By the end of the 1970s, Māori were extremely concerned about the possible loss of their language (Rutene, Candler, & Watson, 2003). The causes for this loss were many and complex. Benton (1996) conducted intensive research in the 1970s on the use of *te reo Māori* in a variety of settings and found

> the century-long exclusion of Māori language from formal education gave rise to a disjunction between the language and new technology and new scientific and social concepts, all of which tended to be transmitted and discussed mainly or even exclusively in English. As the language became less associated with intellectual, technological and philosophical discourse, even 'Māori' knowledge (traditional history, cosmology and genealogical information, for example) became the preserve of those few families which could provide a parallel education in Māori outside the school. Even these, however, had no viable way of adapting the language to the demands of modern life, in an environment in which modernization had become synonymous with Anglicization. (p. 168)

On the other hand, Spolsky (2003) saw the banning of *te reo Māori* at school, since the nineteenth century, as only one factor which combined to bring *te reo Māori* to the brink of extinction. At the same time as the decline in the number of speakers was being noted, academic results for Māori students were acknowledged as being well below that of their *Pākehā* peers.

> [I]in mainstream schools, on almost all measures of educational achievement, the average achievement of Māori students is lower than that of non-Māori students. The disparity in achievement between Māori and non-Māori has been a feature of the Aotearoa New Zealand education system for more than a hundred years. (Rutene et al., 2003, p. 6)

The imminent loss of language combined with a growing anger at poor academic results galvanised Māori to re-evaluate education and schooling. They wanted to investigate how it could be used to support rather than hinder their children's ability to gain what Sir Apirana Ngata outlined as their right in the proverb that opened this chapter. In deciding to, Graham Hingangaroa Smith (2003b) described the impetus for Māori to set up their own Māori-immersion schooling system in the following way:

> A common catch-cry that was used as a justification was that *'we can't do any worse than the system is currently doing – there is only one way to go – upwards'*. (p. 9)

When *kōhanga reo* and *kura kaupapa Māori* emerged in the early and middle 1980s, the Ministry of Education seemed to ignore them, possibly believing them to be a short-lived phenomena. When Māori-immersion education did not falter and disappear, the Ministry remained unsure what to do, and it was not until 1990 that *kura kaupapa Māori* could apply for state funding (Rutene et al., 2003). As the discussion in Chapter 4 on the provision of bilingual examinations reveals, the Ministry of Education continues to be uncertain how to ensure the success of *kura kaupapa Māori.*

Schools that began to teach in *te reo Māori* did so as a result of a grassroots movement rather than as a result of a Ministry of Education initiative. Although Māori-medium education emerged partly from the "bilingual schools movement", bilingual units are excluded in the definition of Māori-medium education. As observed by May and Hill (2005), in other countries, immersion education is regarded as one form of bilingual education and/or located on a continuum. As noted by Hornberger (2002), the Māori-only ideology in schools like *kura kaupapa Māori* is of such integral and foundational importance that the use of two languages, as is suggested by the term *bilingual*, is antithetical to those dedicated to Māori revitalisation. For the purpose of this book, Māori-medium education refers to those schools and immersion units that teach in the medium of Māori 81–100% of the time.

In this chapter, we document the history of Te Kura Kaupapa Māori o Te Koutu (Te Koutu) to illustrate the complexity of influences on its establishment and growth. Within this historical account of the school, we focus on the parents' perspectives, including why they wanted their children to attend Te Koutu. We use interview and meeting data from 1998 to 1999 when the school had been operating for only a few years, and interviews from a decade later. The original data came from a wider study looking at developing a culturally appropriate mathematics curriculum (Meaney, 2001). The second set of data came from interviews, which specifically asked parents about their reasons for sending their children to Te Koutu. Information about the history of Te Koutu comes primarily from Uenuku Fairhall who was involved in the school from its beginnings, first as Chairperson of the Board of Trustees and then as Principal from 1998.

The History of Te Koutu

Te Koutu was one of the schools that formed as a result of the grassroots initiative to revitalise Māori culture, language, identity, and self-determination. In 1992 in the regional town in which they lived, a group of parents whose children attended *kōhanga reo* began meeting. They were fearful that the children would lose fluency in *te reo Māori* if they were not able to continue in Māori-immersion education. They also felt that their older children were not being challenged in the mainstream classrooms. Their Māori language was already strong, and the educational programmes were not designed to match the children's specific needs. The parents wanted the language and cultural experience to be more profound than what they saw their older children receiving in the mainstream, English-medium schools. There was also a sense that, rather than just complain about the situation, they should become more actively involved to ensure that at least their younger children gained the education that they wanted them to have. Smith (2003b) described the shift in Māori parents' conceptions of what they could do in the following way:

> The 'real' revolution of the 1980's was a shift in mindset of large numbers of Māori people – a shift away from waiting for things to be done to them, to doing things for themselves; a shift away from an emphasis on reactive politics to and an emphasis on being more proactive; a shift from negative motivation to positive motivation. These shifts can be described as a move away from talking simplistically about 'de-colonization' (which puts the colonizer at the center of attention) to talking about 'conscientization' or 'consciousness-raising' (which

puts Māori at the center). These ways of thinking illustrate a reawakening of the Māori imagination that had been stifled and diminished by colonization processes. (pp. 1–2)

Like other Māori parents (Reedy, 2000), those who set up Te Koutu desired an education for their children, which built on the language and cultural experiences from *kōhanga reo* as well as providing a strong academic foundation. Many of their younger children were close to school age, and although there was a *kura kaupapa Māori* close by, that school community wanted to remain small in numbers and enrolled only a very few children each year. Many *kura kaupapa Māori* felt that in staying small, they were able to ensure there was as much *whānau*, or family, involvement as possible (Rutene et al., 2003). Each child could be recognised as an individual whilst building relationships (Bull, Brooking, & Campbell, 2008). The involvement of family was seen as one of the key principles for *kura kaupapa Māori* that was likely to contribute to children meeting parental aspirations. When Uenuku was in his first year of being principal at Te Koutu, he suggested the following:

> You are not going to get a lot of support from parents if they feel that they don't make a difference. And Māori statistics aren't going to change no matter what the language is or no matter what the approach you use with their children until such time that education or the striving for education in its formal and informal sense comes more to centre stage in their lives. There were a lot of people who thought that education in Māori would automatically be advancement. The only advancement it's lead to in my way of thinking because it has forced parents to make a choice. To put them [their children] in or not and even by making that small choice meant movement in thought and idea and so that's what we have to do, we have to keep moving and creating choice and realise that parents are making important choices and that they are going to enjoy making those choices. (Uenuku, 16/8/1998)

In the same year, one of the parents expressed his desire to be involved in his children's education at Te Koutu in a similar way:

> I'm pro-Māori but without being anti-anything else, we've just got to look at the stats, the stats will tell us in this country that our people have low academic pass rates. I suppose I can even use myself as an example. It's just, we're not dumb, like we haven't been through these tertiary institutions for one reason or another and it's, you know, hey man, I'm not participating in that and I also believe in something like, which has been involved with a lot of this work which I have being doing now and goes something like this, if you are part of a process, within that ownership you can define how everything is used, how everything is worked. You know, it's just the same as if you think there is a problem. Whoever perceives there's a problem, also diagnoses it ... my aspirations are not any lower than any other person who wants their child to be into things but I want to be part of the process ... I value my time and I am jealous of my time but I also value my children, you know, their time. I'm not saying that my time is running out. No, but I've got to think the same as my parents gave all their time for me, then that's just part of the process. I do want to be in on setting up those, it's seems funny when you say you're setting up boundaries, because I don't want to set boundaries but to set up all those branches to return to. Yep, that's me. (P1: 8/11/1998)

Another parent expressed it more succinctly:

> I know another area [resource in developing the mathematics curriculum] that would help, would be like the desire for these children to succeed. Like there's an area of expertise in the sense that of all these people have made a commitment to their children's education by putting them through an alternative education model. (P2: 8/11/1998)

In 1992, the parents saw the success of the existing *kura kaupapa Māori* in their city and were interested in setting up their own one. Many of the parents had connections to education at different levels and saw that much could be achieved by embedding learning on a platform of Māori values. The two educational movers were Uenuku, who at that time was head of a partial immersion unit at the local high school, and a Ministry of Education officer, specialising in early childhood education as well as in the setting up of schools that moved children on from *kōhanga reo*. Other parents also had an education background. The parents, especially the mothers, desired an involvement in education because they wanted to be part of their own children's learning. For some parents, this meant overcoming their feelings of antipathy towards schools arising from their own poor experiences as students (Smith, 1991). This attitude can be seen in the comment from a mother about being involved in the mathematics curriculum project that was run at the school in 1998 and 1999.

> Just speaking from myself the advantages that we are getting [from being involved in the curriculum development project] are that we can start owning things. We can be in touch with it and then we can start when our kids come back [home]. We are familiar with the contexts, we are familiar with their work and then we can ask them not intelligently, but we can ask them without fear of the maths, so we don't say go away I'm busy. (P3: 8/11/98)

Research into successful home–school partnerships consistently shows that parental involvement in their children's education does result in increased academic outcomes (Bull et al., 2008). Weiss and Edwards (1992) discussed the need for parents and the school to hold "common sets of beliefs, expectations, values, and meanings about achieving a quality education for all the children" (p. 222), but in their research they found that there were no opportunities to negotiate those very beliefs, expectations, values, and meanings.

Being involved in different aspects of the education system meant that many of Te Koutu's parents felt confident to set up a new *kura kaupapa Māori*. It was by having an insider's understanding of the mainstream education sector that the parents had the confidence to spurn it and set up something different. The decision making not only contributed to an increased engagement in their children's education but also ensured that there was an opportunity to negotiate the sort of education that should be provided.

The influence of teacher education on these parents' views about what constituted a good education was profound. In 1998, one parent, who was working in a *kōhanga reo* and studying for an early childhood diploma, described how her thinking had changed about mathematics.

> Actually this is my last year. I'm doing the early childhood diploma. It's changed for me within, from my course. Say three years ago I didn't care about maths. You know it didn't worry me, or it didn't worry for my children, it would have, but, it didn't, maths didn't worry me three years ago but since I've been working, on my course, and implementing a lot of what I'd learnt from my course into *kōhanga [reo]*. A lot of these things have come out and maths has been one of them. I think looking at, we have blocks and different size blocks and my lecturers showed us how maths can come in using the blocks the different size, we've got one long, we've got two. We made, the *tamariki* [children] made two small

> ones, which would fit on one long block, and you know one, concept of maths they can get in *kōhanga* level. Dough play. They can identify sizes whose got more than the others. (P4: 8/11/1998)

However, there was a tension for some parents in that they felt that those who had been trained in mainstream teacher education programmes may be jaundiced in how they viewed the possibilities for children. The following extract from an interview comes from 1998, and similar ideas are discussed in Chapter 12.

> But then the teachers at the school have been grounded in mainstream teaching models, so I mean if it's, yeah their input is somewhat sort of ambiguous isn't it whether or not their input actually is really, really helpful, because obviously it's going to be but there's always that other flipside that it could be sort of tainted, it's not tainted if you really thought about it. (P2: 8/11/1998)

Some of the children of the parents, who were meeting in 1992, were attending the *kōhanga reo* situated on Te Koutu *marae*, a complex that includes a meeting house, dining room, and open courtyard. Shortly before, the *kōhanga reo*, which had been originally in the dining room called Karenga, moved to its own buildings on the *marae*. Consequently, the Te Koutu *marae* trustees were approached to see if they would be agreeable to the new *kura* using the vacated dining room. As Karenga needed some modifications before the school could move in, the school began in the garage at one of the parents' residence from the beginning of 1993. Fourteen children, aged between five and eight years, were initially enrolled. In May 1993, the school moved into the still leaking dining hall. The connection of the *kura* to Te Koutu, both the *marae* and the area, was significant. In 2008, one parent described the importance of this relationship.

> I'm from Koutu and my whānau has lived in Koutu forever. ... No, Koutu born and bred, Koutu's in my heart so I had to say for me it was [a reason for choosing this school for his children]. I would have preferred my children to go to Rotorua Primary because that's the school I went to. It was my wife who wanted our *tamariki* [children] to come here. We only live round the corner and yeah that's the reason we send our *tamariki* here. No, it's cool, most of the teachers here are my relations. (P5: 10/9/2008)

In the beginning, the parents each paid $20 a week for the teacher. This allowed the school to be classified as a private school. As a result, *kura kaupapa Māori* were able to set up "systems of administration, a curriculum, and ways of teaching that were consistent with tikanga Māori (Māori cultural values)" (Rutene et al., 2003, p. 5). However, the number of children at Te Koutu soon rose to the point where a second teacher was needed. This would have been difficult to finance, so the school applied to the other *kura kaupapa Māori* in the city to become a satellite school. Officially, the children became students of the other *kura*, and the teacher was paid as a member of their staff.

From the beginning, parents had high expectations for language proficiency, cultural understanding, and a rich learning environment. The last point gave rise to a lot of discussion about how to implement this. Parents expected the teachers to go beyond what they knew, so that the teachers became learners and contributed everything that they had. In 1998, one of the teachers described how decisions about the teaching were made.

> As a *whānau* [literally family but in relationship to a school means the extended school community], we can come together and say well, as a child I didn't get to learn that so I'd like my child to know about that and, you know, others will say, well, what is that? You know, we never had that when we were a child. So I'd say that when we get together as a *whānau* and we look at, at *kaupapa* [knowledge systems] that we have. Oh, you know, it's really funny because with the different points of views from everybody and of course, you know, we've got one or two who're teachers in other schools and you know, they think well, no I think it's better if we looked at it this way. (T1: 23/8/1998)

The level of involvement by parents was characteristic of the Māori-immersion sector. Irwin and Davies noted the following in 1994:

> The whānau of children in kohanga reo, kura kaupapa Māori, immersion and bilingual programmes are involved in hours and hours of work which is aimed at a fundamentally different level of involvement than is normally the case in other programmes of mainstream schooling or early childhood provision. (p. 80)

In 1998, one of the parents contrasted being involved with her sons in *kura kaupapa Māori* with her experience of being a parent for her daughter in mainstream education.

> At the time that she was these boys' age, I wasn't as active a mother in the school. I didn't ever question what maths was or what science was, I just went to the sports days, I didn't really care that much but that was the environment that those schools offered parents. *Kura kaupapa Māori* is something that we've only just well, we've been in it six years but we've just begun to realise how much of a big commitment it is and how much harder it is but that's the good thing. I get a say, I get to contribute to what's happening with my kids. (P6: 2/12/1998)

After 18 months, Te Koutu decided to request approval to be recognised as a special character school by the Ministry of Education. The Ministry restricted the number of new *kura kaupapa Māori* who could be set up (Reedy, 2000). In a special character school,

> parents are required to provide a statement of aims, purposes, and objectives which explain the ways, other than language, in which the character of the school will differ from ordinary state schools. These are included in the school's charter. The board may refuse to take pupils whose parents do not accept the character of kura kaupapa Māori. In other respects, the legal standing of the kura is the same as for any other state school. (Te Puni Kōkiri, 1993, pp. 6–7)

While the school was on probation for this official recognition by the Ministry of Education, the satellite arrangement stopped. Instead, the Ministry partially funded the school while it prepared its first report for the Education Review Office, who at this time determined whether schools were operating to an expected standard. Uenuku, as chairperson of Te Koutu, and another parent with an education background were in charge of writing the report.

At the beginning of 1996, Te Koutu gained official recognition, whereupon the school began to look for a permanent site. Te Koutu wanted to do things on its own terms. So although advice was provided by the Ministry of Education, it was not always accepted as the community had become used to making their own decisions. Also, by this time, the Ministry of Education had gained some experience with *kura kaupapa Māori*. In 1997, there were 59 *kura kaupapa Māori* with official recognition who were teaching about 4000 children (Reedy, 2000).

The original idea had been to build classrooms on the *marae*, even though the Ministry of Education suggested that this might be logistically problematic. The architectural consultants employed by the Ministry of Education made the parents really think about what they wanted for their school. After much discussion, some parents still wanted to stay at the *marae*, whilst others wanted to look for a new site. This saw the beginning of a school style that was quite different to that of other *kura kaupapa Māori* in the city.

Although all *kura kaupapa Māori* operate in accordance with the guiding document *Te Aho Matua* (Kura Kaupapa Māori Working Group & Katarina Mataira, 1989) (this is described in more detail in Chapter 11), there was a lot of opportunity for difference in interpretation and therefore in the types of education to be provided. As had been the case when the school was set up initially, parents made decisions about their children's education. Some parents moved their children to the other *kura kaupapa Māori*, which had decided at that point to make more places available. Other parents felt that the emphasis on academic achievement at Te Koutu was too strong so they set up another *kura kaupapa Māori*. There was much discussion about the weighting given to different aims, such as language proficiency, cultural competence, and academic achievement. Therefore, it was not surprising to find that parents chose the *kura kaupapa Māori* which best matched what they considered to be appropriate for their children. Although not specifically about the decisions taken at this time, the following quote comes from the parent who had been studying for the early childhood diploma. Her comments show how parents were prepared to put their point of view about their children's education.

> Sometimes I have a bit of a problem with Uenuku myself, you know I say "hey, if you can't teach my girl maths, I want you to go out and find a way that means she's going to enjoy maths". You know we've had this conversation before, him and I. No, you find a way of introducing maths to her in a way that she can learn, you know at her pace, maybe because I always try to say that she's at a different pace than other kids, you know, and his method may not work for her and he may have to go and find other methods. You know, even to go back to hands-on method. You know that's more what I am trying to get at, is a lot of our *tamariki* [children], our Māori *tamariki* need that hands-on, you know. They are not so much want to sit and listen, and more want to touch, yeah. So I would like her to learn first hand experience things, three-dimensional. (P4: 8/11/1998)

One parent, whose children had attended Te Koutu for over ten years by September 2008, stated that she supported the school because it provided the education that she desired for her children. This was the same parent who in 1998 had contrasted her experience of *kura kaupapa Māori* with the mainstream school, which her daughter had attended. Choice about the education they wanted for their children was something that parents viewed as an important right.

> I think Māori as a whole are a bit introspective and tend to focus more on being Māori as New Zealanders instead of Māori on a global scale being able to compete and participate anywhere in the world because that's what I want for my children. I don't want them to think that New Zealand is the be all and end all of the world and I believe that Māori were traditionally explorers and we have forgotten that and New Zealand is the only place we totally feel okay and I don't want that for my children. (P6: Sept. 2008)

At the end of 1996, a site just outside of the traditional area of Te Koutu was offered to the school, but it was rejected because of the close bonds that had developed to Te Koutu. Early the following year, excess railway land that was close to the *marae* and within Te Koutu was chosen as the school site. This land had not been built on previously. It also offered an unbroken view of the mountain, Ngongotahā, a major spiritual landmark of the local tribe, Ngāti Whakaue. Figure 3.1 is a photo from the school grounds looking towards the mountain. The Ministry of Education was petitioned by Te Kura o Te Koutu to purchase the land on behalf of the original owners, Ngāti Whakaue, in order to secure a long-term lease as a school site. The Ministry of Education agreed as the cost of the land was offset by a rent-free period. Fortunately Ngāti Whakaue and the Ministry of Education had a long-standing tradition of cooperation, which greatly facilitated the negotiations.

The new school buildings were opened with 37 children and three classrooms in September 1998. Uenuku had become principal at the beginning of the year, after relinquishing his role as chairperson. Initial plans for the new school worked on the premise that there would be no more that 72 students. This reflected the desire to use the property/staffing formula to maximum advantage. By learning how the Ministry of Education worked, Te Koutu ensured that class sizes remained small, with no more than 20 children per class.

In the late 1990s, an application was made to the Ministry of Education for the school to be extended to a high school or *wharekura*. In 1997, there were only five *wharekura* recognised by the Ministry of Education (Reedy, 2000). Originally, it had been thought that when the children became older they would move to the other *kura kaupapa Māori*, which had already gained status as a high school. When the move was imminent for the first set of Te Koutu children, the other *kura* decided to

Fig. 3.1 From Te Koutu school grounds looking towards Ngongotahā

restrict entry to their own children. The *kura* felt that allowing Te Koutu children to enter at the high school level would disrupt the relationship that their own children had developed between one another and the school community.

As well, the parents of the three oldest children in Year 9 at Te Koutu wanted them to continue there. So, the school applied to the Ministry of Education for secondary school status. They did not realise that the *kura kaupapa Māori* representative body had set up procedures for the establishment of *wharekura* independent of the Ministry. Some other schools felt that Te Koutu had jumped the queue, when it was successful with its application to the Ministry. However, Te Koutu's application came to be viewed as a blueprint by several schools wanting a change in status.

Eventually, the pressure of applications for enrolment and the desire to expand the curriculum base for the secondary school students resulted in a steady increase in student numbers, until there were 207 students in 2010. The Ministry of Education has indicated that they would not like to see the school role rise above 250 students due to the size and configuration of the school site. In 2011, there are 25 staff members of whom 20 are teachers.

Notably, the pressure for an increased enrolment came from the success of the school. In the following extract, one parent described how she moved cities so that her children could attend Te Koutu:

> Tamsin: What I'm asking parents about is why did you send your children to this school?
>
> Mother: Well I was living in Turangi, at the time and they were always going to go to a *kura kaupapa Māori*. I wasn't happy where my kids were because they weren't made to speak Māori over there. I just happened to be looking at the TV and it had Te Koutu kids but it also had other schools kids on there. I was just listening to the way the kids from here spoke their Māori. You could tell it wasn't memorised and it was all natural so I come over here, moved over here. Oh, made sure they got in first and then moved over. (P7: Sept. 2008)

Governance and *Whānau* Involvement in the School

Many *kura kaupapa Māori* have gone through a similar process of expansion. Some have successfully maintained the *whānau* model of governance whereby parents and other caregivers have an extensive say in the school's management, albeit with increased specialisation. The rationale for the *whānau* approach to school management was to prevent parent alienation from the school and the abrogation of their responsibility towards their children's education. Te Koutu still wants to improve this parent–school relationship, but finds it difficult. The following extract comes from an interview with a parent in 1998. It describes in some detail the intensity of commitment that parents made to the governance of the school.

> Yeah. We just have a lot of meetings, a lot of discussions, a lot of things go like, different parts. Like if we're dealing with one subject, one part will go to one *whānau*. You know, everyone will get issued out something and then we bring it back like, yeah, you do, you're given one section and everyone goes out, research it as a *whānau*. We bring it back and discuss it, take the good points and that's just what we do all the time. (P8: 23/8/1998)

The first approach that the school adopted was to assign all parents to one of five areas of governance. Each area then elected its representative, or *māngai,* to Te Ohu Māngai (the committee of representative) who were then endorsed as the Board of Trustees in order to meet the requirements of the Ministry of Education. As time went on, it was found that some of the governance groups operated well, others not so well. So the school opted to vote for its five parental representatives and to dissolve the various governance groups. Internal management currently includes the principal and the heads of the junior, middle, and senior school and the teacher-in-charge of staff development and appraisal.

Even in the early years of the school, involvement with curriculum decisions around subjects, such as mathematics, was seen by many parents as being outside the scope of what they could contribute. They also respected the expertise that the teachers had. This can be seen in the following two comments from parents who were involved in the first project that looked at developing a culturally appropriate mathematics curriculum.

> Tamsin: Is it better just to leave it to the teachers?
> Parent: I don't know how the other parents feel but like I was saying at the beginning it seems to be a bit over my head, so for me personally I'm quite happy to leave it with the teachers. (P9: 20/6/99)

> Well, who is it for us to say what our children need to learn? And why can't our children learn as much as we can give them. I don't know? Umm, I like the ideals and visions Uenuku has for the *kura* so I'm quite happy for him to choose and provide whatever he can for our children. I don't want to say our children only need to learn this because our children can learn anything, so, no that's pretty open-ended for me. (P10: 23/8/1998)

Unlike the economic situation during the establishment of the school, many parents are now in full-time employment, which severely limits the possibilities for involvement in the everyday running of the school. Initially, it was reasonably easy to find Māori-speaking parents who could assist in class or with extra-mural activities, but this is no longer possible. The high levels of commitment to the language and the culture are also not as strong. With the re-emergence of *te reo Māori* in public life in Aotearoa/New Zealand, people have become more complacent about the status of *te reo Māori* than they were in the 1980s. Unfortunately, intergenerational transfer of language has regressed as more and more parents expect Māori-medium education to assume this responsibility.

> So far, 20 years of activity have produced no more than a handful of new speakers who might be expected to ensure natural intergenerational transmission to their own children. It has, however, made it likely that many of the graduates of the immersion and bilingual programs will want their own children to have a chance to learn Māori as their second language. In other words, the institutionalization of schooling in Māori and the establishing of community support (within the Māori community and in national government policy if not yet in non-Māori New Zealand ideology) are starting to set the conditions for continuity. (Spolsky, 2003, p. 569)

Nevertheless, as will be discussed in Chapter 9, the reliance on schooling to provide the transmission of *te reo Māori* falls on the individual teachers and their varying levels of linguistic proficiency (Reedy, 2000). As suggested by Reedy

(2000), the combination of many teachers being second-language learners of *te reo Māori* with the proliferation of new vocabulary to accompany subject areas is likely to give rise to profound language change. Native speakers, who had been so involved in the initial push for *kōhanga reo* and *kura kaupapa Māori*, may no longer recognise or approve of the language that the children are using in these institutions.

As well, the recent release of the report on the Māori language (Waitangi Tribunal, 2010) has indicated a startling drop off in the enrolment of Māori children in *kohanga reo* and Māori-medium primary and secondary education. *Kōhanga reo* do not provide long day-care services as needed by working parents in the current economic climate. If parents cannot send their children to Māori-medium early childhood education that meets their needs, then there will be fewer children with *te reo Māori* skills of a sufficiently high standard to enter *kura kaupapa Māori*. This suggests that the perception that Māori-medium schooling will maintain the language to a reasonable level may no longer be a valid one.

This challenge will soon be at Te Koutu's door. Te Koutu will have to review its priorities and practices to meet these external pressures whilst maintaining its core values and goals. They are currently considering the establishment of a transitional facility for preschool and school-age children to improve their proficiency in *te reo Māori,* so that they can transition smoothly into Te Koutu.

In the early years of Māori-medium education, a lot of parents saw themselves as change agents. Now, they seem to feel that the school is being well run, and so they no longer need to be advocates for their children. Improved Māori prosperity also means that the strong commitment to changing society is no longer there, and parents are more fickle in their commitment to the school. This does not mean that parents are abdicating their right to make choices about their children's education, rather that they see themselves as consumers instead of initiators of alternative schooling options. The following extended interview extract shows how over a decade after the school started, parents were making decisions about sending their children to Te Koutu because they felt that it provided the education that they wanted for their children. What had changed in the intervening years were the constituents of a good education in a *kura kaupapa Māori*. Improved academic results were perceived as having been achieved. Language and culture were still important, but the changing society also meant that education needed to encompass more than what had been identified when the school had first begun.

Parent: We chose to come to Te Kura o te Koutu because of that global aspect of the school because it encompassed what we saw as a future in terms of how kids particularly could get on in the world. What we were looking for is something where they could understand their own culture, number one, where they could learn English and ensure it was learnt properly and it wasn't only a secondary language but it was a language that was taught well and also we wanted a third language which ended up being Spanish because we saw it as a global language. The likeliness of our kids staying at home with us is near on zero.

Tamsin: Why is that?

Parent: Because the world's changing so rapidly that the likeliness that they would end up staying and working in New Zealand is for the beginning stages of their working career is not likely. For me, I saw it as the sheer fact that by giving them their own language they had security in who they were, number one. By ensuring that they had two other languages, English and Spanish, it ensured they could travel and learn more. If they chose to learn another language after that, they had the ability and the skills to do so. If they chose to work in a country that was on the other side of the world they would gain in the same opportunities as everyone else around if not more so. I have always thought New Zealanders as a whole have a very naive opinion of languages be that Māori or any other language so consequently, we allow ourselves to limit our children and I didn't want to do that. When I came down here, Rotorua wasn't the place I was going to come to. I came into the school in the school holidays. I met Uenuku and I was coming. That was the end of it. He had the same thoughts in terms of academia that I wanted that it was not just about our cultural aspects, which are highly important, but it was about ensuring that the kids had a good education across the board in every subject including maths and science and English, which was really important to me, and was the only way I could sell it to my husband. That's the reason we came. (P11: Sept. 2008)

In many ways, the change in perceptions about what the school should offer can be seen in how mathematics learning was discussed by parents in 1998 and in 2008. In the following extract, the parent talked about her concern that mainstream schools were allowing many Māori children to underachieve in mathematics.

I guess because we haven't always seen it [mathematics] as having good results for the children. You know, like I know in mainstream schools a lot of children miss out, they don't achieve, or they don't feel good about themselves through the curriculums and programmes. So, I guess personally, I'd like to see our kids. I think confidence is the most important thing and so, if, if they're learning really happily and they're confident then they're going to learn maths and if they're not then they're not going to learn. I think mainstream school haven't really given children that. Like some of the children get it but not all of them. (P12: 23/8/1998)

The cultural relevance of mathematics also came into the discussions. One of the teachers, who was also a parent and later went on to set up her own school stated the following:

We'd need to focus on things *kaupapa Māori* [Māori knowledge] within *kura kaupapa Māori,* so I would say that the curriculum doesn't totally look at, at that, doesn't look at *kaupapa Māori* in the big way, we would like it to. (T2: 23/8/1998)

This was elaborated on by a grandparent, who was the major caregiver of two children who attended Te Koutu in 1998 and 1999.

Now my kids bring maths homework home and you sort of don't even understand the way they do the addition, times. Now it is different to how we were taught and we're teaching them different, the way we used to, the way we were taught how to do maths and I think that because of the *kura kaupapa Māori* that there's a lot of things our kids could learn through nature, through learning by using objects like in string games, using the stars. And it's just that I think that when they get older it gives them a better understanding. I mean, they'll see things broader when it comes to maths so we can see that the learning of the stars is a big thing for our Māori kids. And any different things through the bush, through nature. That all things comes back to maths too. (P13: 23/8/1998)

This grandparent's focus was on looking back to traditional ways of doing mathematics to gain inspiration for alternatives to how mathematics could be conceptualised and taught. In contrast, parents in 2008 still mentioned these traditional mathematical ideas, but more in terms of Māori always having had mathematical expertise. This positioned Māori as confident mathematical users and learners without restricting them to only being able to learn mathematics in a traditional way. The following extract comes from the parent who in 2008 described her reasons for sending her children to Te Koutu. It was representative of the types of views expressed by parents at this time.

> Maths is not just about numbers that you see or do within a maths class, in a cultural aspect it is about stars, it is about *matariki* [winter solstice] and all those sorts of things. It's about when is the best time to plant the crops, it was all a mathematical equation. It was about gathering food, it was about lots and lots of different things and working out survival techniques. So, from a cultural point of view, maths was highly used, in a way that they [our ancestors] thought was practical. In terms of the current situation, some of those things are still used but maths is used in terms of science if they want to move ahead. My daughter is currently wanting to be a forensic scientist .... (P11: Sept. 2008)

From this parent's viewpoint, Māori people were always good at mathematics because they had used it within many aspects of their lives. Her expectation for her children was that mathematics would have the same role in their lives, both currently and also when they became adults.

Meeting Challenges in Establishing and Operating Te Koutu

Te Koutu began its life within a particular zeitgeist and the demands of that time. The decisions the school community made concerning the school and the status of *te reo Māori* are based on the complex interaction of historical, social, economic, cultural, psychological, legal, linguistic, and attitudinal factors. There have been many challenges along the way for the school community, from maintaining adequate funding to balancing the competing dictates of ensuring that children gained strong *te reo Māori*, as well as cultural awareness, skills and knowledge, and good academic results. Since its inception, Te Koutu's primacy in the revitalisation of *te reo Māori* has changed as the parents' sometimes-competing perceptions of their children's needs have shifted. This is reflected in a shift within a decade, of the primary role of *kura kaupapa Māori* from being the resurrection of Māori pride in their language and culture to that of the need for children to take their cultural heritage into the modern world.

For Māori parents who send their children to Te Koutu the norms about what to expect from sending their children to school have changed. In Kemmis' (2009) diagram, Fig. 1.2, education is perceived as contributing to the good for individuals and also for humankind. Schools, before the setting up of Māori-medium preschools and schools, contributed to the loss of Māori language and culture as well as academic underachievement of students. This did not result in the individual or the society gaining any benefit. The change in perceptions has changed outcomes, including gains in self-determination. However, not all challenges are resolved as changing societal conditions – the sayings, doings, and relatings of the practice architectures – means that providing schooling needs to be a dynamic process. New challenges continue to arise. With the shifting societal conditions, the patterns of collaboration also change. Relationships between government agencies such as the Ministry of Education, parents, and teachers need to be re-established continually so that the challenges first can be recognised and then worked upon. Tensions in discussions about the direction of schools can lead to new possibilities coming into view. When parents resisted the direction that Te Koutu was moving in, it opened up opportunities for the establishment of new schools with different aims. Thus, resistance is not by itself a negative action, but rather can contribute to new opportunities being taken advantage of.

Participating in these evolving understandings about the role of schooling contributed to the politicisation of Māori parents. In a generation, they went from accepting that they had no role in education to ensuring that education provided for the ever-changing needs of their children. In many ways, this politicisation could not have occurred without ongoing cultural negotiation.

> Cultural negotiation is a process that makes schools' hidden values and processes visible to community and school while making the community's knowledge, values, and processes visible to schooling. Schooling then becomes explicit and open to choices – choices that can only be responded to at the local community level as they concern issues of culture, language, and identity. Through an exploration of their own cultural strengths and their particular goals and visions for their children, community and school can construct a curriculum of the possible – creatively devising content and pedagogy. (Lipka, 1994, p. 27)

Over time, the school community of Te Koutu has become more politically conscious. Sometimes this resulted in the core values and aspirations of Te Koutu becoming lost in playing the rules of the game as set by the Ministry of Education. Conforming to measures such as national standards and responding to more centralised compliance agents are an outcome of state funding. These dictates have interfered with Te Koutu's ability to make decisions about its educational priorities. As well, the philosophy of wanting Māori children to live as "citizens of the world" (Durie, 2003) meant that Te Koutu became more open to lots of new ideas, skills, and training opportunities, which other *kura* did not take up. The teaching of Spanish as a second language to all students in the school has led to groups of older students and teachers going to Mexico for three-month periods to take part in Spanish language schooling. However, like playing to the Ministry of Education's rules, some of the foundation ideas, which came from the setting up of the school, got lost in the mix. To maintain language revitalisation and the development of children's *te reo Māori,* proficiency will be an ongoing challenge for Te Koutu.

Chapter 4
It Is Kind of Hard to Develop Ideas When You Can't Understand the Question: Doing Exams Bilingually

For students who complete their schooling at Te Koutu, the final few years are marked by having to do external examinations, which are provided in English and *te reo Māori*. Māori have had to fight with the New Zealand Qualifications Authority (NZQA) for the exams to be provided bilingually. Although at first it seems sensible to provide linguistic clues about what the questions are asking in the students' two main languages, there are inherent inequities in the way the examinations have been constructed. Thus, this issue cannot be considered to be resolved, even though the norm of expecting that examinations would only be in English has been overturned.

New Zealand has a tradition, inherited from the United Kingdom, of assessing students in their final years of high school with external written exams (Jones, McCulloch, Marshall, Smith, & Smith, 1990). The sorting of students by providing them with marks through the examination process assumes that the top students at school will be the top students at university. McKinley (1995) showed that this is not the case, yet universities continue to use this as a basis for accepting students. What is often unrecognised is that the content and the format of the examination questions affect who is seen as successful at school. Clarke (1996) summarised this issue of control as follows:

> The political significance of assessment gives substantial coercive power to those responsible for shaping assessment. What is assessed determines what is taught. The performances valued in our assessment provide a model of the goals of the curriculum. (p. 329)

Nevertheless, as Young-Loveridge (2005) pointed out, there are a number of ways of assessing students, and these will give quite different pictures of how they are doing. She showed that Māori boys' performance in mainstream schools appeared to improve dramatically if they were assessed through diagnostic interviews, as used in the New Zealand Numeracy Project, rather than with paper-and-pencil tests.

The provision of national examinations and the exams' direct relationship to students' possibilities for career choice and further study are a political decision. Although Te Koutu is concerned with the outcomes for their students, it is usually at the wider community level that decisions are made. In this chapter, we provide an

T. Meaney et al., *Collaborating to Meet Language Challenges in Indigenous Mathematics Classrooms*, Mathematics Education Library 52,
DOI 10.1007/978-94-007-1994-1_4, © Springer Science+Business Media B.V. 2012

overview of some of the decisions which have been made and the impact that this has had on students, who then sat the external examinations.

National Certificate of Educational Achievement

In 2002, assessment at the end of high school was radically changed in New Zealand/Aotearoa. In the old system, students had to achieve at least a 50 percent mark to be considered to have passed. The allocation of marks ensured that 50 percent of students in any subject failed. The change in 2002 was from norm-referenced examinations to criterion-referenced assessments. Each discipline area, such as mathematics, was broken down into component parts, with students being tested on their mastery of the different components. Some of these components, Achievement Standards, are assessed internally through NZQA's moderated assessments, whilst others continue to be assessed externally in a formal examination process. Students can gain excellence or merit or achieve levels for each Achievement Standard depending on how they answer increasingly complex questions. The total credits that they gain for the National Certificate in Educational Achievement (NCEA) is the same regardless of the level that they achieved. The introduction of NCEA was seen as one way of ensuring that more students finished school with a qualification.

In order to be admitted to university, students need to have Level 1 numeracy and Level 2 literacy credits, as well as a good performance in other subjects at Level 3. Mathematics Achievement Standards, therefore, are important for students who wish to go on to university. Level 1 Achievement Standards usually are gained when students are in Year 11 or in the third-to-last year of high school.

Making the Exams Bilingual

The movement to NCEA formalised the provision of assessments in *te reo Māori*. This had taken a long time to become normal practice, and without it schools' willingness to teach high school subjects in *te reo Māori* had been restricted. The unwillingness of NZQA to provide examinations in both languages exacerbated the tension experienced by parents and teachers about what was best for the language and what was best for the students' future careers.

There had been ongoing requests for exams to be bilingual from 1992, when Uenuku Fairhall was teaching mathematics in *te reo Māori* at a Rotorua high school (Barton & Fairhall, 1995a, 1995b). In 1993, his students had been able to respond in *te reo Māori* to the School Certificate[1] examination written in English, with a special marker being employed to do the marking. However, at this time NZQA would not support the translation of the exam into *te reo Māori,* because, for so few students, it was considered an excessive cost. The parents and Uenuku, as the teacher, did not

[1] The exam prior to 2002 that was completed by students in their third final year of school.

consider this appropriate. The following extract comes from a parent and is part of a discussion that occurred in 1999 about the lack of support for bilingual exams:

> My son went to primary school in English and then he did Māori at high school and Uenuku taught him that whole year in Māori, fifth form maths [third last year of high school] and then they sat the exam in English and he passed quite well. I was lucky that I had a good experience but I remember parents of that year were really angry because their kids had to do that. They had to translate as well as sit the exam. I think most of the girls passed that year, it was that first class at [high school] and two of the boys passed. But there was a huge uproar from the parents that they learnt in Māori and then they had to sit in English. We were told that they were going to have the exam in Māori but it didn't happen.

Consequently, the school decided not to continue teaching the final years of mathematics in *te reo Māori* as long as the examinations remained in English.

In 1999, end-of-high-school exams were still held in English only. Uenuku was now principal at Te Koutu. Although the school did not have students doing School Certificate or Bursary[2] examinations, it had just been granted permission to teach high school subjects. In the meetings held at Te Koutu at this time, there was discussion about whether mathematics should be taught in English at the high school level of the school. Similar to the earlier discussion at the local high school, some of the parents felt that their children would be disadvantaged if they were taught in *te reo Māori* but had to read and then respond to the examinations in English. However, Uenuku had discussed again with NZQA the possibility of the exams being bilingual. The following extract comes from a transcription of a meeting held on 15th August 1999 between parents and teachers.

Tamsin: Assessment but there was that thing about School Certificate and having to sit and do it in English, and then discussions about how they could answer in Māori and that came into it as well. That discussion.

Parent1: Somebody said that the paper was in English but that it could be given to the children in Māori, was that you Uenuku?

Uenuku: Yes, the exam was...

P1: Written in English, eh?

Uenuku: Yeah, but it will provide translation facilities. But what I've always asked for is that it's in both languages because if you get someone who's used to a certain type of mathematical language in one school, because you know we're not all, whether we like it or not, we're developing separately, the various schools, because there's not the networking available so the language can sometimes

[2] The exam prior to 2002 that was completed by students in their final year of school.

be quite, quite different. So in many cases have at least an offering in English, so that the children have got two sets of registers to model, understand what the question is about.

JuniorPrimary Teacher: So what do you propose? For instance?

Uenuku: The exam is already in English. What I think is going to happen in this year and the upcoming years is that when they do the computer templates. The beauty of computers, everything is already on the computer. What they'll give to the translator, is that and there will be room to put in, in italics, in a smaller lettering is the Māori, so that's what, the kids will be looking at, the italics all the time but if they're having any trouble because whoever the translator is, wherever the translator is from the school is from Auckland and they're using particular terms. It's not a tribal thing, it's more, it's the maths register that they're actually using and they've got another clue in the English. If they put it all in Māori they probably wouldn't . . .

SeniorPrimary Teacher: And there seems to be some problems even with, not just *iwi* [tribe], but the understanding of the maths in itself. I mean, mostly what you're teaching our kids, what is for another school totally different, even when it's written the same, what's on the paper is actually depends on who's doing it. Say if it was you, you'd be put down what our kids would know. So there is a conflict on the other side and then there's a general, how the. . .

SeniorPrimary Teacher: It will take a while.

T1: Yeah, yeah.

P1: So that means that each *kura*, their presentation of curriculum, teaching curriculum for the *tamariki* [children], you would take your own *matua* [teacher] to teach your *tamariki* [children].

The discussions held at Te Koutu in 1999 were probably similar to discussions happening in other Māori-medium education facilities around the country. Certainly McKinley and Keegan (2008) reiterated that parents at a *kura kaupapa Māori* in the early 2000s were more worried about their children's job prospects than insisting that their children were taught in *te reo Māori*. The beginning teacher, who had recently started teaching science, was concerned that an exam written only in *te reo Māori* would not be accessible to the students because of the large amount of new vocabulary they would have to learn. Although revitalisation of *te reo Māori* was a major focus for Māori-medium education, so also was improving student academic outcomes. Without an appropriate resolution these two aims were in conflict, with one aim having to be prioritised over the other.

Having the exams provided bilingually was important because it signalled that *te reo Māori* could and should be used to discuss subjects at a higher level. Part of Te Koutu's discussion was that students would be disadvantaged if they were to learn mathematics in one language but be expected to do the examinations in another language. If the school responded to students' needs by teaching high school mathematics in English, they would be going against the commitment of *kura kaupapa Māori* to revitalise the language (see Chapter 2). However, providing exams in *te reo Māori* only was not considered appropriate, because individual differences in the mathematics register between different schools may have meant that an exam written by a Māori speaker from one school could not be understood by students from another school. The potential for the exam to be incomprehensible to students, if it was only provided in *te reo Māori,* was seen as something that would disadvantage students. The most appropriate solution was the one that was adopted, which was to have the questions presented in both English and *te reo Māori.* During the last few years before the introduction of NCEA, some exams had been translated "under limited arrangements made between NZQA and individual *wharekura* [high school]" (Stewart, 2007, p. 5).

With the introduction of NCEA, when schools enter students for the external assessments in April, they can state that they wish students to complete the exams in *te reo Māori*. If schools do not request an exam in *te reo Māori*, then it will not be translated. One consequence of this has been that there have been limited practice exams for students doing Achievement Standards in *te reo Māori* for the first time. Previous exams are publicly available, but if no other student had attempted a particular Achievement Standard in *te reo Māori,* then there would be no practice examinations.

NZQA contracts translators to produce the bilingual exam booklets. The different language versions are at opposite ends of the booklets, thus making them twice the size of the English-only exam. The exam questions are always written first in English and then translated into *te reo Māori* (Stewart, 2007). The consequences of this decision are many, and some are described later in the chapter.

Results from Bilingual NCEA Examinations

Although the NCEA external assessments are provided bilingually, the knowledge that they assess is Western knowledge based on the Ministry of Education's curricula. Thus in regard to students' achievement in mathematics, although the language may be Māori, the content is not. As Stewart (2007) wrote with regard to science:

> Just as the Pūtaiao [science] curriculum document 'is not considered to be a Māori curriculum' ([McKinley, 1995], p. 55), neither can these translated examinations be considered distinctively Māori science assessments. The assumption is that the content knowledge to be assessed is exactly the same in wharekura [high school] as in mainstream schools: indeed,

it is reasonable to suggest this system constitutes added motivation to teach a mainstream science programme, since wharekura [high school] wish to maximise student achievement. (p. 5)

Results for students completing NCEA in Māori-medium situations are not always readily available. On the whole, Māori students who were enrolled in Māori-medium education appear to be more likely to achieve in NCEA at the expected year level than their peers at English-medium education. This can be seen in Fig. 4.1 from the Ministry of Education's *Senior Secondary Students' Achievement at Māori-Medium Schools – 2004–2006 Fact Sheet* (Wang & Harkess, 2007).

In Fig. 4.1, each column represents a year, whilst the different shadings outline the percentage of students who achieved credits at the different year levels. The percentage of students not covered in the graph (when the column does not reach 100 percent) represents the percentage of students who were enrolled but who gained no credits at any level in that year. It is expected that most students enrolled in Year 11 would be doing NCEA Level 1, in Year 12 NCEA Level 2, and in Year 13 NCEA Level 3. Although this is the general trend in the graph, it can be seen that in all years some students are doing credits above or below what is expected from their year level. This flexibility has been one of the main advantages of ensuring that more students complete high school with at least some qualifications.

However, Fig. 4.1 does not provide details about whether the Achievement Standards were for internal or external assessments or details about the subjects in which the Achievement Standards were gained. The authors also warned that the number of students in Māori-medium education completing end-of-high-school assessments was small, and this could result in a bias when the results are presented as percentages. For example in 2004, there were only 154 Year 13 students

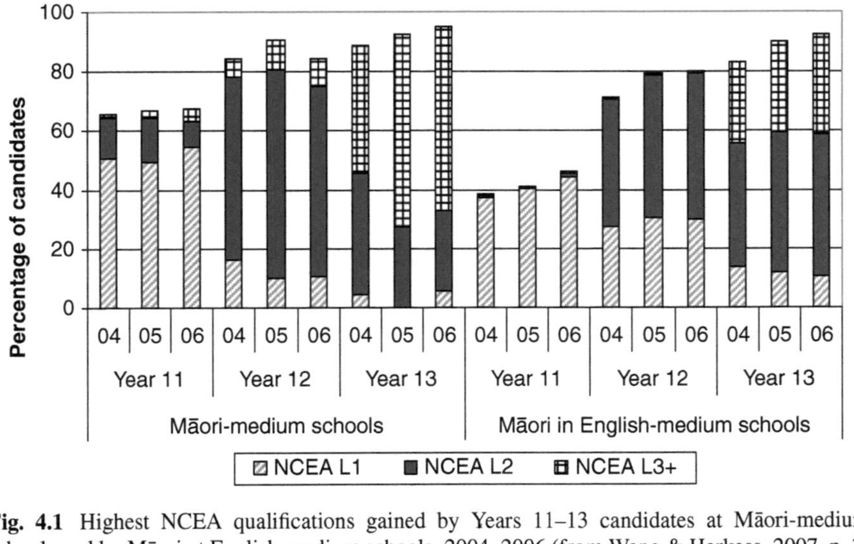

Fig. 4.1 Highest NCEA qualifications gained by Years 11–13 candidates at Māori-medium schools and by Māori at English-medium schools, 2004–2006 (from Wang & Harkess, 2007, p. 3)

in Māori-medium education. By 2006, this had risen to 253, but these numbers were very small when compared to the number of Māori students in English-medium education, which number in the tens of thousands.

When the mathematics results are isolated, the figures show that Māori studying in *te reo Māori* performed significantly worse than their non-Māori peers as well as their Māori peers studying in English. Stewart (2007) combined NCEA Level 1 results from 2002 and the following two years to produce Tables 4.1 and 4.2.

Table 4.1 Number of mathematics exam papers completed by Māori-immersion students between 2002 and 2004

	Number of papers	Not achieved	Achieved	Merit	Excellence
All Pāngarau [mathematics]	941	693 (73.5)	210 (22.3)	38 (4.0)	0 (0)

Adapted from Stewart (2007, p. 8)

Table 4.2 Individual Achievement Standards with non-Māori, English-medium Māori students and Māori-medium students' results combined for 2002–2004

Achievement standard no. (level 1 credits) title	Cohort	Achieved results (%)	Merit and excellence results (%)
90147 (4)	Non-Māori	61.6	28.1
Use straightforward algebraic methods	Mainstream Māori	40.1	10.9
and solve equations	Te Reo	20.6	2.5
90148 (3)	Non-Māori	56.2	15.9
Sketch and interpret linear or quadratic	Mainstream Māori	34.3	5.2
graphs	Te Reo	12.7	2.0
90151 (3)	Non-Māori	76.2	37.3
Solve straightforward number problems in	Mainstream Māori	58.2	18.7
context	Te Reo	31.4	4.6
90152 (2)	Non-Māori	67.7	29.3
Solve right-angled triangle problems	Mainstream Māori	47.2	13.6
	Te Reo	42.9	7.9
90153 (2)	Non-Māori	75.4	29.0
Use geometric reasoning to solve	Mainstream Māori	58.4	14.2
problems	Te Reo	39.1	7.3
90194 (2)	Non-Māori	58.3	24.8
Calculate relative frequencies and	Mainstream Māori	37.6	9.5
theoretical probabilities	Te Reo	13.3	0.6

Adapted from Stewart (2007, p. 7)

The results in Table 4.1 reveal that almost three quarters of exam papers attempted by Māori-medium students between 2002 and 2004 were not successfully completed. Each Achievement Standard noted in Table 4.2 would have a separate exam paper, which it is expected to take students half an hour to complete. So while in Table 4.1, 941 exam papers may appear to be a lot, students would sit up to six papers over a three-hour exam period.

Table 4.2 shows the percentage of students who achieved in each of the Achievement Standards that were assessed externally between 2002 and 2004. Although in all cases the percentage of students in Māori-medium situations who achieved the standards was lower than that of their peers, there were differences between the different standards. The two Achievement Standards that Māori-medium students did particularly poorly at were 90148: Sketch and interpret linear or quadratic graphs, and 90194: Calculate relative frequencies and theoretical probabilities.

The results in Table 4.2 show that far fewer Māori-medium students gained an excellence or merit in Achievement Standards assessed externally compared to their peers. One reason may be that some students in New Zealand focus on credit collection rather than on the quality of the assessments that they gain (Burkhill & Bye, 2005). It is unclear if this was why Māori-medium students did not gain higher levels and why they might be affected more than other students.

Another suggestion is related to students' command of the subject register in *te reo Māori*. Stewart (2007) suggested that it may not be possible for Māori-immersion students to gain Excellence in science because they need to be able to use the science register extremely well in order to provide appropriate responses to complex reflective, evaluative questions. Teachers of other subjects in English-medium situations also stated that they struggled with finding sufficient time to prepare students to answer the more complicated Excellence questions (Burkhill & Bye, 2005). As well, students complained about not having the differences between the levels of questions explained to them, making it difficult to know how to respond appropriately. However, the complexity would be increased in order to answer highly reflective questions in what for most Māori-medium students is a second language. The results in Table 4.2 suggest that whatever the reasons, students in Māori-medium education do not achieve at the same levels as their English-medium peers.

All students who want to enter university must gain 14 credits in Level 1 mathematics. Given that most Achievement Standards provide only 2 or 3 credits each, students must gain several different Achievement Standards to make up their 14 credits. Comparing the percentage of students who gain this minimal requirement in Fig. 4.2 with the percentage of students who achieved external credits in Table 4.2 suggests that many students rely heavily on internal mathematics Achievement Standards to meet the minimal requirement. Figure 4.2 also indicates that almost 70 percent of students gain the numeracy requirement in the Māori-immersion situation. This is much higher than could be expected from the external assessment results. However, ten percent more Māori students in English-immersion schools gain the numeracy requirement than Māori-medium peers. Poor results in the external assessments mean that Māori-immersion students must gain their credits from

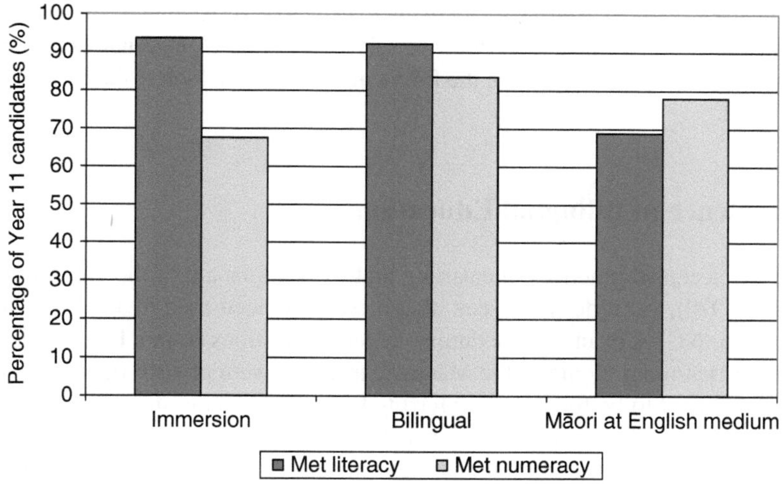

Fig. 4.2 2005 Year 11 candidates – percentage to meet the literacy requirement, percentage to meet the numeracy requirement (from Murray, 2007)

completing internal assessments, which places pressure on schools to provide these opportunities as often as possible during their teaching year. If it was possible for Māori-immersion students to perform as well on the external assessments as they do on the internal assessments, then it is more likely that a greater percentage of these students would achieve the university entry requirements.

On the other hand, Cathy Dewes, principal of Te Kura Kaupapa Māori o Ruamata, suggested that internal assessment is better suited to the philosophy of *kura kaupapa Māori*.

> I believe internal assessment takes into account the way Māori students best learn; both the learning and assessment can be hands-on and interactive. We've moved away from a total dependence on the traditional pencil and paper assessment. It's great to see the progress that Aotearoa [New Zealand] has made, away from the English colonial method of assessment. Paolo Freire calls it 'transmission education' you cram the information in during the year and you expect perfect regurgitation in a three hour exam at the end of the year. (NZQA, 2001)

This chapter does not debate the rights or wrongs of external high-stakes assessment. Rather, the discussion is about the issues concerned with the provision of bilingual examinations in English and *te reo Māori*. This is not because the other debate does not exist in New Zealand. For example, Cathy Dewes also raised the possibility that in the future, Māori-medium education should develop its own more culturally appropriate assessments that would have equivalent status with those of NCEA (NZQA, 2001). McKinley and Keegan (2008) also felt that traditional Māori understandings of science were being lost with the present concentration on teaching and assessing of Western science.

Translating examinations into *te reo Māori* has not resolved all of the issues, as had been hoped by the parents advocating for it in the 1990s. Students sitting these examinations in *te reo Māori* are less likely to do well compared with students taking the ones in English. In this section, we raised the issue of external examinations being culturally inappropriate. In the following section, we look at the quality of the translation.

Equivalence in Bilingual Education

There are accepted practices in ensuring that translations are of the highest quality (Brislin, 1970), yet it does not seem that these have been used to any great degree in regard to NCEA exams. We examine in detail questions from a Level 1 algebra[3] external assessment. Some of the Māori translations were of such poor quality that students seemed to be forced into referring to the English question in order to produce an answer. The weaknesses that were inherent in the English versions of the problems were exacerbated by the translation, and this was likely to result in extra demands being placed on students who were completing the exams in *te reo Māori*.

Pollitt, Marriott, and Ahmed (2000) highlighted three main concerns that question writers should take into account when producing assessments. Their research was into the difficulties of answering assessment questions experienced by second-language learners of English (ESL), who had a strong mathematical background. Although the situation for Māori-medium students is not the same as those for ESL learners, the concerns still appear relevant. The concerns were as follows:

> Linguistic:
> The uses of 'ordinary' English words with special meanings can cause unexpected problems for L2 students;
>
> Contextual:
> 'Real world' contexts can so complicate the task, if the context is not familiar, that comprehension and task solution are prevented;
>
> Cultural:
> Language and context may interact in subtle ways such that apparently easy questions become impossible for culturally disadvantaged students. (Pollitt et al., 2000, p. 1)

We do not believe that any Māori student is culturally disadvantaged. However, we do acknowledge that there can be a mismatch between the cultural experiences of students and those expected by mainstream schooling. When the cultural situation is unfamiliar and this is combined with other issues, Māori can be made to feel that mathematics is something that can be done only by English speakers.

Most external examinations consist of a series of word problems situated in a context (see Fig. 4.3). In discussing the cognitive demands on English language learners in solving mathematical problems, Campbell, Adams, and Davis (2007) stated that "[l]ife experiences, language, cognitive processes, and knowledge of and the ability to apply mathematical content all interact in the solution process"

---

[3] Achievement Standard 90147, *Te whakamahi tikanga taurangi māmā me te whakaoti whārite* (Using straightforward algebraic methods and solve equations).

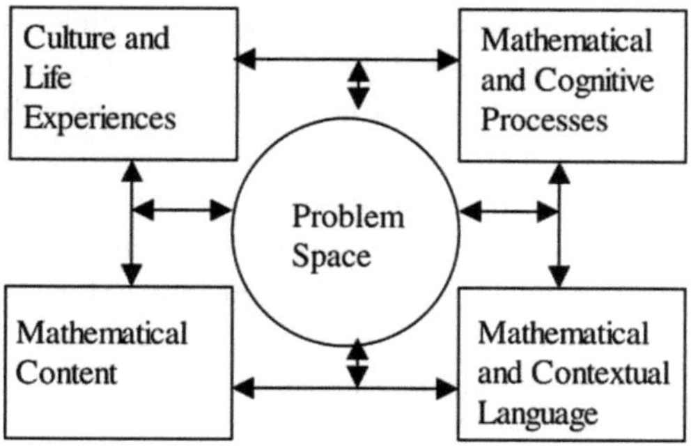

Fig. 4.3 Meta-reflective interactions in the problem-solving space (from Campbell et al., 2007, p. 9)

(p. 8). When the solving process demands the retrieval and processing of too much information, a student suffers from cognitive overload, which results in the student being unable to solve the problem. The interaction of the different components is illustrated in Fig. 4.3.

The work of Pollitt et al. (2000) and Campbell et al. (2007) suggests that it is possible to identify potential cognitive demands faced by students and whether these would contribute to them being unable to respond to the questions. After seven years of the provision of NCEA examinations in Māori, the translated instructions, questions, and problems still present anomalies. We analyse several questions and problems from the Level 1, 2008, examination for algebra – one of the mathematics Achievement Standards – to demonstrate such anomalies and to consider the cognitive demands made on students. The 2008 exams are the latest ones that we have available for analysis. Given the relative newness of NCEA, it was decided to choose the latest exam, because it was less likely to exhibit the teething problems identified by Stewart (2007) in earlier exams.

The first example is Question 2 of the examination. The English version of the question is as follows:

> Pam sends Christmas cards to her friends.
> The stamps cost 50 cents for each friend.
> The cards cost $2.75 for each friend.
> She spends a total of $68.25.
> The equation for the amount she spends is:
> $0.50f + 2.75f = 68.25$
> where f is the number of friends she sends Christmas cards to.
> Solve this equation to find how many friends she sent Christmas cards to.

Translations often reveal weaknesses in the original text. In relationship to the cultural and life experiences of the students, the context for the problem has some improbable circumstances. Normally the purchase of stamps would take place after the cards (and envelopes) were purchased and inscribed. Moreover, it would be more natural to buy according to the number of cards to be sent, even if that required nothing more than matching them to the number of friends. However, in this case, the given equation indicates that the number of friends is the unknown. This may seem pedantic but as suggested by Pollitt et al. (2000), there is a possibility that improbable contexts would lead students to misunderstand the question, thus jeopardising their ability to show that they could solve such an equation.

The translation in *te reo Māori* is given below and it indicates other potential problems, perhaps because the translator has tried to make the question as close to a literal translation as possible. The consequence is a loss in some of the transparency of the context in the Māori version. Thus the mathematical and contextual language, rather than clarifying, can contribute to the students' confusion.

> Tuku kāri Kirihimete ai a Pam ki ōna hoa.
> E 50c te utu mō te pane kuīni mō ia hoa.
> E $2.75 te utu mō ia kāri mō ia hoa.
> E $68.25 te tapeke ka whakapauria e ia.
> Ko te whārite mō te nuinga e whakapauria ana e ia ko te:
> $0.50f + 2.75f = 68.25$
> arā, ko *f* te tokomaha o ngā hoa e tuku kāri Kirihimete ai ia.
> Whakaotihia tēnei whārite kia mōhio ai ki te tokomaha o ngā hoa i tukuna kāri Kirihimete ai e ia.

The translator translated "sends" (*Tuku kari .... ai*) with a Māori structure that implied an activity that occurred regularly. Although the first sentence in English also implied this, the English in the rest of the problem soon narrowed the activity down to a one-off occasion. It is highly unlikely that the prices quoted and the number of friends would be the same from year to year, and therefore that the activity could have happened more than once.

The translator also made several other decisions, which emphasised the Pākehā [European] aspects of the problem. He or she used the same name, rather than a transliteration of "Pam" or an authentic Māori name. This in itself is not problematic, but the names used in the other questions of the examination – Anne, Garry, David, and Jim – are all Pākehā. Sheffield School is the only location given. Such monocultural nomenclature is surprising in modern New Zealand/Aotearoa. For most Māori students, sending Christmas cards would not be a family practice. Therefore, perhaps the name "Pam" would seem more likely. Given Pollitt et al.'s (2000) concerns about how difficulties in interpreting the context combined with being miscued by the language, it is possible that some students would struggle with understanding what they were required to do mathematically. This may also contribute to students believing that mathematics is part of the Pākehā world.

The translator used the same letter for the variable "*f*" that was in the English original. This is surprising as the letter "*f*" was chosen as a cue because it was the first letter of "friends" and represents the number of friends in the equation. This cueing was not replicated in the Māori translation. The use of letters drawn from the nouns in algebraic word problems has been shown to be problematic for students' understanding (Reedy, 1999). Furthermore, "f" is not even a letter in the Māori alphabet, thus making it doubly confusing for students doing this problem in *te reo Māori*. Student expectations in reading questions can often support them to adopt the appropriate approach in answering a question (Sweiry, Crisp, Ahmed, & Pollitt, 2002). In this case, the students would have had to first spend time trying to work out what the Māori version of the question was asking, before turning to the English version.

Another difficulty inherent in translation from English to Māori is the relative usage of the passive voice. This is manifested in a variety of ways in Māori. In the case of this problem, the culminating question would be better served by an active clause using *e* ... and the anaphoric particle, *ai*. The use of *e* ... *ai* indicates the present or future tense. The anaphoric particle allows the object of the main clause to be the subject of the sub-clause, thus connecting related ideas or clauses. These particles indicate that a sub-clause follows and help to clarify the sentence. However, many of the questions do not include these. The following is a version of the question written in what we consider to be more appropriate *te reo Māori*.

Kei te tono kāri Kirihimete a Pam ki ōna hoa katoa.
E $2.75 te utu mō tētahi kāri.
E 50 heneti te utu mō tētahi panekuīni.
E $68.25 ka pau i a ia i te hoko kāri me te hoko panekuīni.
Ko te whārite mō te moni ka pau i a ia ko te:
$0.5h + 2.75h = 68.25$
e tohu ai te h i te tokomaha o ngā hoa.
Whakaotia taua whārite nei hei tātai i te tokomaha o ōna hoa e tono kāri ai ia.

The contexts for the questions in this external examination – sending Christmas cards, building up a savings account, using a garden irrigation system, constructing a concrete path, and transporting students in vans – may be seen as culturally neutral, yet some are likely to be beyond many students' experience. This implied that the examiners believed that they had provided contexts that all students would know about, but perhaps they had not considered that some Māori would not have ever experienced them. Christmas cards are something that Māori may know about, but not something that they would ever actually experience. It is unlikely that Māori students could see themselves or other Māori in such contexts. On the other hand, there are no activities that would specifically invoke Māori participants' cultural knowledge. If the students generally were to see Pākehā as belonging to these contexts, would this result in Māori students developing the notion that mathematics was an activity only undertaken by others, not by themselves? The following quote from Campbell et al. (2007), though relating to ESL students, could just as well be referring to Māori-medium students.

> Problems and other mathematical materials are often written using implicit assumptions about the typical student who would use such materials at a particular developmental level or grade. For ESL students, these assumptions may be incorrect. These assumptions, in effect, increase what Paas, Renkl, et al. (2003) termed "extraneous cognitive load" and require students to "search for referents in an explanation". (p. 2)

In the rest of the questions, contexts often are not explained or supported well. Question 7 is based on a water tank that is connected to "drippers" that irrigate a garden. The word "drippers" is written in the original question with quotation marks, highlighting its unusualness as a lexical item. There are no diagrams or photographs to illustrate the context, and so the "extraneous cognitive load" is heavy.

The Māori translation also uses quotation marks for the chosen equivalent, *pū turu-turu*. This phase is impossible to decipher without reference to the English version. Exacerbating the difficulty is the use of *kura* for a "water tank", a word more commonly used to mean "school". This latter meaning initially seems possible and so further problematises the context.

There are other examples where the choice of constructions and vocabulary are baffling. Question 4 reads, "*I kiia a Anne ko te (x – 2) tētahi tauwehe o te* (x^2 *+ 48x – 100)*". This translates as "Anne was said to be $(x - 2)$ a factor of $(x^2 + 48x - 100)$", whereas the original began with "Anne was told that...". The translation would be better started with "*I kiia atu ki a Anne ko te ...*"

Another question in Māori, number 8, contains the information "*He rawaka te raima tō Jim ki te hanga i tētahi ara hīkoi, ko te 9m*^{2} *te horahanga katoa*". The word *rawaka* means "sufficient" as in the English version, but it is a rarely used term. A more common term for "sufficient" is *rahi* or *nui*. It is unclear why such an obscure word was used because most, if not all, students would be obliged to refer to the English original. There would be an increase in cognitive load for students who try to make sense of the Māori version. We suggest that a better translation would be "*He rahi te raima a Hēmi hei hanga i te ara hīkoi, e 9m*^{2} *nei te horahanga*". There are various changes besides the word *rawaka*. We felt that the sentence was better served by using different constructions, plus a change from the possessive "*tō*" to "*a*" as *raima*, in this case, falls into the "*a*" category of possessives. In Māori there are two categories of possession expressed by the base words "*a*" or "*o*". For example, the phrase "Rangi's story" could be translated in Māori as *te kōrero a Rangi* or *te kōrero o Rangi*. However, the first strictly translates as "the story composed (or told) by Rangi", whilst the latter translates as "the story about Rangi". The *raima* in this case was concrete to be used by Jim. The changes we are suggesting overcome, to some degree, the cognitive load from the interactions between the mathematical and contextual language and the culture and life experiences.

In other instances, the Māori translation seems unintelligible. The final instruction of Question 9 is translated as "*Me tuhi mai i tētahi whārite kotahi i te itinga rawa e whakamahi ana koe ki te whakaoti i te rapanga*". The sentence on its own makes no sense, leading to several possible understandings, or none at all. Again, this incomprehension forces the students to refer to the English version.

Improving the Quality of the *Te Reo Māori* Examinations

What, then, is the status of the translation? How does this affect the students' trust in the translation as a "definitive" version? Do the implied culture, language, and ethnicity of the actors in the questions affect the students' valuation of the examination and, by interpolation, their self-efficacy as students of mathematics? Can the Māori students have faith that the translation is a semantically and mathematically sound version of a semantically and mathematically sound original? It is likely that students will lose confidence in the translation over time and refer to the English original even if they think they have understood the Māori translation, or not even bother to read it at all.

We offer four suggestions that could overcome the anomalies found in the 2008 NCEA external assessment. First, the translated Māori version needs to be independently back-translated into English. This would verify the robustness of the initial translation (Brislin, 1970). A follow-up should involve students in the auditing process by soliciting their understandings and opinions the following year. This extended process will also better inform the writers of examinations about which items function well as contextualised word problems and which are better left as simple questions to be solved. Secondly, the Māori translations need to be vetted for low-frequency contextual vocabulary, grammatical and syntactical errors, and unnecessary obtuseness. Thirdly, half of the questions should be written in *te reo Māori* and then translated to English, to enable all stakeholders to appreciate the difficulties and challenges involved.

The fourth suggestion involves a lot more thought and time. The language of word problems, indeed of any mathematics question, that include words outside of the mathematics register has yet to be standardised in Māori. Word problems written in Māori are too new to have yet developed a specific structure. The expectations that students have when reading mathematics problems in English about how to gain the relevant information to solve the problems are not yet built into the way mathematics problems are constructed in *te reo Māori*. Translations, far too often, have to rely on seldom-used vocabulary, or try too hard to capture how the English presents the situation rather than what it purports to present. A metaphorical rather than a literal translation would be more useful for Māori-immersion students.

The context in which these suggestions are provided includes consideration of the time factor in the development of these ideas. In the Welsh-medium education system, translation of assessment items from English to Welsh is done within a two-year cycle, allowing for independent testing of the equivalence of the test items (Evans, 2006). In New Zealand, translation of NCEA exams does not begin until NZQA is informed in April that there will be students sitting the exam in November. This timeline is significantly shorter and, with the small number of Māori-medium students who sit the exams, it is unlikely that there would be a population who could be pre-tested to ensure the equivalence of items.

It would seem that once *te reo Māori* versions of the questions were added to the exams, NZQA felt they had fulfilled their commitment to Māori-immersion students. Given that it is likely that dialectical differences and non-standard use of some mathematical terms mean that students will need to refer to the English version at some point during the exam, NZQA are not obligated to provide a "good" Māori version of the paper. However, the consequence for Māori students of a "poor" Māori version being presented to them is that they could suffer from cognitive overload, as they try to resolve differences between the two versions. Students at the very least will have wasted a lot of their precious exam time trying to make sense of the Māori version before being forced to then decipher the English version. Consequently, Māori-immersion students do much more than what is expected of English-medium students in these exams, and this is likely to be a major contributor to their poor results in the mathematics external assessments.

Students' Responses to Doing Exams Bilingually

In 2005 and 2007, some students who had recently completed external examinations were interviewed about their experiences. Evans (2006) had noted that it was difficult to really understand language-based differences in test performances without having an understanding of how students work out their answers. Therefore, we asked students how they used each of the languages when answering questions, so we could better understand what was most likely to be causing them difficulties.

Before looking at students' opinions about the bilingual exams, we present the accumulated results from 2003 to 2008 at Te Koutu for the external NCEA Level 1 Achievement Standards in Table 4.3. Te Koutu did not do NCEA in 2002 because there were no students who were at Level 1. Although we compare the results with those presented in Table 4.2, the differences in the years as well as the small numbers mean that it can only be a very general comparison. For each achievement result, the accumulated number of students varied between a maximum of 56 and a minimum of 36. Up to 2005, all students at Level 1 completed at least some of the external exams. However in 2006, Te Koutu made available more opportunities for students to complete internal assessments and to gain the necessary numeracy requirements for university in this way. This has become a common practice throughout New Zealand, with most schools putting only those students who are expected to gain external assessments in for those assessments.

The results for Te Koutu are varied. In all Achievement Standards, they outperform the average for Māori-medium schools in Table 4.2. In most cases, the results are better than those for Māori students in mainstream schools. The two exceptions

Table 4.3 Individual achievement standards with Te Koutu students' results combined for 2003–2008

Achievement standard no. (L1 credits) title	Achieved results (%)	Merit and excellence results (%)
90147 (4) Use straightforward algebraic methods and solve equations	61	0
90148 (3) Sketch and interpret linear or quadratic graphs	39	2
90151 (3) Solve straightforward number problems in context	44	0
90152 (2) Solve right-angled triangle problems	77	12.5
90153 (2) Use geometric reasoning to solve problems	80	9
90194 (2) Calculate relative frequencies and theoretical probabilities	26	3

are "Solve straightforward number problems in context" and "Calculate relative and theoretical probabilities". As discussed in Chapter 7, the teachers found probability hard to teach because it requires the students to be able to explain themselves in written sentences requiring a high level of linguistic ability. The straightforward number problems also require very good reading skills and an ability to connect to the contexts for the different problems. However, in other Achievement Standards Te Koutu's results were above the average for non-Māori in mainstream schools. These were "Solve right-angled triangle problems" and "Use geometric reasoning to solve problems." Both these Achievement Standards require students to use symbolic mathematics to show their reasoning, with their reading and writing of sentences not so much an issue. Therefore, it is valuable to find out from students how and why they used the two languages when completing the exams.

On the whole, students switched between the languages to ensure that they gained the most meaning from the question, and thus gave themselves the best opportunity to answer appropriately. As one of the students said, "*nō reira he āhua uaua ki te – ki te um whakaputa i ngā whakaaro ina kāre e mārama ki ngā patai. Engari ka tarai*" (therefore it's kind of hard to develop ideas when you couldn't understand the question. Still, you've got to try). The following extract provides information from another student that summarises some of the information provided by several of his peers. The student had done Level 2 exams, in 2007, and so had completed quite a few external assessments by this time, not just in mathematics, but in other subjects as well.

Interviewer	Ina mahi koe i aua whakamātautau, ka pēhea tō whakamahi i *te reo Māori?*	If you did that exam, how did you use *te reo Māori*?
Student 1	Ā – ōrite ki te reo Ingarihi na... – āe, ōrite ki te reo Ingarihi na te mea i te wā ka haere ki te... ki te whakamātautau, he ō... he whakamātautau i roto i te reo Ingarihi, ā, he mea ōrite i roto i te reo Māori na reira ka mahi te mahi ōrite mō ngā mea e rua.	Same as English. Yes, same as English because when you sit the exam it's in English, same as Māori therefore I use the two languages in the same way.
Interviewer	Nā tērā ka pānui reo Māori nei?	Do you read the Māori?
Student 1	Āe, āe, ka pānui reo Māori, ka tuhi reo Māori.	Yes I read and write in Māori.
Interviewer	I te wā ka whakamahi koe i te whakamātautau, ka whakaaro Māori i a koe e whakautu ana i ngā pātai?	When you are doing the exam, do you think in Māori while you are answering the questions?
Student 1	Āe, na te mea kua ako i te reo Māori – nā reira kei te whakaaro Māori na te mea kua ako i te Tātai i roto i te reo Māori mō ngā tau e hia kē nei, nā reira ka haere ngā whakaaro i roto i te reo Māori.	Yes, because I have learnt in *te reo Māori* – therefore I think in Māori because I have learnt in Māori for how many years now. Therefore I think in Māori.
Interviewer	Ka pēhea tō whakamahi i te reo Ingarihi i roto i aua whakamātautau?	How do you use English in those exams?

Student 1	Ā, he āhua..., ētahi wā he āhua uaua i te mea ko ētahi o ngā kupu kāre i te mārama i te mea kua mā... kua mārama kē ki ngā kupu Māori, nō reira ētehi o ngā kupu Ingarihi kāore anō kia ako, nā reira ētahi wā ka uaua mena ka pērā.	Sometimes it's hard because I do not understand the [English] word. I understand the Māori language one. Therefore some of the English words I have not yet learnt. Therefore sometimes it's difficult if it's like that.
Interviewer	Ka whakawhiti koe i aua reo e rua?	Do you switch between the two languages?
Student 1	Āe, i ētahi wā... i ētahi wā ka um ina kāre au i te mārama i te reo Māori, ka huri atu ki te reo Ingarihi ina kāre i te mārama ki te reo Ingarihi, ka huri atu ki te reo Māori, ā, ka whakangāwaritia i nga mahi. He āhua – āe, ki ō... ki ōku whakaaro kia whakangāwari te whakamātautau, na, pai ake ngā reo e rua i te reo kotahi.	Yes, sometimes. Sometimes I do not understand the Māori, so I switch to English. If I do not understand the English I switch to Māori – this makes things easier. I think this makes the exam easier. It's better in two languages than the one.
Interviewer	Ina he whiringa, ka hiahia koe ki te mahi i aua whakamātautau i te reo kotahi?	If you had a choice, do you want to do the exam in the one language?
Student 1	Kāo. Pai... pai ake te mahi i ngā reo e rua.	Better to use the two languages.
Interviewer	He aha ai?	Why?
Student 1	Pai ake te reo Māori me te reo Ingarihi i te reo Ingarihi anake. Ko te mea e rua ō whakamāramatanga, nā reira, āe, ka ngāwari i reira.	It's better in Māori and English rather than only English. The thing is that you have two explanations, therefore it makes it easier.

Students were able to supplement their dominant language by using words from the other language to help clarify the question. For most students, the dominant language for discussing mathematics was *te reo Māori*, and so they would just check an unfamiliar word by looking at the English version. *Te reo Māori* was their language for thinking about mathematics because this was the language that they had studied most of their mathematics in. For this student, having two languages gave him two understandings, and he really valued being bilingual.

The type of words also indicated to the students which language they should use. Contextual words seem to be easier to understand in English, whilst mathematical vocabulary is handled more easily in *te reo Māori*. This distinction can be seen in the following extract from an interview with a student in 2005.

Student 2	Ko te reo, i te mea... ā, kāore au i te tino um mārama ki ngā pātai, arā, kāre... kāre he uaua mōku ki te um mōhio... ō... ki te mōhio he aha te hiahia o te pātai.	The language ... I do not really understand the questions. It's difficult for me to understand the intent of the question.
Uenuku	Āe. [nods]	Yes.
Student 2	Kei te rapu i te aha, āe, engari, āe.	What is it trying to find out.
Uenuku	Tēhea reo?	Which language?

Student 2	Ngā reo e rua, te reo Māori me te reo Pākehā.	Both languages, Māori and English.
Uenuku	He aha te huarahi ka whai koe kia... kia mārama ai koe ki te hiahia o te pātai?	What strategy do you follow so that you understand the intent of the question?
Student 2	Ka pānui au i te pātai i roto i te reo Māori i te tuatahi, anā, ki te kore au e mārama ka pā... ā, ka huri au ki te pātai i roto i te reo Pākehā. Anā, um... i te mea i te parakatihi, i te mahi kē ngā um i roto i te reo Pākehā, nō reira i āhua um... um hoki ngā mahara ki te hiahia o taua pātai ka oh... nō reira koira tōku huarahi hei mōhio ki te kimi i te aha.	I read the question in Māori first. If I do not understand the question, I switch to the question in English. Because the practice paper was in English I can remember the intent of the question. Therefore that's my strategy to know how to find whatever.
Uenuku	Ka hoki ki te rerenga Māori i ētahi wā?	Do you return to the Māori sentences sometimes?
Student 2	Āe.	Yes.
Uenuku	He aha anō ngā... he aha anō ngā take e hoki ai ki te mea Māori? He aha anō te mea ngāwari atu i tērā take?	For what other reasons do you return to the Māori? What other things make it easier?
Student 2	Ko ngā kupu pērā ki te taunga me ngā pūwāhi me ērā mea, i te mea kua waia kē ki aua kupu. Kāore au e waia ki roto i te reo pākehā. [nods]	The [Māori] words like *taunga* [co-ordinate] and *pūwāhi* [point] and those things because I am familiar with them. I am not familiar with [these] words in English.

This student relied on the predictability from answering practice English papers to help her work out what the questions really were. In 2005, there would only have been three years of previous exams in *te reo Māori* that could be used for practice. Figure 4.4 provides an example from 2007 of a student's attempt at completing a practice exam in *te reo Māori*. There were many more practice English exams that could be purchased by students. Therefore, it was not surprising that students had used the English practice exam papers to gain an insight into what would be expected of them when they sat their own exam. The lack of resources in *te reo Māori* is an ongoing issue in Māori-immersion education.

Another student doing Level 1 in 2005 started by looking at the questions in *te reo Māori*, but found that the way that the sentences were set out made them difficult to understand. This student struggled with mathematics, and so the unfamiliarity of the grammatical expressions only added to her uncertainty about what was required of her. Campbell et al.'s (2007) model suggests that the combination of mathematical and contextual language with the mathematical content was placing the student in a situation of cognitive overload. This student was thankful that her class had looked at the way that questions were expressed in English from past exams. The grammar of the English questions provided a predictability that *te reo Māori* questions did not have.

PĀTAI TUATORU

Mā te Kaimāka anahe

He tākaro tēnehi tonu a Tom, Dick me Harry.

Ka tākaro a Tom rāua ko Dick, ka toa i a Tom $\frac{1}{4}$ o ngā kēmu.

Ka tākaro a Tom rāua ko Harry, ka toa i a Tom $\frac{4}{5}$ o ngā kēmu.

Kotahi te kēmu tēnehi i tākaro a Tom ki a Dick, ā, kotahi hoki ki a Harry.

E whakaaturia ana ētahi o ngā kōrero ki te hoahoa i raro nei:

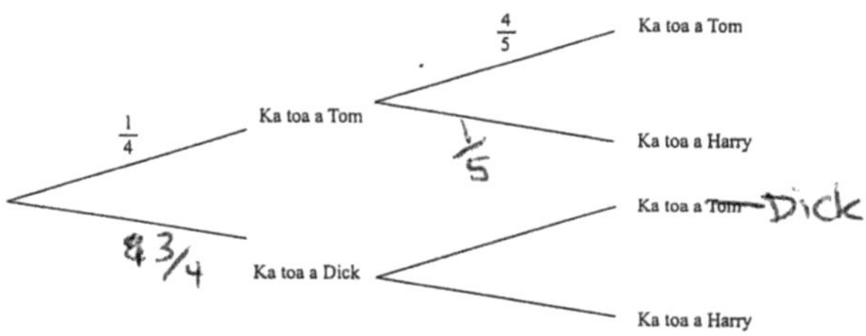

(a) He aha te tūponotanga ka toa a Tom ki ngā kēmu e rua tahi?

$\frac{1}{4} \times \frac{4}{5} = \frac{4}{20} = \frac{2}{10} = \frac{1}{5}$

Te tūponotanga: $\frac{1}{5}$

(b) He aha te tūponotanga ka kotahi pū o ngā kēmu e rua e toa ai a Tom?

T H D T

$\frac{1}{4} \times \frac{1}{5} = \frac{1}{20}$ $\frac{3}{4} \times \frac{4}{5} = \frac{12}{20}$ $\frac{6.5}{20} = \frac{13}{40}$ $1+12 = \frac{13}{2}$

Te tūponotanga: $\frac{13}{40}$

(c) Kia hauhau te rā, kāore e tino pai te tākaro a Tom i a Harry.
Ko te tūponotanga ka toa ia ina tākaro ia i a Harry i te rā hauhau he $\frac{2}{5}$ noa iho.
Te tūponotanga ka tū tā rāua kēmu whai mai i te rā hauhau he 0.3.

He aha te tūponotanga ka toa i a Tom te kēmu?

$\frac{3}{10} \times \frac{2}{5} = \frac{6}{50} = \frac{3}{25}$

Te tūponotanga: $\frac{3}{25}$.

Pāngarau 90194, 2006

Fig. 4.4 Student attempt at completing a practice exam question for Achievement Standard 90194: Calculate relative frequencies and theoretical probabilities

Some of the issues raised in the analysis of exam questions were also signalled by the students. Although only a small number of students were interviewed, two in 2007 and five in 2005, their answers were similar. Pollitt and Ahmed (1999) suggested that in reading exam questions, students are looking for prompts that indicate what they should do. In the interviews, students clearly stated their need to work out what they had to do by reading the question carefully. Unfamiliarity with the terms in either language meant that the students used the other version to check their understanding. Although the students did not comment on the contexts as being difficult, there would be less reason to switch to English if the contexts for the problems were familiar to the students. Switching between languages takes time, and NZQA could consider allowing Māori-immersion students more time to complete the exams so that they are not disadvantaged by having to do this. This is a practice used by some universities where students are required to do exams bilingually. For example, at the University of Auckland, ten minutes extra reading time is provided to students who are sitting exams in Māori with English translations.

Meeting the Challenge of Doing Exams Bilingually

As a practice in Schatzchi's (2005) sense, completing examinations is complex. The history that positions examinations as a neutral way of determining academic merit was connected to the social and economic needs of a society that could be said to no longer exist. To some degree, the change from a norm-referenced assessment to a criteria-referenced assessment reflects the change in the needs of a society that no longer wants half of its students to fail at the end of their eleventh year of high school. The majority of jobs that are now available require students to be able to deal with a range of complex and abstract ideas. The provision of exams in *te reo Māori* is part of this wider discussion about assessment, but in many ways it reflects some of the wider concerns. In dealing with the complexity of providing appropriate exams for Māori-medium students, there are tensions between providing an education for the good of each individual and providing an education for the good of the wider society (Kemmis, 2009). For the parents and teachers, the issue is the balance between the needs of the language's revitalisation process and those of the students, giving them the best opportunity to gain good jobs. This ongoing tension is a feature of the cultural-discursive, material-economic, and social-political dimensions that shape responses to the issue.

Parents and teachers are torn between wanting what is best for the language and what is best for their children and students. This can be seen in their repeated requests in the 1990s for exams to include *te reo Māori* as well as English. Once Māori-medium education was recognised by the state, parents and teachers saw it as being obligated to provide bilingual exams. Even though it took ten years for this to be achieved, parents and teachers continued to operate with this belief. Parents and community members concerned with the revitalisation of *te reo Māori* ensured the provision of bilingual exams by continually reminding NZQA of the contradiction in educational aims if exams remained monolingual. Māori parents and teachers were able to disrupt power relations so that their expectations of what should be provided to their students were fulfilled.

However, the fact that after seven years the bilingual exams are still full of errors means that New Zealand society through NZQA has not yet accepted the importance of Māori-immersion students being provided with the same opportunities to show what they can do as their English-medium peers. The current rates of poor results in external mathematics Achievement Standards will continue unless the level of commitment to these students is changed. Such a change is likely to be difficult for parents and teachers to achieve because of the care that must be taken in raising the issue into general consciousness.

Although NZQA would not acknowledge that they are resisting the challenge, their lack of quality assurance for the exam suggests that this is the case. In some ways, the readily available statistics that are provided, and which show that Māori-immersion students are doing better than their Māori peers in mainstream schools, limit discussion about these same students' poor results in mathematics and science. There is a tension here because, if discussion about Māori-medium schooling revolves around poor results, national discussion can revert easily to one about this schooling system being a failure. On the other hand, not having a discussion about the difficulties facing Māori-medium students in completing external achievement standards in mathematics and science restricts the ability of parents, teachers, and NZQA to recognise and deal with the problem. Consequently, it is important that the way the discussion is phrased illustrates the inconsistencies between the internal and external achievement standard results. It may be that in so doing, new opportunities, such as Cathy Dewes's suggestion for a completely different assessment scheme, become part of the discussion not just for Māori-immersion students but for English-medium students as well.

Part II
Meeting Mathematical Challenges

Māui

Taranga kept dragging the boy until they were well out of earshot. "Go on, continue with your tale".

Defiantly, the boy stared directly at her before saying, "It is no tale". He carried on nevertheless, "Thinking me dead, my mother cut off her topknot. She carefully wrapped me in the hair and saying a few prayers, she lowered me onto the sea's surface to be carried away by the outgoing tide".

Taranga wondered how the boy knew all this. Though the details differed, she was suddenly swept back to the horror of that awful night. The sudden, terrible pain and the anguish and terror of seeing the incredibly small inert baby lying on the blood-soaked sand. Her wailing was due more to fear, fear that the malevolent spirit of this miscarried, misshapen child would haunt her for years to come, as the unhappy unborn were known to do. She had cut off her hair in contrition, hoping her sacrifice would be recognised. She had hardly looked down as she wrapped the hair around the bloody body of her tiny, tiny son. She had then hurriedly muttered some prayers – for herself rather than for the baby – before she threw it as far as she could beyond the crashing waves, not even looking to see where it landed. And here it was, the spirit of her last-born that had finally found her. She could hardly breathe as she waited.

The boy's face softened somewhat. "I am no spirit. I was barely alive when I was born, but I was alive. The prayers you said must have been powerful, for I never sank below the surface, though I floated for days and days. Jellyfish and kelp became wrapped around your hair, protecting me from the sun and seawater".

Taranga said nothing, not daring to believe the child's fantastic story.

He continued, "Finally I was washed ashore again. The stench of the dissolving jellyfish and rotting kelp drew swarms of flies and hungry seagulls. Luckily, I was seen by my ancestor, Tamarangi. He shooed away the gulls and flies and them pulled away the kelp, jellyfish, and hair. He took me back to his house and put me in a basket that hung from the rafters above the fire. The heat dried me out, and the smoke made me sneeze, making the phlegm gush from my nose and mouth. I was alive!"

Taranga had started to shake. She knew all about Tamarangi and his powers.

Māui-pōtiki went on. "He brought me up. He taught me all that he knew, both practical and spiritual. He also told me who I was and where I came from. Now that I am old enough he sent me this way, telling me how to find you and my brothers and sister. Like I said, *nāu anō au*".

This part revolves around issues of using language to support mathematical thinking. When Māui was born, his life was unrecognised. When the mathematics register was developed, it was done for entirely practical purposes in relation to teaching it in classrooms, and the value of the language for thinking mathematically went unrecognised. In considering the orders and arrangements that affect how practices operate, the social-discursive influences of doing mathematics is our concern here (Kemmis & Grootenboer, 2008). The mathematical register within the natural language facilitates but also constrains the way that mathematical thinking can be discussed and used. However, like Tamarangi's recognition of Māui's life force, the strength that is within the language cannot be utilised if it is not recognised.

Thinking mathematically needs to be encouraged by using the resources within the language that the teachers and students use for communication. In recent years, it has been recognised that thinking is something that does not occur exclusively within the confines of an individual's mind (Radford, 2003a). The community in which the thinking takes places is considered to make a significant contribution to how the thinking is shaped. There is a major role for language in the development of understanding through this joint negotiation in the learning community. Radford considered that it is the cultural/historical connection between words and actions that enables thinking to occur: "In this line of thought, cognition appears as definitely consubstantial with the various forms of social organization, types of production and cultural modes knowing." (Radford, 2003a, p. 125). Therefore, supporting students to think in mathematics in *te reo Māori* is not a simple matter of changing the language of communication. There are a number of wider considerations that are related to fulfilling the aims of *kura kaupapa Māori*, but also related to the students' ongoing lives as Māori living in a Westernised country. All of the challenges outlined in this part could be considered as being at the stage of finding appropriate ways forward. In meeting the challenges, "normal" becomes redefined, but this process is not completed as some issues relating to these challenges are still to be worked through.

Kura kaupapa Māori were established to both revive *te reo Māori* and ensure that students improved their academic results. Accompanying the elaboration of the mathematics register, there was the need for students to use it effectively to think mathematically. Even with all the *aroha*, or love, that went into the development of *kura kaupapa Māori* and the mathematics register, it could have been all too easy for mathematics in *te reo Māori* to be stillborn. Without a caring environment in which it was nurtured, it may have turned out to be a lifeless imitation of its English-medium counterpart. Like Māui's search for himself, thinking mathematically from a Māori worldview means identifying the valuable resources within mathematics and the language, thereby identifying familial context. Without considering how mathematics intersects with the Māori language, it is difficult to see how thinking mathematically can be done coherently. This is a challenge for all *kura kaupapa Māori*.

The role of language in supporting students to think mathematically has been emphasised for some time including in the English and Māori-medium mathematics curriculum statements (Ministry of Education, 1992, 1996). In describing the mathematical processes that support learning, the Māori-medium curriculum states the need for students to "*Whakaputa whakaaro pāngarau* [to express mathematical ideas]" (Ministry of Education, 2006, p. 14). Although teachers are required to show this aspect in their planning, the curriculum does not explain what this means in practice.

The importance of language in helping children make sense of their world is supported by Campbell and Rowan (1997) who assert that "language has the power to help children organize and link their partial understandings as they integrate and develop mathematical concepts" (p. 64). Language does not do the thinking for people but rather it supports the organisation of that thinking and acts as a viaduct to other knowledge. The historical and cultural development of the language will have an impact on the organisational structures that connect new thinking to already understood knowledge. Roberts (1998) discussed the relationship between language and worldview to show how a person's purpose for noticing something is affected by his or her own perspective, which draws on the knowledge, including cultural understandings, which they already have.

The teaching approaches recommended by the national curriculum statement (Ministry of Education, 1992) are, at heart, constructivist. The constructivist view is that people make "sense of the world in ways that are always historically and culturally specific" (Begg, 1999, p. 5). Developing a shared meaning of mathematical ideas is a key process within constructivist learning theory (Good & Brophy, 1990). This means that children should have the ability to verbalise someone else's understandings to themselves so that they can re-organise external language into an "inner language" or "internalised thought".

> The learning of mathematics is fundamentally a matter of constructing mathematical meaning. The environment of the mathematics classroom provides experiences which stimulate this process of construction. While the mathematical knowledge of school children will incorporate visual imagery, both at the level of iconic thought and more elaborate visual representations (geometrical, graphical), mathematical meaning requires a language for its internalization within the learner's cognitive framework and for its articulation in the learner's interactions with others. (Clarke, Waywood, & Stephens, 1993, p. 235)

Mousley (1999) reported that many mathematics educators believe that students who understand mathematical concepts are likely to be able to talk about them, explain them, make links between them, apply them across contexts, and represent them.

However, Radford (2008) problematised the constructivist view of learning as not being explicit enough in recognising the role of culture in the way that learning is achieved. Consequently, he considered learning as the relationship between objectification and subjectification which was "a process of knowing but also becoming" (p. 225).

> According to the theory of knowledge objectification, learning does not consist in constructing or reconstructing a piece of knowledge. It is a matter of actively and imaginatively endowing the conceptual objects that the student finds in his/her culture with meaning. It is what we will later call a process of objectification. (Radford, 2008, p. 223)

In the process of objectification new knowledge, a person reflects both on what he or she already knows as well as on how this affects who he or she is. This reflective subjectification process is mediated by the cultural-historical understandings that are embedded in the knowledge object and the person himself/herself. Consequently, language is just one resource that is available during the objectification/subjectification process.

> This led us to envisage a broader context large enough to conceive of tools, things, gestures, speech, writing, signs, and so forth, in relation to the individuals' activities and their intentional goals. In this broader context, we called semiotic means of objectification the whole arsenal of intentional resources that individuals mobilize in the pursuit of their activities and emphasized their social nature: The semiotic means of objectification appear embedded in socio–psycho–semiotic meaning-making processes framed by cultural modes of knowing that encourage and legitimize particular forms of sign and tool use whereas discarding others. (Radford, 2003b, pp. 43–44)

In *kura kaupapa Māori*, the cultural and historical ways of objectifying knowledge will not be the same as in English-medium schooling even when the mathematical knowledge is considered to be the same. The way that meaning is given to mathematical object will be different because of the organising structures within Māori culture including the language. Brown (1997) discussed how in the process of engaging with a mathematical task, he would describe the mathematical ideas in a way that reflected his individual understanding, but in using the language of the society in which he lived, his membership to that society was reinforced. One of the challenges in trying to understand how students can develop mathematical thinking in *kura kaupapa Māori* has been to identify what are the similarities and what are the differences, in how the languages are used in doing mathematics.

Within the wider field of mathematics education, there seem to be a number of issues relating to the forms of communication expected of students. Underlying these issues is the expectation that children need to communicate effectively. Sfard, Nesher, Streefland, Cobb, and Mason (1998) stress the importance of developing children's communication skills but question how this will be done and what should actually be taught. They commented that this issue has not been given much thought by the mathematics education community. They argued that children need to be taught how to communicate with their peers and teachers so that there is a baseline of shared understanding.

In this part, we consider how speaking and writing in Māori is connected to thinking mathematically. In particular, we look at the resources within the language that can support children to think mathematically and some of the challenges involved such as how written mathematical communication can be used to support fluency in what was traditionally an oral language. We also explore how teachers make students aware of the transparency of some of the mathematical terms in *te reo Māori*, with specific reference to probability. Like Māui's fruitful search for his family, we feel that the search for the uniqueness of the features of *te reo Māori* has real value in supporting Māori students to think mathematically.

Chapter 5
The Resources in *Te Reo Māori* for Students to Think Mathematically

Te reo Māori, like all languages, contains features that can be used to support thinking in mathematics. Some features exist traditionally within the language, such as logical connectives, and others have become newly available with the development of the mathematics register. The challenge continues to be one of identifying these features so that they can be used in such a way that the integrity of the language is maintained and that the benefit to students when doing mathematics is realised. For second-language learners of *te reo Māori*, such as most teachers and students in *kura kaupapa Māori*, the influence of English often makes it difficult for them to appreciate the features in Māori which could contribute to mathematical thinking. Once the features have been identified, there are further challenges in being able to understand why some terms are difficult to learn. The ultimate aim is to support students to think mathematically through explaining and justifying what they know.

Thinking mathematically is about using mathematical understandings to create mathematical solutions to problems. Using a symbolic interaction perspective, Erna Yackel (2001) observed that

> [s]tudents and the teacher give mathematical explanations to clarify aspects of their mathematical thinking that they think might not be readily apparent to others. They give mathematical justifications in response to challenges to apparent violations of normative mathematical activity. (p. 14)

Language, including diagrams and symbolic equations, is more than just the vehicle for the thinking. The linguistic features of a language support or constrain the way that ideas are discussed. Halliday (1978) summarised how languages both reflect and shape different worldviews of people from different cultures:

> Languages have different patterns of meaning – different 'semantic structures', in the terminology of linguistics. These are significant for the ways their speakers interact with one another; not in the sense that they determine the ways in which the members of the community *perceive* the world around them, but in the sense that they determine what the members of the community *attend to*. (p. 198)

However, recent work suggests that even when a language has limited counting words, speakers can still complete enumeration activities (Butterworth & Reeve, 2008). This reinforces the fact that language can constrain but not predetermine

T. Meaney et al., *Collaborating to Meet Language Challenges in Indigenous Mathematics Classrooms*, Mathematics Education Library 52,
DOI 10.1007/978-94-007-1994-1_5, © Springer Science+Business Media B.V. 2012

what can be seen and acted upon. Thinking mathematically involves being able to perceive a situation and recognise how mathematics could be utilised to resolve an issue within that situation. Burton and Morgan (2000) stated that "[t]he language used in mathematical practices, both in and out of school, shapes the ways of being a mathematician and the conceptions of the nature of mathematical knowledge and learning that are possible within those practices" (p. 445).

English-medium mathematics education research has suggested for some time that language has a considerable impact on the teaching and learning of mathematics (Cocking & Mestre, 1988; Ellerton & Clements, 1990; Durkin & Shire, 1991). Although the focus of much early research was on the specific vocabulary terms (Love & Tahta, 1991), this was replaced by an interest in the features of English that are significant in explanations and justifications in mathematics, which support the solving of problems. For example, Bills (2002) highlighted certain word-level features as being useful indicators of students' mathematical thinking. Personal pronouns ("you" and "I"), present tense, and logical connectives such as "because", "so", and "if" were more likely to be found in appropriate answers to mental arithmetic questions (Bills, 2002).

Using slightly different resources to those in English, the mathematics register in *te reo Māori* supports mathematical thinking in a different manner. In the next section, extracts from transcripts of lessons and staff meetings illustrate how the mathematics register in *te reo Māori* is used for thinking mathematically. Although we concentrate on spoken language in this chapter, mathematics is often done in conjunction with some form of written text, and we have included these when relevant. Chapter 6 focuses more on how writing in *te reo Māori* contributes to students' thinking mathematically.

Identifying relevant features in the mathematics register that support mathematical thinking needs to be done in conjunction with fulfilling the aims of *kura kaupapa Māori*. Thus, the use of the mathematics register should help students achieve academically, but also support the revival of *te reo Māori* by using it to fulfil a range of different functions. Nevertheless, there is still room for improvement in how the students use the mathematics register, and this is an ongoing challenge.

Resources in *Te Reo Māori*

Given that Māori-immersion education was set up to reverse the decline in Māori language (Spolsky, 2003), it has been recognised that there is a need to ensure that "the authenticity of the language is maintained" (Christensen, 2003, p. 12). Māori mathematical discourse has several distinct characteristics that are similar to those found in the English discourse. It is conceptually dense and jargon-filled (Halliday, 1978; Pimm, 1987; Dale & Cueras, 1987). There are also linguistic characteristics specific to *te reo Māori* which can be used to discuss mathematics and, when continuously used, can enhance the learning of *te reo Māori* (Barton et al., 1998). For example, a very important construction in Māori, and one which is used more frequently than its English equivalent, is the passive tense (Harlow, 2001). A feature of mathematics is that there is an inherent requirement to perform certain actions – to add, to multiply, to increase, to find out, to solve, and so on. In *te reo Māori*, when an action is required, in most cases a passive tense is used. Māori passives have a variety of suffixes, and there are some restrictions on their use. Therefore, learning mathematics in the medium of Māori supports the learning of this very important linguistic construction. Similarly, Māori verbal numerical markers do not have English language counterparts and differ according to function of the grammatical expression (Trinick, 1999). For example, the verbal particle *ka* is used when counting, *e* when quantifying, and *kia* when expressing a need for a certain number of things. In *te reo Māori*, numbers are preceded by a range of particles depending on the function and context (Barton et al., 1998).

Concerns have been raised about the possible implications for *te reo Māori* as a consequence of its use for discussing mathematics (Barton et al., 1998). For example, it would be a great pity if the grammatical structures used in English to discuss mathematics were superimposed onto *te reo Māori*, so that the language became a Māorified version of English. Whilst *te reo Māori* is traditionally characterised by the liberal use of passive verbs, some writers argue that many contemporary speakers and learners of *te reo Māori* have an inability to make use of these passives (Barton et al., 1998; Harlow, 2001). English is much more likely to use the active tense in situations, where native speakers of *te reo Māori* would apply passives. Christensen (2003) found that the Māori-immersion teachers who had learnt mathematics in English resisted discussing mathematics in the passive voice.

> Difficulties were experienced because in Māori the words do not follow the sequence of the written symbols, as they do in English. English was also seen to be more concise than Māori. For this reason, many teachers and students simply follow the linguistic structure of English, using Māori words. For example, an addition problem is written in symbols as *3 + 2 = 5*. In English it is most common to say this as it is written, symbol for word, *three plus two equals five*. In Māori it is linguistically correct to begin with the verb *tāpirihia te toru me te rua, ka rima*. However, many teachers and students have adopted the English structure, saying *toru tāpiri rua ka rima*. While it may be pragmatic to accept this borrowed linguistic structure as an example of language change resulting from contact between English and Māori, it is unclear whether such a borrowed structure used specifically for pāngarau [mathematics] could transfer across to general language use. (Christensen, 2003, p. 37)

Therefore, the problem may not necessarily be the inability to use the passive voice, but rather the inability to choose when it is appropriate to apply it. In English, mathematics is often discussed without reference to an active participant as the action has been included in a nominalisation, and the verb identifies the type of relationship involved (Meaney, 2005a, 2005b). The natural use of the passive voice in *te reo Māori* may well support the same conceptualisation more easily, so it may be valuable for students to learn how to use it in *te reo Māori* from an early age.

Rather than have to make Māori sound like English in order to discuss mathematics, we argue that the authentic resources within contemporary *te reo Māori* can provide students with resources to think mathematically. In the following sections, we outline some of these resources.

Linguistic Markers

One beneficial resource is the linguistic markers within *te reo Māori* that forewarn listeners about the type of information that is to follow. These markers assist listeners' thinking, because they add meta-level information about the importance of what they are receiving. Although English has some ways of forewarning listeners about the type of information that is to follow, there does not seem to be the diversity that is available in *te reo Māori*. One of these markers, *kē,* tells the listener that what is to follow is unexpected. Another, *arā*, is used to emphasise that an elaboration is following.

Y6Teacher: Ānei tētahi o ngā ahutoru, arā, te koeko tapawhā, mahara? (2005 lesson)	Y6T: Here is one of the 3D shapes, [namely] the square pyramid, remember?

This utterance began the first of this teacher's filmed lessons at Te Koutu in 2005, and referred to material covered in the previous lesson. The teacher highlighted one term *ahutoru* (three-dimensional shape) as the word that needed to be recognised and understood by the students. *Arā* then emphasised that an elaboration was coming. Although the term was used in a previous lesson, the teacher assumed that many of the students still needed to have it highlighted.

In the next extract, which also comes from the same teacher's 2005 set of lessons, the teacher's language suggests that she expected some students to struggle to follow the logic in the argument being presented. She used words and commands to ensure that they paid attention to the important sections.

Y6Teacher: Tekau ngā tapa, tekau ngā mata me ngā akitu, tekau mā rua ngā tapa, tāpirihia kia rua, ā, ka tekau mā rua kē tērā. Heoi anō, i mutu i te karaehe, i kā mai kē tētahi; "Whaea, kei te hē tētahi o ngā mahi, me kā, ngā kaute, kua hē tētahi o ngā wāhanga." Ko [Ākonga 1] tērā, he aha tāu i kite ai?	Y6T: 10 sides, 10 faces and vertices, 12 sides add another 2, that's 12. However, at the end of the class, someone said "Whaea, some of the working out is incorrect, according to calculations one side is incorrect". That's [Student1], what did you discover?
Student1: E waru ngā tapa?	Student1: Eight sides.
Y6T: E hia?	Y6T: How many?
Student1: E waru ngā tapa.	Student1: Eight sides
Y6T: Me whai kē mehemea kei te tika ia. Tahi, rua, toru, whā, rima, ono, whitu, waru, nā reira, kāore ko te tekau. Nā reira, kei te tika te maha o ngā mata me ngā akitu?	Y6T: Follow along to see if he's got it right. One, two, three, four, five, six, seven, eight. Therefore it's not ten. Therefore, is the number of sides and vertices, correct?
Students: Āe!	Students: Yes!
Y6T: Āta whakaaro koa!	Y6T: Please think carefully

Students: Āe!	Students: Yes!
Y6T: Āe, i te mea he aha tētahi atu huarahi i kite kē?	Y6T: Yes, because what other way did you discover to solve the problem?
Students: Tāpirihia te rima ki te rima?	Students: Add five to five?
Y6T: Nā reira, kei te kōrero, i rongo koe, koutou? I a ia e kā ana? Kōrero mai anō koa, tama.	Y6T: Therefore, you're talking, did you, all of you, hear what he was saying? Tell us again, please.
Students: Tāpirihia te rima ki te rima?	Students: Add five to five?
Y6T: Tāpirihia te maha o ngā mata ki te maha o ngā akitu, kua puta kē ko te tekau, nēhā? Te maha o ngā tapa, me kā, waru ināianei.	Y6T: Add the number of faces to the vertices and it will be 10. The number of the sides, we need to say, is eight.

This was part of a discussion of Euler's rule (Vertices + Faces – Edges = 2) applied to a pyramid and how some of the previous day's work had been incorrect. The *kē* highlighted for the listeners that they should notice and be surprised by what followed. It acted as a scaffolding device for students' listening so that they could understand the differences between what had been said on the two days. This was further emphasised by the teacher with the command *Āta whakaaro koa* (Please think carefully) which occurred a few turns later. Once the student had responded to the initial question, the teacher re-emphasised the need to listen. She then had the student repeat what he had said. These examples suggest that the teacher was confident that the students would understand what was being discussed but, because of its complexity, she needed to remind them to be careful so that they would not miss valuable information.

Linguistic markers, such as *arā* and *kē*, which forewarn listeners that important information is to follow, can help students to focus on what the speaker feels they should be paying attention to. These markers were used by the teachers and did not appear to the same degree in students' oral explanations and justifications. If students could learn to make use of these resources, they would not only be showing a rich command of *te reo Māori* but also a more in-depth understanding of the mathematics that they were describing.

Transparency Within Terms

As is discussed in Chapter 2, the development of terms for the mathematics register in *te reo Māori* was done in such a way that the meaning of the terms should be transparent to the learner. In this section we look at how the teachers at Te Koutu made students aware of this transparency.

In the following extract, the teacher explicitly made the students aware of how the label for a "square" in *te reo Māori* provided them with the clues about its definition. The word for square in Māori is *tapawhā rite*, which literally means "four equal sides". She emphasised the features of a square through words, symbols, and diagrams.

I kā koe i mua he tapāwha rite. He aha te tikanga o tērā? He pai ngā ingoa Māori no te mea ka whakamārama i te āhua i roto i te ingoa, nē? He tapawhā rite. He aha te tikanga o te rite? [draws shape on board] He ōrite te aha? He rite ngā taha. Mehemea ka whakamahia au taku rūri. . . he rite ia taha? Nā reira he tapawhā. . .. He tapawhā. . . he tapawhā rite, na te mea he ōrite ngā taha.	You said earlier it was a square. What does that mean? Māori names are good because the shape is explained in the name, isn't it? A square. What is the meaning of "same"? [draws shape on board] What is the same? The sides are the same. If I use my ruler are the sides the same? Therefore, it's a quadrilateral . . . a quadrilateral. . . It's a square, because all the sides are the same.

If students are not familiar with, or do not use the cognates of mathematical terms in their conversational language, they are unlikely to benefit from these everyday meanings when the words are introduced into a mathematical setting. For instance, if students do not have *horahanga* in their conversational language, which means a "spreading out [of food]", then they are unlikely to see a connection with its mathematical meaning of "area".

The transparency of the mathematical meaning of the terms has the potential to support students in thinking mathematically. Yet, this is unlikely to happen without instruction. The teachers at Te Koutu felt that the students had to learn the conversational meaning for such terms as *mua* ("before" or "in front of") and *muri* ("after" or "behind") before these terms could be used for talking about "the number before" or the "number after". Christensen (2003) noted that the teachers in the Poutama Tau – a professional development project on numeracy – also struggled with these terms. In the diagnostic interview that teachers gave students as part of Poutama Tau, they had to ask:

> Kia tatau whakamuri mai i te 23. (Count backward from 23.)
> Kia tatau whakamua koe, atu i te 10. (Count forwards from 10.)
> Some teachers recognised that the use of these words in Māori is different from their equivalent use in English and that this may be one of the reasons for confusion, especially for teachers and students whose stronger language is English. (p. 37)

Although the transparency of some terms such as *tapawhā rite* (square) supports students' understanding and thinking in mathematics, not all terms that were chosen for transparency turned out to be as transparent. As described in Chapter 9, not all of the teachers at Te Koutu were aware of how the terms had been constructed. Consequently, they were still grappling with how best to support students to gain the vocabulary to achieve academically, and to use *te reo Māori* fluently in a range of contexts. This challenge will be ongoing.

Logical Connectives

Western mathematics utilises many types of relationships at different levels. At one level is the nature and origin of mathematical objects and their relationship with language. For example, numbers are related to other numbers by such relations as "greater than" (*nui ake*), "less than" (*iti iho*), and "equal to" (*ōrite ki*). Additionally, a "relation" in Western mathematics can be defined as a set of ordered pairs {(1,3), (2,6), (3,9)...}. In this relation, the ordered pair is connected by a mathematical relationship of multiplying by three.

The syntax of the language describes these mathematical relationships (Carrasquillo & Rodriguez, 1996). One syntactical device is the logical connectives; these are words or expressions between clauses or sentences that are used to join or connect ideas that have a logical relationship (Dawe, 1983). The types of relationships indicated by these expressions include time and space, enumeration and exemplification, amplification and contrast, inference and summation, cause and effect, etc. Within each relationship category, the logical connectives, which join the ideas or clauses, are used differently, with different grammar and punctuation.

Logical connectives determine what can be inferred from these relationships, and mathematical reasoning relies heavily on their use. Research has shown that when students read mathematics text and/or engage in mathematical conversations in English, they must be able to recognise logical connectors, and the situations in which they appear (Spanos, Rhodes, Dale, & Crandall, 1988). Solomon and O'Neill (1998) argued that "[m]athematics cannot be narrative for it is structured around logical and not temporal relations" (p. 217). Generally in narratives, cohesion is achieved by placing a series of events in a timeline. In mathematics, cohesion is achieved by logically joining separate ideas together. For example, in problem solving "[a] convincing argument makes a clear connection, using reasoning, between what is known about a problem and the suggested solution" (Meaney, 2007, p. 683). Logical relationships are timeless and, although time markers are common in recounts, they are inappropriate in discussing mathematics. For English speakers, Esty (1992) stressed the importance of "five key logical connectives: 'and', 'or', 'not', 'if ... then' and 'if and only if' ", which provided mathematics students with an understanding of when equations were true and, therefore, provided the limits of their generalisations.

For Māori, relations are also important and vary to suit different contexts. The word "relation" can be translated as either *whanaunga* or *pānga*. However, both these words are context specific. *Whanaunga* is a generic term applied to kin of both sexes related by marriage, adoption, and or descent. This word implies some human kinship relation. *Whānau* terms are considered inappropriate to use when describing "relationships between mathematical objects" (Trinick, 1999); it is more appropriate to use terms like *pānga* (a connection) or *tūhono* (join), for non-kinship/human relations.

Te reo Māori has an abundance of logical connectives that illustrate the range of possible relations. Table 5.1 is a sample of *te reo Māori* connectives.

Table 5.1 Logical connectives in *te reo Māori*

Relationship category	Logical connectives	English translations
Time	*kia*	when, until – used for future time
	rā anō	right to, as far as, since long ago
	i	while, during
	ina	for, since, inasmuch as, when, if, and when.
	muri	after, afterwards, the time after, the sequel – often modified by *mai, iho*, or *atu*.
	ka ... ana	when, whenever
	tonu & rawa	"as soon as" and "by the time"
Mathematics example		
I a koe e whakaroa ana nga taha ka aha?		While you were making the sides longer, what happened?
Ina tango te rima ka ...?		If [you] subtract the five, then ... [what happens]?
Tāpiri tonu te whitu ka tau tōrunga.		As soon as [you] add the seven [it] becomes positive.
Causal (Reason and Purpose, Cause and Effect)	*kia....ai*	so that
	e....ai	in order, whereupon ... that
	na te mea	because
	nō reira	therefore, thereby, that's why, so, consequently, for that reason, hence, thus, accordingly.
Mathematics example		
Nō reira kei hea pea tona tuaka hangarite?		Therefore, where perhaps is the line of symmetry?
Kia tuhia te rārangi e hono ai ēnei kotinga e rua.		Draw the line in order to join these two bisectors.
Tāpirihia kia rua kia nui atu ai te roa.		Add two so that the length is longer.
Adversative (unexpected result, contrast, opposition)	*ahakoa tonu*	even though, even so
	ahakoa	although, notwithstanding, despite, even though,
	kē	whatever, no matter, in spite of, nevertheless
		indicate difference or unexpectedness.

Table 5.1 (continued)

Relationship category	Logical connectives	English translations
Mathematics example		
Kua roa kē tēnei i tēnā.		This has become longer that that.
Ahakoa he roa atu he nui atu te horahanga o tēnei.		Although it's longer, the area of this is greater.
Condition	*mehemea*	If
	ki te	Condition about the future
Mathematics example		
Ki te tāpiri i te rua ka waru.		If [you] add two [you] get eight.

Logical connectors in *te reo Māori*, as in any other language, are acquired and mastered by children as part of their language development in and out of school. An examination of Te Koutu teacher and student talk shows that the basic and more frequent connectors are acquired early in this development, such as *anō* (again) or *engari* (but). Other connectors are mastered much later, if at all, and only after students have being exposed to a variety of language learning situations. For example, Uenuku, who teaches the older students, frequently uses the particle *ai* in his mathematics talk. *Ai* is a particle of great use, particularly in the older generation of speakers of *te reo Māori*. It mainly represents the English "who", "which", and "what", and has reference to the time, place, manner, cause, means, intention, and so on of an action (Harlow, 2001). This connector is almost absent in the talk of teachers of younger students. It is unlikely that these students will learn how to use this particle appropriately without modelling from their teachers.

Linguistic Complexity

Even when the features in *te reo Māori*, which would be useful in thinking mathematically, have been identified, there can be difficulties in learning them because of their complexity. An extended debate in English-medium mathematics education has focused on what features of the mathematics register are difficult for students to learn. Was it the difficulty of the mathematical concept, which made the language hard to acquire, or was it the mathematical language itself which contributed to the problems in understanding the mathematical concept? In this debate, there is an awareness of how the contexts in which the mathematical language was acquired contributed to the ease with which it was learnt.

In an early study, Knight and Hargis (1977) posited that since mathematics is a study of relationships, comparative structures are an essential and recurring component of mathematical language. Nevertheless, they also argued that comparative structures are difficult for many students to master. In contrast, Walkerdine (1988) suggested that rather than some terms being more conceptually difficult for children to master, it was the context in which terms were learnt that contributed to children's difficulties. For example, small children often exhibited much more difficulty with the concept of "less" than with the concept of "more". This had led to suggestions that it was cognitively more difficult to master. However, when children's lives were examined, there were few instances when children asked for less, but many instances of children asking for more. Consequently there is a need to consider how language is used in children's lives, both in and out of school, to better understand what aspects of the mathematics register are the most difficult to master.

In Māori-medium mathematics education, there has been less research into the linguistic features that may be difficult for students to learn. Hereafter we outline some suggestions about the features, which may cause problems for Māori-medium students. We include our reasons as well as some examples of these features from recorded lessons and meetings.

1. Some of the features in *te reo Māori*, particularly the particles, are not always semantically transparent and can have a variety of meanings. This can prove a challenge to teachers and students alike. For example, the word *ki* has many major functions, and many different grammatical constructions. In modern *te reo Māori* mathematics register, the word *ki* has taken on heightened significance, more so than in common daily usage. This is because of its role in introducing an instrumental phase, that is, the thing by means of which some action is carried out. For example, *Whakareatia te 5 ki te 4* means that the 4 acts on (replicates/multiples) the 5 because of the position of the word *ki*. However, *ki* as a preposition also means: "motion towards a place", or "on to", or "in the event of", or "according to".

Kua haere ***ki*** *Rotorua.*	They have gone *to* Rotorua.
Kua paea te waka ***ki*** *te ākau.*	The boat is stranded *upon* the beach.
Ki *te puia he uka, he aha ngā putanga e taea ana?*	*If* a coin is tossed, what results are possible?
Ki *a Uenuku, he nui atu tēnei.*	*According* to Uenuku, this is bigger.

Another example of connectors that can prove challenging is the set of particles common in teachers' mathematics talk: *nei, nā*, and *rā*. Their basic meaning is as a locative particle to indicate position near the speaker *(nei)*, position near the person being spoken to *(nā)*, and position distant from both, *(rā)*.

E hia ngā mata nei?	How many [geometric] faces are there?
He nui atu te koki rā i tēnei?	Is this angle [over there] bigger than this angle [by us]?

As well as these spatial relationships, *nei* and *rā* can be used to imply "nearness to" and "distance from" the present time.

Nō mua atu rā *I te rā nei*	some time ago, some time before today

These particles are also often used with pronouns and personal nouns to strengthen and emphasise the reference to "me", "us", or "you".

Ki ahau nei	in my view/opinion

Additionally, *nā* has several different functions. Without a macron to indicate a lengthened vowel sound, *na* is used at the beginning of a narrative, to call attention or to introduce some new element or emphatic statement, to which special attention needs to be drawn. This is a device teachers use frequently to signal to students that they are going to introduce a new or additional idea.

Na, ko te rīrapa, anā, he momo ara, nē?

Now, the maze, well, it's a sort of pathway, isn't it?

Na, ka haere atu koe ki tatahi, nēhā? Ka kite i ngā anga ma e toru nēhā.

Now, you go to the beach, yes? You will see three white shells, yes.

Very often, in teacher mathematical talk, a question is asked inviting the students to agree with, and/or to support a particular statement. In questions, which serve this purpose, it is very common to use a device called a tag at the end of the sentence (Harlow, 2001). In the two earlier examples, the tag used is *nē* or *nēhā*. The root word is *nē* and can be followed by *rā* or *hā*, depending on the dialect of the speaker.

2. Many logical connectors in *te reo Māori* have comparable, yet different variants in low- and high-frequency use, such as "if" which can be *mehemea* in high-frequency use, and *ki te* in low-frequency use. There are a number of different words, which translate to the English word "if", and some care is needed in selecting the form to use on any particular occasion, since they are not all equivalent in meaning (Harlow, 2001).
3. Some logical connectors are polysemic in structure and are made up of two or three words together such as *ki te* previously described. Others commonly used in teacher talk include *heoi anō* (however, so much for that) or *nā reira* (therefore, that's why, so, consequently, for that reason). The individual words have their own meaning, but the new multi-lexical form has a new function.
4. As noted earlier, the syntax of mathematics is seen commonly as the language that describes relationships. Often, mathematics discussion involves understanding a number of related ideas in one sentence. A challenge for students is to master the correct word order to illustrate the desired relationship between the main idea (contained in the main clause) and the modifying or supporting idea(s) (subordinate clause). A simple sentence consists of a single clause, for example, *tāpiritia te 5 ki te 2* (add 5 to 2). However, mathematics discussion involves much more than simple sentences and often requires the joining of a number of ideas to create complex sentences. For example, *I te tuatahi, me tāpiri te 5 ki te 2 kia kimi ai te otinga* (First, add 5 to 2 to find the answer). "To find the answer" is

a subordinate phrase. "First" as the logical connective joins the ideas in "add 5 to 2" in a logical manner "to find the answer" in order for the sentence to make sense. Syntactically, some of the relationships between the main clause and the subordinate clause are linguistically more difficult to master than others. These include the following:

Clauses of purpose
Kia tāpiri ngā tau matitahi i te tuatahi kia ngāwari ai te kimi i te whakaotinga.
Add the single digits first **so that** it is easier to work out the answer.
Particular relative clauses
These are clauses whose function is to qualify the noun.
Haere ki te whare e tū nei te hui.
Go to the house [where the meeting is taking place].

Learning How to Give Spoken Explanations

In the next chapter we look extensively at students' written explanations and justifications. Students usually begin to explain and justify their reasoning through speaking before they put their thoughts into writing. As a research area, speaking about mathematics came to the fore in the late 1980s with the publication of David Pimm's 1987 book *Speaking Mathematically.*

Since then, research, with English as the language of instruction, has tended to focus on dialogical structures in mathematics classrooms, and their contribution to students' mathematical understanding (see Nathan & Knuth, 2003; Bill, Leer, Reams, & Resnick, 1992; Moskal & Magone, 2000; White, 2003; Tanner & Jones, 2000).

The role of the teacher in supporting students to talk about the mathematics they were engaged with has been a focus in this research. Early on this research identified the typical teacher–student exchange known as the IRF (initiation – response – feedback) exchange (Mehan, 1979). The teacher asks a question and sometimes leaves a sentence incomplete. The students are expected to provide a response, and then the teacher gives either explicit feedback, through affirmation or negation of the response, or indirect feedback by asking a new question. Nathan and Knuth (2003) discussed the difficulties that teachers had in reconciling the need to accept all students' responses (social scaffolding) and the need to ensure that mathematical ideas were central in these responses (analytical scaffolding). In order for teachers to persuade children to take risks and put forward their ideas, teachers sometimes accept all of the students' responses. However, Khisty and Chevl (2002) showed that unless the talk within the classroom focused on developing mathematical understandings, then students were unlikely to gain anything from the talking.

There have been a number of critiques of the IRF exchange, which suggest that it is unlikely to lead to improved mathematical understandings. Wood (1998) criticised the use of leading questions where the student simply provided a one-word answer to questions because they did not push students to think mathematically. She stated, "[A]lthough the teacher may intend that the child uses strategies and learns about the relationship between numbers, the students need only to respond to the surface linguistic patterns to derive the correct answers" (p. 172). She suggested an alternative pattern, that she labelled "focusing", would be more effective in promoting learning: "A high level of interaction between the teacher and students creates opportunities for children to reflect on their own thinking and on the reasoning of others" (Wood, 1998, p. 172).

For students to give explanations and justifications, they need to understand how they are constituted and to see them as essential components of doing and learning mathematics. "The understanding *that* students are expected to explain their solution is a social norm, whereas the understanding of *what counts* as an acceptable mathematical explanation is a sociomathematical norm" (Yackel, 2001, p. 14). Students need to learn how to phrase explanations and justifications, but teachers also need to expect that children will provide these as part of each mathematics lesson. Gibbons (1998) in studying students' acquisition of the English register for a science topic found that "as the discourse progresses ... , individual utterances become longer and more explicit, and this occurs as the students begin to formulate explanations for what they see" (p. 109). Gibbons suggested that teacher requests for explanations were what triggered students to move from the "doing" to the "thinking" in their learning.

At Te Koutu, teachers recognised that children needed to explain their understandings as a normal part of a mathematics lesson. This awareness was linked to the teachers in the primary section of the school being involved in a New Zealand–wide professional development program on numeracy, Poutama Tau. In this program, teachers learnt about the need to have children explain their strategies when solving arithmetic problems.

Y4 Teacher: It was all the little words too that they got mussed up on like *atu* (away), *mai* (towards), *i* and *ki*. Yeah, and I noticed with Student1, he is a good mathematician but his language lacks and when it came to the actual explaining of how he did it he couldn't really explain but he can do it in his head but he can't explain because his language is quite poor, actually, I was talking to somebody about. I think it is *te reo* in the home too isn't that strong.

Tamsin: Because I was also talking to Y7 Teacher today and your kids now, you have been forcing them to explain themselves for a while.

Y7 Teacher: Yep

Tamsin: Could you talk a little bit about the consequences of that just to...

Y7 Teacher: Yeah, forcing them to explain everything that they do, it doesn't matter what it is, whether it is number, we are looking at algebra and statistics. So it doesn't matter what they are doing, they have to explain every answer that they ever get, the same sort of scenario. For some, it is quite easy to explain it in words. For some, the language is just not good enough for them to do that.

So in that sense, although they are writing stuff, they are also explaining it. Now those ones that are pretty good at explaining themselves in writing, it is quite easy for them to explain it talking as well, verbally, I mean. Whereas those ones that are a bit slower, they have to read what they have written and in same cases that is an opportune time to fix up what they have actually said.

Y1 Teacher: We do that in *Poutama Tau* anyway. We always ask them how they got that answer, "*pēhea koe e mōhio ai*" [how do you know?] things like that. And it's also getting them to, because they will only give you the straight answer like the basic answer but you get them to repeat it. Like you say "*kei hea te tūru*?" [where is the chair?] and they will say "*kei kō*" [over there] and you go "*whakamārama mai*" [explain it to me] or "*kei te taha o te tēpu*" [at the side of the table], you know. You are just getting them to use it even with the little ones as they are going up, so they do get to the higher levels. "*Kei te mārama, kei te pai pea tō reo*" [Do you understand, is your language okay]? Yeah, to *whakahāngai* [relate] to them too. When I do lessons with them I also think how I am going to *whakahāngai ngā kōrero ki a rātau* [relate what is being discussed to them], to relate to them, how it is going to relate to them? Like shapes, naming all the shapes in the classroom, . . . using the language because my ones can't write either. They are not, some of them are starting to write, but it is getting them to talk about it.

Tony: These are great you know, some of the language teaching stuff in this is pretty good. (Meeting Sept. 2008)

The teachers could see the potential in having the students explain their thinking. However, it is clear that students' lack of exposure to *te reo Māori* outside of the classroom was a challenge that teachers had to address. This is discussed further in Chapter 10. An example of a classroom exchange can be seen in the following extract from a lesson that was recorded in 2006.

Y6Teacher: Kotahi rau, rima tekau mā ono, anā, āe, ngāwari tērā! Nō reira he mea ngāwari tērā Student1?	Y6T: One hundred, fifty six, that's easy, therefore that's easy.
Student1: Āe!	Student1: Yes!
Y6T: He aha ai?	Y6T: Why?
Student1: Nō te mea ka mōhio ko te tekau whakarau tekau mā toru ko te 'tahi rau, toru tekau, anā ka mōhio ko te rua whakarau tekau mā toru ko te rua tekau mā ono, anā, me tāpiri noa iho i te 'tahi rau toru tekau ki te rua tekau mā ono.	Student1: Because you know that ten times thirteen is one hundred and thirty, you know that two times thirteen is twenty six. You only need to add one hundred and thirty to twenty six.
Y6T: Āe, ka pai, [I] kite au i tērā!	Y6T: Yes that's good, I see it now!

This child was able to give an explanation about how to do the calculation. Gradually, the teachers were beginning to expect students to give these explanations, and the students were beginning to know that they had to give them. When the teachers started requiring students to provide this information, students often gave answers to questions about what had they done as "I just knew it" or "I just guessed". Nevertheless, even in 2008, the teachers were still the most dominant speakers in most mathematics lessons. They did find that it was easier to have students give verbal explanations and justifications around regular writing activities, and this is discussed in more detail in the next chapter.

Kanikani Pāngarau – Dancing Mathematics

As well as using writing to support students giving explanations and justifications, the teachers in the junior end of the school involved students in thinking mathematically through an activity known as Kanikani Pāngarau (mathematical dancing). This activity was taken from the New Zealand television programme *Toro Pikopiko* and was initiated by the teacher Horomona Horo.

Students learnt a series of movements for each of the numbers from zero to ten. They also learnt symbols for the four operations (+, –, ×, and ÷) and for the equals sign. Children were then given problems by the teacher through movement and asked to provide an answer by also using movements themselves. Figure 5.1 shows Horomona illustrating a problem with children watching and then writing down their answers.

Addison and Te Whare (n.d.), the originators of Kanikani Pāngarau, explained that it was based on the principles of *kapa haka* where specific words were represented by specific actions. *Kapa haka* is a traditional team dance that is often performed competitively (Murray, 2000). In the *haka*, actions emphasise the sung or chanted words (Matthews, 2004). However, as Matthews noted, "[I]t was the body that was the instrument and vessel of delivery" (p. 9). In Kanikani Pāngarau, the actions must carry all the meaning. The audience is expected to respond in kind, making the expression of meaning paramount. The children at Te Koutu loved their involvement in Kanikani Pāngarau, and although it was only concerned with basic facts, it did seem to resonate with the children's cultural background.

Kinaesthetic involvement is believed to support students' understanding and was labelled by Howard Gardener as an intelligence-kinaesthetic intelligence (Touval & Westreich, 2003). Sellarés and Toussaint (2003), in considering why some algorithms in computational geometry failed, found that these were based on kinaesthetic heuristics rather than logico-mathematical ones. However, the kinaesthetic-based algorithms were much faster, even when they were incorrect, thus suggesting that they are more computationally efficient. Sellarés and Toussaint suggested that there is a need for algorithm designers to bridge the gap between the two types of heuristics in order for efficiency to be combined with accuracy. Although Kanikani Pāngarau deals with much simpler mathematics, there is potential for it to be a support for students to think mathematically if it is further developed.

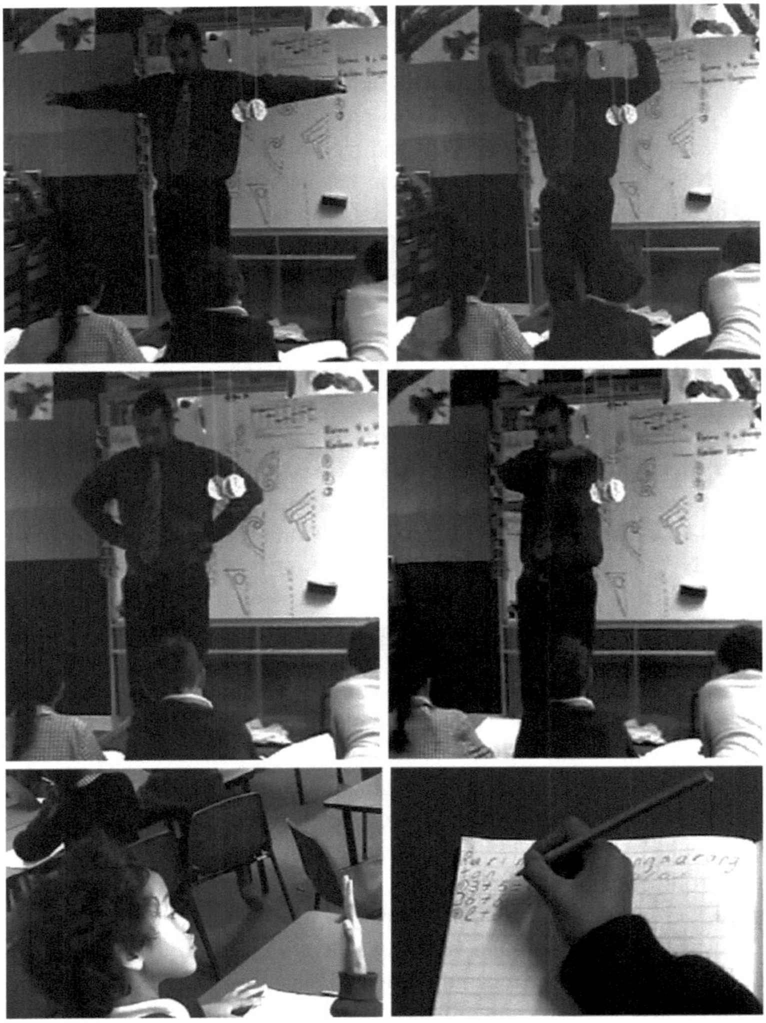

Fig. 5.1 *Kanikani Pāngarau*

Kinaesthetic activities related to gestures that accompany speaking have begun to be researched (see Roth, 2001; Radford, 2003a, 2003b). This research into gestures has concentrated on how extra meaning is added to oral descriptions of mathematical ideas and how it supports students to understand what they are learning. Using culturally appropriate movements to provide extra layers of meaning may well contribute to students being able to think mathematically. However, much more work needs to be done to identify the actions that teachers and students are already using to support mathematical thinking in classrooms, not just those used in *Kanikani Pāngarau*. By working with teachers and students, it would be possible to determine the most effective way that these actions could be used to support students.

Such a challenge is only just being recognised as having potential for improving students' understanding.

Meeting Challenges Around Thinking Mathematically

Thinking mathematically in *te reo Māori* has not yet been fully explored. Different aspects of this are at different stages in the "overcoming challenges cycle" described in Chapter 1.

The mathematics register in *te reo Māori* has features that have the potential to be a useful tool in supporting students to think mathematically. Linguistic markers and logical connectives that are in *te reo Māori* can be useful for linking ideas logically. There may well be other supportive features, which are yet to be identified, but the issue of using such features is a challenge that has been recognised.

As well, there remains a challenge to have students realise the potential from using these features when thinking mathematically. This involves determining potential difficulties in learning aspects of the mathematics register and then looking at how these can be overcome. We are still at an early stage in meeting this challenge. As Christensen (2003) described, there is some resistance to using traditional features of *te reo Māori* with some teachers using English grammatical structures with Māori words. Yet as teachers begin to insist on students explaining and justifying their mathematical understandings, the need for the features within *te reo Māori* may become more self-evident.

The interactions between teachers and students indicate that these traditional features to some extent are being utilised already. However, until the education system as a whole recognises this utility, many will continue to see English as the more appropriate language for thinking about mathematics. Thus, this challenge is far from being met.

Other aspects of Māori language and culture are only now being identified at Te Koutu as having potential, but are yet to be explored fully. They offer openings, which, if followed up, can provide unexpected solutions in meeting the challenge of both supporting students to think mathematically and ensuring the integrity of the language. There is a need for more research to support a better understanding of their efficacy. For example, the use of actions that are part of cultural activities such as *kapa haka* may also have a greater use beyond Kanikani Pāngarau. Unless further work is done to explore this area, it is unlikely that its potential will be realised. This is something that needs further consideration in the coming years.

Chapter 6
Writing to Help Students Think Mathematically

In this chapter, we consider students' and teachers' perceptions of the contribution that writing makes to students' understanding of mathematics. We also outline the genres or text types identified in the mathematics classrooms at Te Koutu. The teachers believed that students' written explanations and justifications supported them to gain a deep understanding of mathematics. The students described how they perceived writing in mathematics to be primarily for themselves and only secondarily for their teachers. This suggests that writing supported their mathematical thinking.

Learning how to support students to increase the quality and quantity of their writing in mathematics was a challenge that the teachers at Te Koutu took upon themselves. Earlier work at Te Koutu had found that teachers provided very few opportunities for children to write explanations and justifications in mathematics. At a meeting in 2006 the teachers, in the secondary section, described their experiences of students not wanting to elaborate on their explanations, but rather keeping them short and only using the simple language they felt most comfortable with. Mathematical language was used only because students believed that teachers expected them to use it. All the teachers agreed that it was in the junior classes that students should start developing these skills, and that a coherent approach across the school was necessary.

The teachers were also interested in exploring whether mathematics could be a vehicle for improving students' *te reo Māori* language skills including increasing their use of logical connectives and the passive voice. The teachers were committed to the revival of the language, and wanted the students to use the authentic resources within the language. Writing can more easily be used in explicit language learning, as it can be referred to time and time again. On the other hand, concerns were also expressed about how a concentration on writing might result in a devaluation of speaking.

The Role of Literacy Within a Traditionally Oral Culture

Te reo Māori has an extensive oral tradition, and the decision at Te Koutu to concentrate on writing was made within this context. Western beliefs about the value of writing in improving students' thinking processes, as propagated by researchers

T. Meaney et al., *Collaborating to Meet Language Challenges in Indigenous Mathematics Classrooms*, Mathematics Education Library 52,
DOI 10.1007/978-94-007-1994-1_6, © Springer Science+Business Media B.V. 2012

such as Vygotsky and Luria (Gee, 1989), cannot take precedence over issues relating to the immediate situation of a *kura kaupapa Māori*. Although literacy is believed to have a role in the regeneration of a language (Hohepa, 2006), concern has also been raised about the possible imposition that writing can have on Indigenous communities' worldviews (Cavalcanti, 2004; Street, 1995). Gee (1989) stated that:

> Discourse practices are always embedded in the particular world view of a particular social group; they are tied to a set of values and norms. In learning new discourse practices, a student partakes of this set of values and norms, this world view. Furthermore, in acquiring a new set of discourse practices, a student may be acquiring a new identity, one that at various points may conflict with the student's initial acculturation and socialization. (p. 59)

As a discursive practice, mathematical writing has an impact on students' identities. Identities formed around mathematical writing can come to be in conflict with the students' Māori identities, depending on how the discursive practices are taught. In *kura kaupapa Māori*, it is important that the role of writing in the teaching and learning of mathematics is problematised. With *te reo Māori* still in a process of regeneration, the teaching and learning of writing in mathematics need to be done in a culturally appropriate way. In discussing the role of reading in the home, Hohepa (2006) wrote:

> A significant issue in the context of language regeneration concerns how language practices both reflect and construct cultural concepts and values. One way to address this issue is to ensure that ways of carrying out an activity such as book reading do not undervalue or replace existing cultural ways but are added to family repertoires (McNaughton, 1995). Also, ways of participating in the activity which are not inconsistent with the specific literacy goals, but which are consistent with culturally preferred ways of participating can be promoted. (p. 299)

Hopeha's warning is also relevant for mathematical writing. The advantages gained for students' mathematical thinking from writing about mathematics had to be considered in relationship to other priorities of the school and its local community. Trying to achieve an appropriate balance between different, often conflicting, aims is one of the major challenges facing Te Koutu, not just in regards to writing in mathematics, but to the learning of Western mathematics generally.

Writing to Support Reflection

In English-medium education, writing in mathematics is not seen as problematic, but rather as an under-utilised resource for supporting the reflection process (Albert, 2000). It has been shown that students who wrote descriptions of their thinking were significantly better able to solve mathematical problems than those who only verbalised their thinking processes (Pugalee, 2004). The National Council for Teachers of Mathematics (NCTM) (2000) suggested:

> Writing is a valuable way of reflecting on and solidifying what one knows, and several kinds of exercises can serve this purpose. For example, teachers can ask students to write down what they have learned about a particular topic or to put together a study guide for a student who was absent and needs to know what is important about the topic. A student

> who has done a major project or worked on a substantial long-range problem can be asked to compare some of their early work with later work and explain how the later work reflects greater understanding. In these ways, teachers can help students develop skills in mathematical communication that will serve them well both inside and outside the classroom. Using these skills will in turn help students to develop deeper understandings of the mathematical ideas about which they speak, hear, read, and write. (p. 352)

The importance of students being able to explain their thinking process is also valued in New Zealand's end-of-high-school assessments. As was discussed in Chapter 4, the National Certificate in Educational Achievement (NCEA) in mathematics requires students to be able to write explanations and justifications (Meaney, 2002). Hipkins and Neill (2006) wrote about the impact of NCEA on high school mathematics teachers' awareness about language:

> both mathematics and science teachers give a relatively high priority to the need to develop language and literacy practices associated with each discipline. In at least two cases the teachers' awareness of these issues has been sharpened by participation in school-wide literacy initiatives. (p. 63)

Students' written explanations and justifications also provide teachers with more information than what is gained from simply listening to students (Drake & Amspaugh, 1994). Moskal and Magone (2000) stated that "students' written explanations to well-designed tasks can provide robust accounts of their mathematical reasoning" (p. 313), and so can be used by teachers to assess students' knowledge. In Māori-medium education, Christensen (2003) found that "[n]one of the teachers recognised that allowing students to develop ways of recording their [numeracy] strategy use might help their thinking, their own and teacher review of strategy use, and their communicating of mental processes" (p. 36). He recommended that this be emphasised more in the Poutama Tau professional development programme.

At Te Koutu, there were many meetings, which discussed the writing in mathematics project. In Chapter 10, we describe some of the approaches that the teachers adopted to support students in acquiring mathematical writing. The meetings also illustrate some of the teachers' reasoning for having students write in mathematics. At a meeting in November 2007, one teacher described what he had been doing with the students.

Y6 Teacher: I found I was finally getting them to restate the question in a different way was a couple of lessons at least. To answer it as well, rather than just saying yes or no. And then to justify their answer and to attach an example onto that as well of what they actually did if they were going to do some *mahi hanga* [geometry work] whatever. Getting them to answer the question, what they exactly did and why did they believe that the answer, to the question, they had given was right. Then they draw their *tauira* [example] for me as well within the answer. That's even after playing around with it before actually starting to answer the question. We've sat on that for a long time. And now they all have to stand up and give their answer.

Tamsin: So they are orally presenting it? And then do they write?

Y6T: No, they are orally presenting their writing because if there are questionable answers that are produced and they've obviously heard everyone else's then they go maybe, maybe I have to redo mine. Plus I don't mind telling them that's not quite right, start again.

...

Y6T: I tried to tell the kids that mathematics is not just a series of numbers, it involves writing and we are fortunate enough that we are doing *āhuahanga* [geometry] at the moment. I would like to try a times table equation and see if they can give me that same sort of format answer to explain how they got their answer. We try to say to them that as they get further up towards where *Matua* [senior teacher] is, you'll notice that you do more explaining, more and more writing so you need to justify your answer even if it is wrong, but you don't know that. You justify as far as you can then the teacher will tell you what parts are wrong, once you've justified it pretty well. How you got to where you are, explaining what actual process they took. Because we looked at car logos in the car park, so they looked at the Mitsubishi and when they looked at it, straight away they thought the whole star series was what they were transforming but when they had to re-look at it again and it was only one diamond that had been rotated three times, in a third of a circle so the writing about it they realised oh, yeah, you're right. I'm not turning the whole star I am turning just one diamond around.

Tamsin: So it was the process of writing that forced their thinking?

Y6T: Some of them got it wrong but they justified their answer. How come they ended up with it? Then at the end we added another bit where they look at it and say what they thought of it. I thought why not have a go and see how it ends up. Have a dive in and have a look. I was noticing if I asked them how they got their answer, the answer had been "I just know", "I just did it" and "it came out like this". Now they justify everything they've done.

Making students aware of the usefulness of writing in the mathematics class has not been simple. In her research on implementing a writing programme in a middle school mathematics class, Johanning (2000) found that it took time, partly because students needed "to learn to focus on what they were doing mathematically and why" (p. 156). The teacher at Te Koutu provided the students with reasons for writing in mathematics and introduced a structured approach to writing explanations. The oral presentations also reinforced to the students the need to be aware of their readers' needs. This is likely to have an impact on the students' mathematical thinking, but it is a slow process.

The same Year 6 teacher had no qualms about telling students when they were wrong. This may have been because he had been teaching the same mathematics group for the previous two years. For him, the important part was the students "having a go". Making mistakes was part of the learning process, rather than something that should cause the students' anxiety. Although many teachers in English-medium situations might find this direct intervention problematic (see Nathan & Knuth, 2003), the relationships between teachers and students in *kura kaupapa Māori* are different to those in English-medium situations. As was the case with Yup'ik teachers (Lipka, Mohatt, & Cisulistet Group, 1998), Māori teachers need to find their own ways of interacting with their students that reflect their cultural protocols, rather than blindly adopt Western teaching models. For *kura kaupapa Māori* teachers, it is important that teaching writing in mathematics is done in a way that respects the relationship between students and teachers and thus remains culturally appropriate. This is discussed further in Chapter 11.

Figure 6.1 provides an example written by a student from the above Year 6 class. This student still has some way to go in providing a coherent explanation. Nevertheless, it is possible to see how the student has combined the diagram with the words and has started to provide details, so that a non-participant could follow what he had done. As a starting point, it provides information to the teacher about what the student knows and provides an opportunity to discuss how it could be improved.

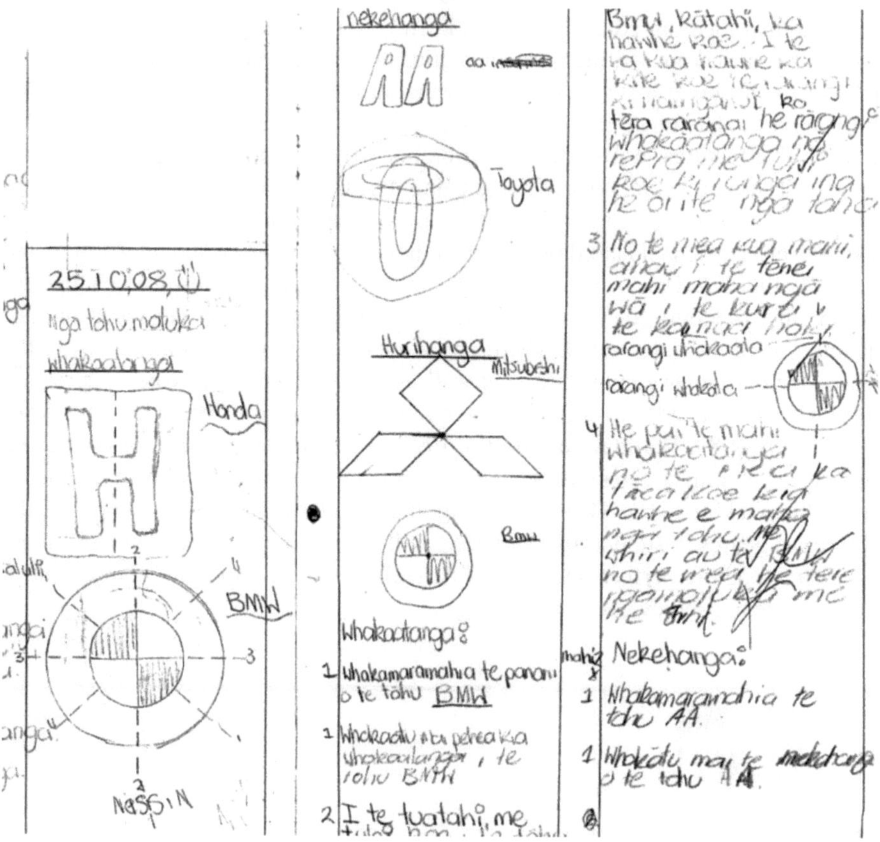

Fig. 6.1 A student's explanation of the reflection used in different car signs

Types of Writing in Mathematics

Often writing is introduced to record how ideas have been manipulated (see Burns, 2005) and in so doing links students' understanding from a concrete experience to an abstract concept. Figure 6.2 suggests that the development of the ideas about shapes is a continuum from recognising them in the environment to being able to manipulate abstract ideas such as the relationship between a net and its solid. The drawing of the triangle with its measurements in the centre of the continuum may have been copied from using a concrete triangle, but was more likely constructed from written instructions. At every stage, the markings on the paper form an iconic representation, which has some resemblance to the actual object being represented (Roth, 2001). As the ideas about shapes develop, the immediate relationship to concrete items that need to be viewed and manipulated by students becomes less important. This continuum can be considered as another way of representing how children's everyday language is developed into official mathematics language that was discussed by Herbel-Eisenmann (2002). As the forms of writing progress, the marks on the paper become more abstract and the relationship to actual manipulation of concrete objects less transparent. Students gradually shift to manipulating abstract concepts without the need for concrete materials at all.

In English-medium education, it has been recognised that in order for students to gain the most from writing in mathematics, they need to understand the conventions of the different text types or genres that are used. A genre is a text type that fulfils a particular function within a communicative interaction and thus has some recognisable features. Any text, whether oral or written, is influenced by three components. These are as follows: What is being discussed; who is involved in the text (producing it or interpreting it); and the form of the communication (written, oral, gestures, etc.). Michael Halliday described these components as the field, tenor, and mode (Halliday & Hasan, 1985). Changes to any of them will result in changes to the text that is produced. To structure texts in unexpected ways can result in their meaning being misinterpreted. Therefore, students need to learn how to write the genres for particular content areas as well as knowing how and when genres are useful (Unsworth, 2001). As Pimm and Wagner (2003) wrote "[m]uch of this work (e.g. Martin, 1989; Halliday and Martin, 1993) is also rooted in questions of school systems developing greater equity by means of students gaining access to linguistic-cultural capital" (p. 162). Having a strong understanding of how to manipulate the features of genres to support their thinking will be invaluable to students in increasing their academic performances. This is likely to be an important component in Māori academic achievement improving.

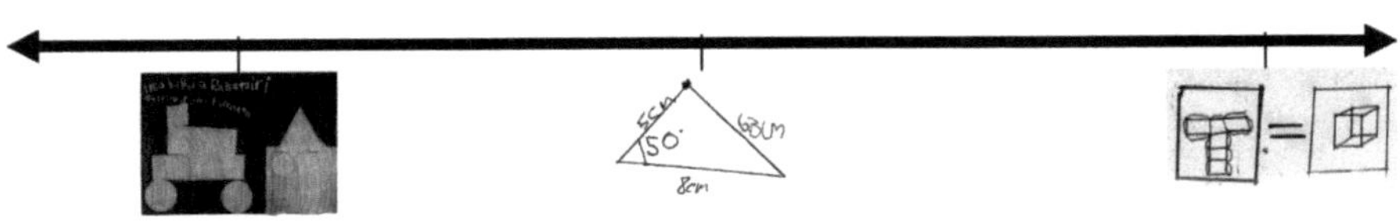

Fig. 6.2 Continuum from diagrams reliant on viewing or manipulating concrete objects to those drawn from abstract knowledge

Although genres have received significant attention since the 1980s, especially in Australia and United Kingdom (Unsworth, 2001), little research has been done in regard to those typically found in mathematics classrooms, especially in non-English-medium situations. Researchers, such as Morgan (1998) and Marks and Mousley (1990), identified several genres that mathematicians use and that, therefore, should be included in students' repertoire of mathematical writing. These genres were as follows: procedural, description, report, explanation, and exposition. However, when Marks and Mousley investigated the genres used in 11 classrooms (seven primary and four secondary), they found many instances of recounts, incorporating symbols, and visual representations, but very few examples of other genres. Recounts described what a student had done during a mathematical activity and was generally expressed as a narrative. As recounts relate ideas chronologically rather than logically, they are of limited use in helping students to think mathematically (Solomon & O'Neill, 1998; see also Chapter 5).

Although the purposes for writing are different for mathematicians than they are for students, by the time that students are in their final years of high school they need to be able to produce a range of mathematical genres. Some genres, or early versions of these genres, need to be present in mathematics classrooms for all year levels. At Te Koutu, identifying the types of writing, or genres, was a starting point for teachers in considering how to improve the quality and quantity of students' mathematical writing.

Writing in Mathematics at Te Koutu

The first step of the project involved identifying the writing that was being done already in mathematics lessons and then classifying it. At a meeting in March 2007, the teachers in pairs sorted samples of students' writing into categories. The samples came from different ages of students and were from a range of topics, and mostly had been collected at the end of 2006. The quality of the writing varied. Two pairs then shared their categories and decided on a joint set. Once the groups of paired teachers had agreed on a set of categories, there was a combined discussion about the genres. The teachers had grouped the writing samples by their function. They identified the primary purpose of each piece of writing and then looked at the structure within each group. Three genres and an initial set of mathematical modes were identified. The three main functions identified were as follows: "described", "explained", and "justified". Thus the genres were labelled in the following manner: *whakaahua* (descriptions), *whakamārama* (explanation), and *parahau* (justification).

Figure 6.3 is one representation of how field, tenor, and mode in each of the genres could be connected. It illustrates the differences between the three genres according to function, the relationship between writer and reader, and the modes used.

Whakaahua refer to mathematical objects or facts. Although they can contain many details, these are additive rather than causal. Causal relationships are more commonly found in the other two genres. *Whakaahua* belong to what Unsworth (2001) described as *recognition literacy* in that they support the "learning to

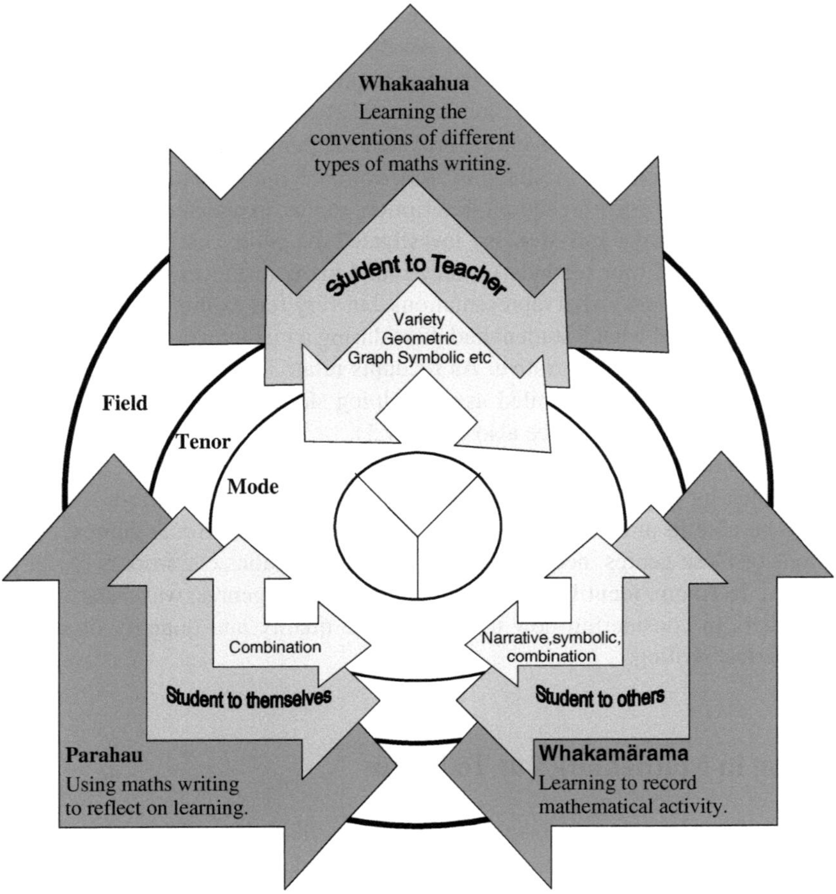

Fig. 6.3 Context of situation for the three genres of mathematical writing

recognize and produce the verbal, visual and electronic codes that are used to construct and communicate meaning" (p. 14). Without this literacy knowledge, *whakamārama* and *parahau* cannot be produced. Figure 6.3 differs from the ideas of Unsworth in that he linked recognition literacy to common experiences of everyday life, whereas in a mathematics classroom, descriptions are closely aligned with students using the mathematics register.

W*hakamārama* provides explanation of a mathematical event or phenomena, and thus is about using mathematics and recording what is done. An explanation is commonly for the benefit of someone other than the writer, such as peers or parents who were not involved in the actual mathematical activity. On the other hand, *parahau* is about explaining why something has been done in a particular manner and is helpful for the writer in working through the choices that were made. Both *whakamārama* and *parahau* are useful in thinking mathematically because they involve different kinds of reflection about the knowledge of mathematics already imparted. These genres require particular linguistic constructions, including the use of logical connectors.

Other researchers have considered what we have labelled as modes to be genres. For example, Solomon and O'Neill (1998) stated that "[i]n so far as genre shapes and constrains the nature of a text, then graphs, equations, proofs and algorithms can be considered as expressions of genre" (pp. 217–218). Our definition of genre was based on the function that it performed. Therefore, the channel through which the function is delivered was considered to be the mathematical writing mode. Ben-Chaim, Lappan, and Houang (1989) described three modes that were used by students to describe an object made from cubes taped together. These modes were as follows: verbal, graphic, and mixed mode. The verbal mode occurred when the student's message was carried by words. A diagram could accompany the words but did not add any more meaning. A graphic mode used diagrams with only labels, at the most, to accompany it. A mixed mode used both diagrams and words to convey meaning.

Whilst in Ben-Chaim et al.'s (1989) study, the graphic mode was believed to be the one more successful at accurately conveying information about the object, the teachers at Te Koutu felt that the explanation and justification genres commonly require a combination of modes rather than being exclusively of one kind. This is supported by O'Halloran (2000):

> Mathematics is not construed solely through linguistic means. Rather, mathematics is construed through the use of the semiotic resources of mathematical symbolism, visual display in the form of graphs and diagrams, and language. In both written mathematical texts and classroom discourse, these codes alternate as the primary resource for meaning, and also interact with each other to construct meaning. Thus, the analysis of "mathematical language" must be undertaken within the context in which it occurs; that is, in relation to its co-deployment with mathematical symbolism and visual display. (p. 360)

Figure 6.4 provides an example of a student's explanation that includes several modes. The student began by using diagrams to describe the various combinations

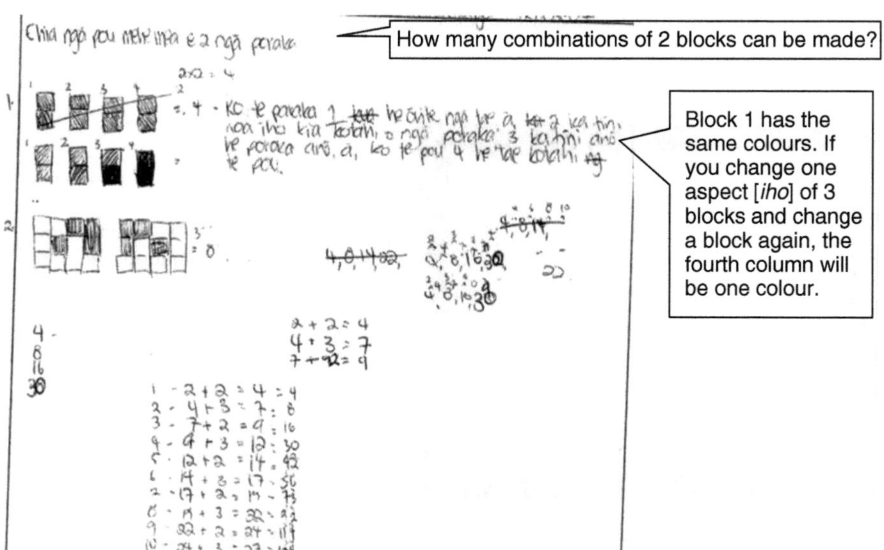

Fig. 6.4 Writing sample showing a combination of description and explanation

that different amounts and colours of blocks could form. She then gave an explanation in words of how to determine the number of combinations formed from two blocks. Symbols were used to try to determine the pattern that related the number of different coloured blocks to the possible combinations. Although it included some descriptive elements, this example was classified as an explanation as it used a variety of different modes – symbolic, verbal, and iconic – to carry its meaning.

Over the course of 2007, more than 2000 pieces of writing were collected. All writing samples were scanned and classified according to genre and mathematical mode and then named and filed in a database. The next sections provide more detail about the three genres.

Whakaahua

Whakaahua refers to the description of a mathematical object or situation, usually by using only one mode. Marks and Mousley (1990) separated what we called *whakaahua* into descriptions and reports. Wallace and Ellerton (2004) stated that "[d]escription and report genres provide the nature of individual things and the nature of classes of things, respectively" (p. 9). In our data there did not seem to be such a clear-cut distinction. Figure 6.5 provides an example of *whakaahua* that was produced for public display by a Year 5 student.

Figure 6.5 includes both diagrams and words in its description of translations, but they are provided as separate entities. There is mention of the squares used to contain each diagram in the sentences, but the connection is not explicit. The diagram shows two translations through the use of pictures, whilst the sentences provide a description in words. Without an interaction between the two modes, it is difficult for the two modes to work together to form an explanation.

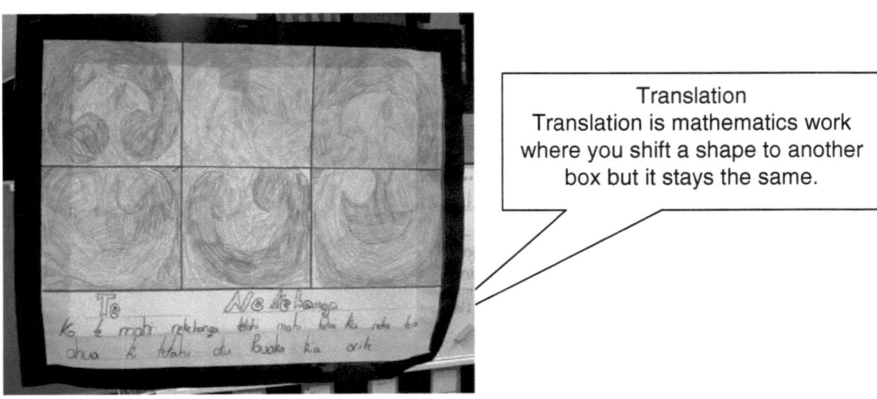

Fig. 6.5 A description using words and diagrams to illustrate translation

Writing *whakaahua* involves students in learning the mathematical writing conventions. At the very elementary levels of school, this involved students in learning to form the numbers or shapes. At later stages, they learnt the appropriate way to write a number sentence or produce a graph.

Whakamārama

Whakamārama, or explanations, often include a series of steps to illustrate how something came to be, and thus can be connected to solving problems. They can include verbal explanations of how to turn a net into a 3-dimensional shape as well as multi-step equations. Marks and Mousley's (1990) equivalent genre was labelled as a "procedure". However, procedures usually imply a lock-step process. For example, Unsworth (2001) listed the stages in a procedural text as: "goal; materials; steps" (p. 123). The pieces of writing that we identified as *whakamārama* explain how something has been done, but do not always provide the steps in a set order. An example of this can be seen in Fig. 6.6.

The different number stories and written sentences combine to give an explanation of how first 12 and then 18 can be calculated. This is different from merely working out what the answer to the calculation would be. For example, $6 + 6 = 12$ on its own would have been considered a description. However, such an explanation is quite simple. In a more elaborated explanation, there would have been a comment about the relationship between 6×2 and 2×6. Later in the chapter, we describe how the teachers worked through identifying features of a high-quality piece of writing.

Whakamārama use a restricted number of modes. The most common modes are the verbal mode, written sentences, and the symbolic mode, where explanations are provided entirely in symbols. Combinations of these modes, such as in Fig. 6.6, also occur as do some other combinations such as graphs with written sentences.

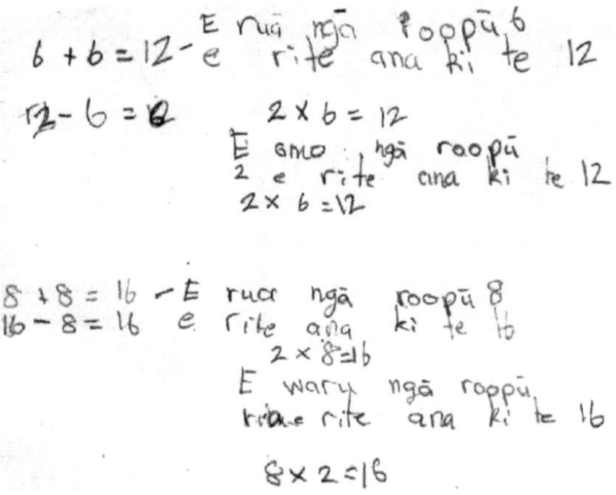

Two groups of 6 is the same as 12
2 x 6 = 12

Six groups of 2 are the same as 12
2 x 6= 12

Two groups of 8 are the same as 16
2 x 8 = 16

Eight groups of 2 are the same as 16.
8 x 2 = 16

Fig. 6.6 An explanation provided in a non-lockstep manner

Parahau

Over the course of 2007, only a few *parahau* or justifications were collected, and this remains an area of concern. The primary purpose of *parahau* is to provide information about why something is done and so are more reflective. In much of mathematics, students need to evaluate different courses of action in order to choose an appropriate one. Justifications provide information on this evaluation and, therefore, commonly involve a combination of several modes, for example, as shown in Fig. 6.7.

Figure 6.7 was produced after Uenuku challenged his Year 9 class to produce the dimensions for a cylinder that would hold 1 litre of water. The students had to justify what they did. A number of different approaches were taken. The student who drew Fig. 6.7 started by deciding that the *whitianga* (diameter) was 10 cm and thus the *pūtoro* (radius) was 5 cm. The student determined the area of the circle at the top of the cylinder. Using this area, and that 1 litre was equivalent to 1000 cm^3, she calculated the height of the cylinder. All the calculations are done using symbols, numbers, and algebraic variables. The justifications for what was done are provided in sentences or phrases connected to the calculations with lines. Not every decision is provided with a justification. Why the student began with a diameter of 10 cm and other justifications could be elaborated. Inclusion of this extra information would have made the justification stronger. Although the calculations proceed from the top to the bottom of the page in a logical order, the addition of justifications disturbs this

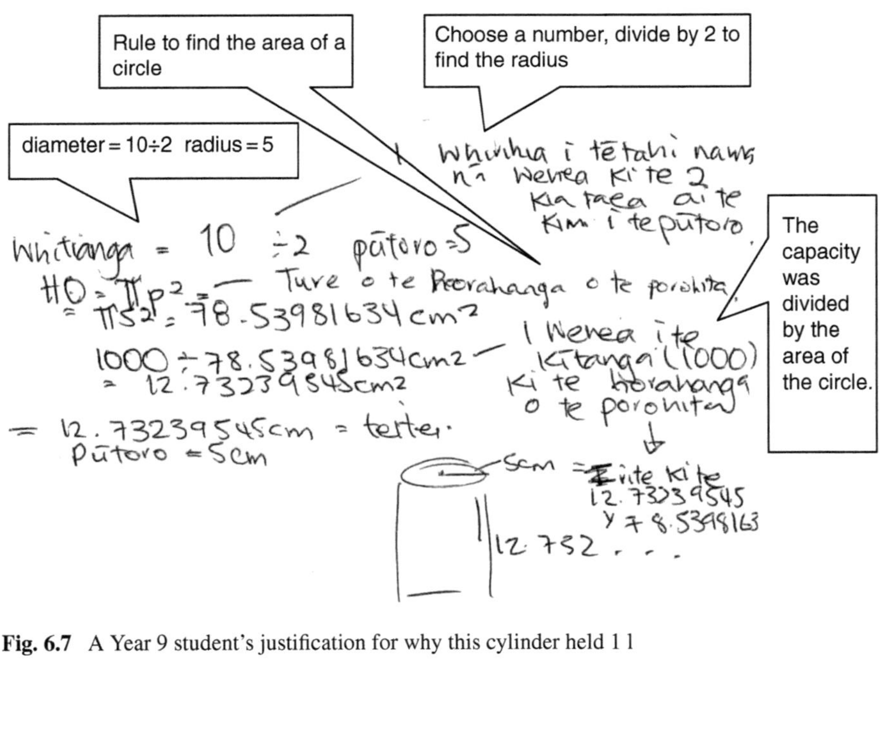

Fig. 6.7 A Year 9 student's justification for why this cylinder held 1 l

flow, as it requires the reader to follow the lines upwards, downwards, and across. Most of the justifications written in sentences appear on the left-hand side of the page.

Judging the Quality of Mathematical Writing

For teachers to improve the quantity and quality of students' writing, there was a need not just to understand the features of explanations and justifications but also to have a clear understanding of what constituted a "good piece of writing". Identifying the features of an exceptional versus an average, or unacceptable piece of mathematical writing is something that is rarely discussed in curriculum documents such as National Council of Teachers of Mathematics (NCTM) (2000) *Standards*. Instead, there seems to be an assumption that teachers and their students are already aware of the criteria for judging mathematical writing. However, this is rarely the case. As one of the teachers in Doerr and Chandler-Olcott's (2009) study stated, "[T]here was no discussion at all about writing, what makes it good, what makes it acceptable, and what makes it mathematically correct" (p. 292). With little research to guide them, the teachers at Te Koutu had to determine for themselves what constituted a quality piece of writing for each of the different genres, and whether it would differ across year levels. This was another internal challenge that Te Koutu took on and is still one that is being worked through.

Many factors, not just the need to convey meaning, influence the "readability" of a piece of writing. For example, demographic characteristics such as gender and ethnicity may influence how students choose to express themselves mathematically (Meaney, 2005b, 2006a). Meaney (2005b) found that senior high school students embedded their algebraic responses within a narrative depending upon whether they answered correctly, their gender, and the socio-economic background of their school, as well as on the actual question asked. Simply producing a formulaic list of features for students to follow may restrict their ability to express themselves fluently, because it does not take into consideration individual preferences. Therefore, the task of teachers in providing guidance to students about good mathematical writing is complex.

Doerr and Chandler-Olcott (2009) working with middle school teachers of mathematics identified the features of good mathematical writing, which are listed in Table 6.1.

This list is quite explicit. However, it could well be that if the word "math" was replaced with another subject such as social studies, then the list would be very similar.

Yet, general features of writing such as students' poor use of grammar did have an impact on their opportunities to take part in writing in mathematics. The Year 8 teacher described her students' poor writing about probability:

> It is something, the subject of probability, what they do is more or less explain what they saw from the data and what I had seen was the writing was erratic. They wanted to put every word they could think of on paper and who cares about grammar. (Year 8 Teacher, Interview/November 07)

Table 6.1 Teacher-generated description of good mathematical writing

Characteristics of good math writing	
• Contains examples/drawings	• Labels diagrams, examples, and numbers
• Uses math vocabulary	• Addresses all parts of the question
• Restates the question	• Addresses the key concepts
• Answers the question	• Is clear and legible
• Is edited	• Has complete sentences and appropriate grammar
• Responses are organized/sequential	
• Explains examples	
• Includes formulas where appropriate	

From Doerr and Chandler-Olcott (2009, p. 293)

The Year 7 teacher in 2007 found it hard to let students write in his mathematics class because he was conscious of how much their *te reo Māori* was influenced by English. He felt that having them write would reinforce inappropriate expressions in *te reo Māori*.

As a consequence of these concerns, time was put aside in the project to discuss what constituted good mathematical writing. In a staff meeting in September 2007, each teacher examined a set of their own students' writing. The teachers then shared their responses to the following questions:

- What is a good piece of writing?
- What is a poor piece of writing?
- What are the features that make it a good piece of writing?
- What strategies can be used to improve students' writing?

Uenuku, like that of many of the other teachers, found it difficult to determine a good piece. All of the students' examples provided an appropriate answer. However there was much variation: Sometimes the mathematics was better; sometimes the mathematical diagrams used for the explanation were better; sometimes the measurements were added (mm squared or not); and sometimes the written explanations were better although the maths was not as detailed as some of the others. Using the work of Uenuku's students, four examples are provided in Fig. 6.8 to illustrate this complexity.

The original activity had required the students to:

> Write for someone who doesn't know how to do it.
> Write why you should do it like that.
> At the end write anything else you know about Pythagoras.

The examples in Fig. 6.8 reveal that students were often good in one area of mathematics but not necessarily in another. The first example has a confused written explanation, although the calculations are clear and the diagram does provide some help to the written explanation. In the second example, the calculations are correct and the diagram provides more details, but the written explanation is mathematically incorrect. The use of non-mathematical terms such as *pouaka* (box) also reduces the clarity of the explanation. The third example also provides a correct answer through the calculations, but with no explicit link to the Pythagoras' theorem as there was in the other three examples. The answer is reduced to two decimal places, which is a sensible response in this situation, and something not done in

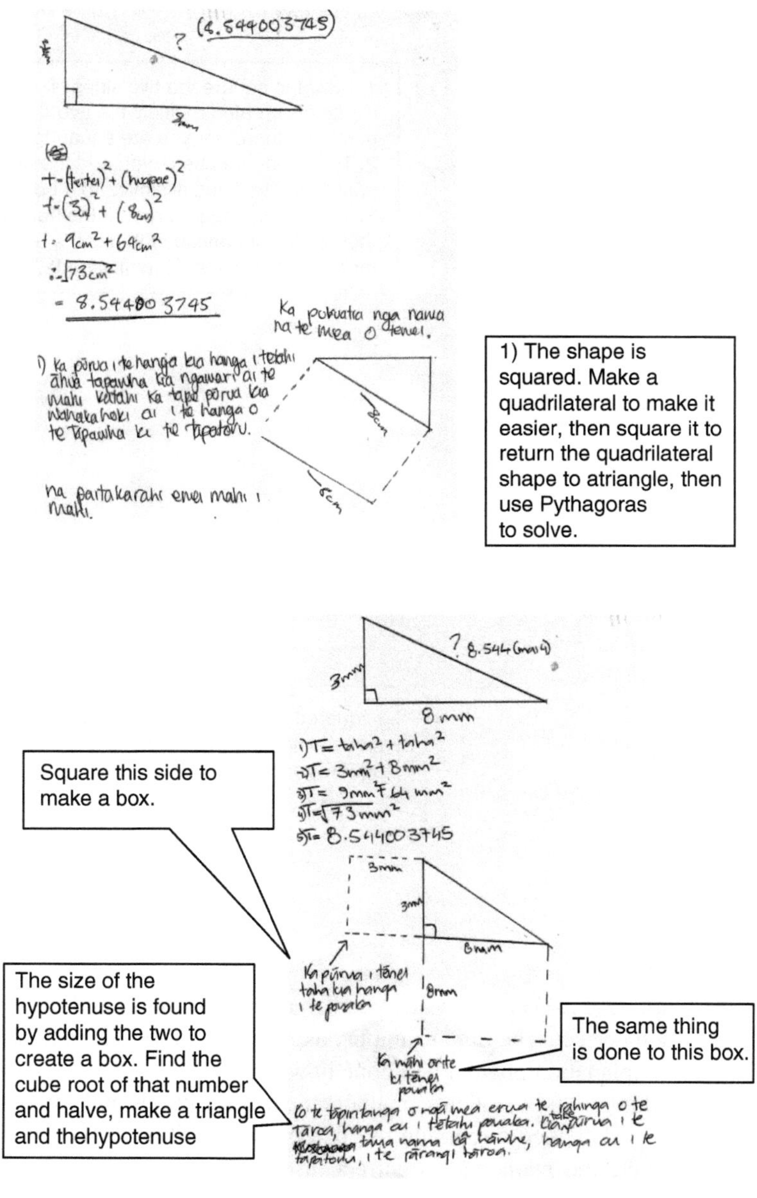

Fig. 6.8 Samples of Year 10 students' explanations of how they found the length of the hypotenuse

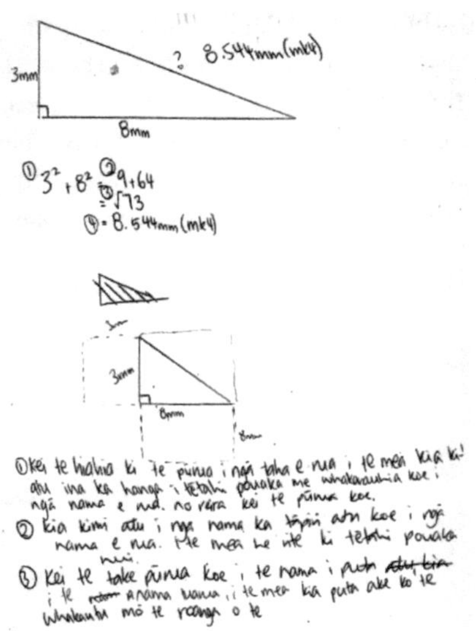

1. I want to square the two sides. To see if a box is created multiply the two numbers, therefore you are squaring
2. To find the numbers, you add the two numbers. The thing is similar to a box
3. You find the square root of the number that comes out because that is the answer for the length of the … (unfinished)

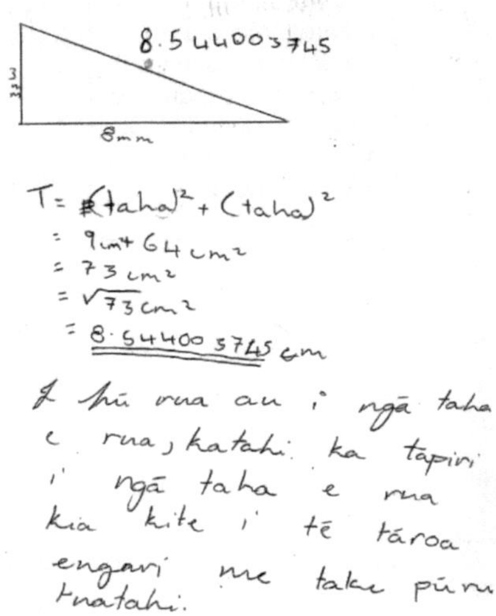

I squared the two sides, added the two sides to find the hypotenuses, but find the square root first.

Fig. 6.8 (continued)

the other examples. However, the written explanation is confused and the diagram mathematically incorrect. The final example has no explanatory diagram, although there is a good explanation of what was done in writing.

To improve this group of students' mathematical descriptions and explanations, all areas would need explicit discussion. A good piece of writing about Pythagoras' theorem must have the mathematics correct, use diagrams and text clearly, and be able to integrate these concisely into a coherent whole. An explicit discussion about these features would contribute to all the students becoming aware of what constitutes good mathematical writing. Students' mathematical thinking is also likely to improve as they gain more clarity about what they had done and why.

As a result of our exploration, we have combined the concerns of the Te Koutu teachers with ideas of Doerr and Chandler-Olcott (2009) to suggest the following criteria for judging the quality of mathematical writing:

- mathematical: The writer needs to show a thorough awareness of the mathematical ideas and how they relate to what is being discussed
- integration of modes: The writer may need to use a range of different modes to increase the clarity of the ideas they were presenting
- stylistic: The writer needs to show an awareness of mathematical, stylistic writing conventions such as syntax, verbal competence, text organisation, cohesion, awareness of reader, and appropriateness of text (adapted from Wilkinson, Barnsley, Hanna, & Swan, 1980).

Some of the teachers felt that sometimes a student knew more than they were able to express on paper and that forcing them to write might be limiting their thinking. On the other hand, being able to clearly justify their use of mathematical concepts was seen as being valuable for students' further learning, and if it was not achieved through writing, then it still needed to be achieved. One of the teachers' reasons for concentrating on writing was that it was easy to see when students were able to show their ideas clearly because it was an object that could be repeatedly referred to.

Students' Views About Writing in Mathematics

The teachers felt that having students write in mathematics would support their mathematical thinking, but we also surveyed and interviewed students to find out their views. Unless the students could see some point in what they were doing, it was likely that they would resist efforts to make them write.

In November 2007, students from all year levels completed a survey about writing in mathematics. As the students were aged from 5 to 18, the survey predominantly used pictures and multiple-choice questions. Up to 102 students or approximately half the total student population completed the surveys. Not all students completed each question, and so the totals rarely equalled 102. However, the students who failed to answer were different for each question. Previously, in September 2007, two children from each class from Year 0 to Year 10 were also interviewed about their experiences of writing in mathematics. A total of 17 children were interviewed. All the interviews were carried out in *te reo Māori* and were based around the following series of questions:

- Do you remember doing this writing? [students were requested to bring some writing with them]
- Can you tell me something about it?
- Do you remember why you did it?

- Do you ever go back and read through anything you wrote earlier in the year in maths?
- What do you think about writing in mathematics?
- What kinds of writing do you do in mathematics?
- What kinds of writing do you like doing? Why?
- What kinds of writing do you not like doing? Why?
- Why might you have to do writing in mathematics?

The interview responses were categorised into five main themes. These were feelings, descriptions, explanations, difficulties, and purpose. Many of the students' comments showed that they perceived these themes as being interrelated. When the primary school students' (Year 1–6) responses were separated from those of students in intermediate and high school (Year 7–13), the same themes were present. However, the number of connections between the themes was much greater in the comments of the older students.

The theme of feelings related to whether students liked or disliked different types of mathematical writing. Further information on this can be seen in the graph of the survey results of students' favourite type of writing in Fig. 6.9. It mirrored the graph of students' dislikes.

Types of writing were also connected to the content that students were engaged in. In the interviews, a Year 2 student stated that she did not like drawing clocks. One Year 3 student also did not like drawing clocks because it was "too easy", whilst her peer felt it was "too hard". Although students disagreed about whether particular topics were easy or hard, they were clear that if the activity was too challenging or too easy, they did not enjoy it. Ensuring that students are appropriately challenged in mathematics classrooms is linked to students being involved in higher levels of reasoning (Anthony & Walshaw, 2007).

The theme of purpose was linked to whom the students saw the writing being for. Morgan (1998), in considering the literature on school writing across the curriculum,

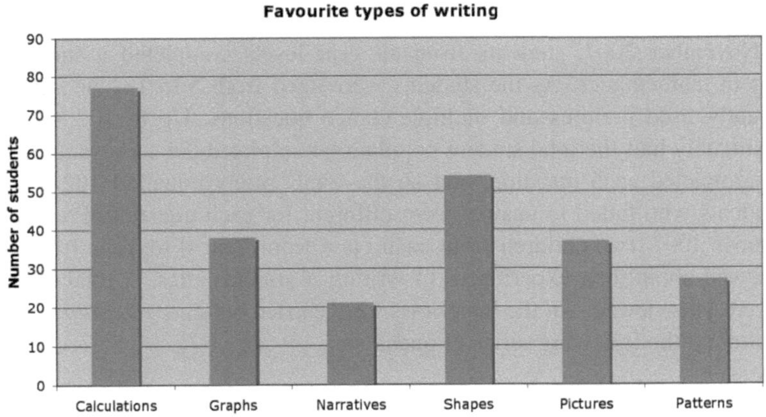

Fig. 6.9 Students' favourite types of writing

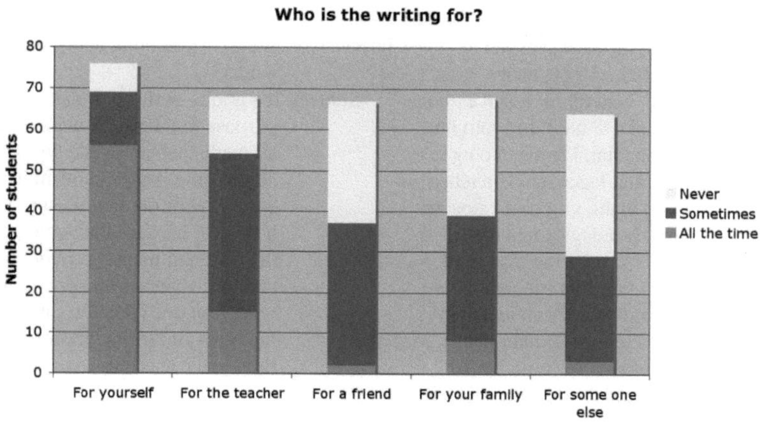

Fig. 6.10 Audience for students' mathematical writing

suggested that "one of the roots of students' difficulties and lack of motivation in their development as writers" was the fact that the teacher as examiner was the audience on most occasions (p. 43). Figure 6.10 shows that students predominantly felt that the writing that they did was for themselves. While many Te Koutu students also felt that they sometimes wrote for the teacher, mostly they wrote for themselves. In the interviews, some students talked about needing a purpose for their writing. If students saw mathematical writing as primarily being for themselves, then it is logical that they would need to understand the purpose for doing it.

The following extract comes from the interview with a Year 11 student. She talks about what she saw as being the purpose of writing in mathematics. It begins with her talking about some of her written work on trigonometry.

Student: Ko te take i whirihia i te mahi nei, he rahi nei nga momo nuka tā te mahi nei. Nō reira, he pai ki au ki te kimi huarahi rerekē – ki te kimi otinga. Koirā te mea pai o te tātai i nga wā katoa, ka kite koe i nga momo huarahi rerekē, He pai ki a koe i te wā ka kimi i te otinga, me te huarahi i tae ai koe ki taua otinga – he pai. Ko ētahi o nga mahi, ko te kimi i nga koki, ko te kimi i te rahinga me te tāroa. Ka hoatu te pātai e pā ana ki ētahi roanga me kii o nga taha – āna, ko tāu, ko te kimi i te huarahi kia kimi i te inenga o te toenga o nga taha. ...	Student: The reason I chose this work is because it involves lots of strategies. Therefore, I enjoy finding different pathways – to find solutions. That's the thing [which is] good about maths – all the time to see different pathways. You feel good when you find that solution – and the pathway to reach that solution – it's good. Some of the work are to find the angles, to find the size of the angles and tangent. The question is provided in regard to find the length of a side. It's for you to find a pathway to find the measurement of the remainder of the sides. ...

Interviewer: He aha ōu whakaaro mo te tuhituhi i roto i te tātai?	Interviewer: What do you think about writing in maths?
Student: He pai. Nga tuhinga. – Ka tino whai take te tuhituhi i roto i te karaihe tātai, i te mea ko nga mahi tātai katoa, ahakoa tēhea momo mahi, i nga wā katoa he momo hononga a nga mahi katoa. Nō reira me mau tonu koe, i te mea ka rite nga momo nuka o tērā, o tērā o nga mahi, ahakoa te momo mahi tētai e mahi ai koe. Nō reira he pai nga momo tuhinga i te mea , i nga wā katoa me hoki atu koe hei te wā – Āe.	Student: It's good – writing. It has real purpose in the maths class. Because all maths, whatever the type of work, all the time it has connections to all work. Therefore you need to keep hold of it because the strategies of the different maths activities are similar, whatever the type of maths work you do. Therefore, it's good – the types of writing because all the time you can return to it.
...	...
Int: He aha nga momo tuhinga e pai ana ki a koe?	Interviewer: What types of writing do you like?
Student: Nga mahi pai ki ahau – nga mahi whāritenga. Nā te mea i nga wā ka mahi whēritenga, i nga wā katoa me kimi nuka, me whai nuka rerekē ki te kimi i nga – pērā ki nga mahi "r" me te "t" – I nga wā katoa me kimi he aha te tikanga o tērā, o tērā o nga reta. Ana, pai noa iho ki au i te mea i nga wā katoa, he uaua te mahi, he pai ki au te mahi uaua i te mea he mea hou, nō reira, e tino ngana ana ki te tangata ki te kimi i te otinga.	Student: The work I like – the equations. Because at the time you do equations, you find strategies, to follow different strategies for example to find "r" and "t". All the time you seek the value of the different letters. I like it because the work is challenging, because it's new, therefore it feels good to find the answer.

For this student, mathematics was about connections and having access to different strategies. Her writing about mathematics enabled her to revisit ideas discussed earlier in the year and to remind herself about these connections. She enjoyed working hard to find an answer, and this gave her a great feeling when she succeeded. She had a strong sense of her own learning needs and valued persevering to achieve an answer. The joy of learning came from feeling good when the answer was found. Writing for her was very much about thinking mathematically.

At the beginning of the chapter, we mentioned the need for students to see mathematics and writing in mathematics as being connected to their identities, including their identities as Māori. For this student, being able to write in mathematics contributed to being able to do mathematics, and this had a positive impact on what she gained from being in a mathematics lesson. Although she did not explicitly mention her Māori identity, there is no sense that doing mathematics involved her leaving her Māori identity at the door.

Students' opinions about writing in mathematics surprised us and opened up unforeseen opportunities to meet the challenges involved in improving students' writing. The teachers were aware that the students resented having to write sentences in mathematics, but had not realised that students saw themselves as the major audience for their writing as being for themselves. As explanations and justifications seem to require some written sentences to carry their meaning, it may be possible for teachers to use students' awareness of their own needs to encourage them to write more in sentences. We are still working on trying to utilise this idea.

Challenges in Writing to Support Mathematical Thinking

Having students think mathematically is an aim for many teachers of mathematics. However, in a Māori immersion school, this must be done in conjunction with the other aims for the school such as supporting the revitalisation of the Māori language. Having students write sentences in mathematics had the potential to improve their *te reo Māori* skills because it can be reviewed many times. Yet there is a need for improvements in written *te reo Māori* to flow back into students' oral language. Without this flow, the oral tradition of the language is at risk of being lost.

In the previous chapter, features of *te reo Māori* were identified as being valuable in supporting students to think mathematically. Supporting students to learn to use these features to improve their mathematical thinking is not simple. In the younger classes, the teachers agreed on specific terms and then channelled students into using these terms. In older classes, students repeatedly wrote sentences using the correct sentence structures to support them to use oral language correctly. However, teachers had to value these approaches if they were to implement them. When a new teacher for the senior students began teaching at the school in 2008, he was worried about covering all the content in time for the exams at the end of the year and initially saw work on writing in mathematics as an unnecessary use of students' class time. Thus the challenge of using writing in mathematics was complex and needs ongoing discussion if it is to benefit students' academic performances and efforts to revitalise *te reo Māori*.

By deciding to increase the quality and quantity of writing in mathematics in 2007 the teachers accepted a challenge on their own initiative. However, the lack of other research in this area has meant that they have had to work with researchers to identify what student writing was already going on and what kind of writing they felt would be most beneficial for students to engage in. In identifying these kinds of writing, we are still learning how features, such as logical connectives, can be used effectively by students. This process contributes to an understanding of what constitutes a quality piece of writing.

By recognising the value in having students write in mathematics, this challenge is on the way to being met. Although there is still some way to go in working with students to produce revised rather than just first drafts of written work, there has been a shift. Writing in mathematics is now perceived as a normal part of mathematics lessons. There is also an increasing expectation across the *kura* that students will explain and justify their answers through writing. Changing students' understanding that mathematics involves writing may take an equally long time to that of changing teachers' beliefs about what they should be expecting from their students in mathematics. As staff come and go, mathematics teaching will need to be reviewed on a regular basis, but Te Koutu is on its way to redefining what are considered "normal" practices.

Chapter 7
The Case of Probability

In this chapter, we explore how students use speaking and writing to think mathematically, or in this case probabilistically. The teachers at Te Koutu found probability to be one of the most difficult areas to teach, and at times the terminology in *te reo Māori* contributed to these difficulties. There is a growing body of research, which investigates the teaching of probability in English, but there is almost nothing that looks at how the ideas develop across year levels in a particular school. One of the strengths of Te Koutu is that mathematics teaching is discussed by all the teachers across the year levels and issues are worked through together.

Probability is perceived as not being a simple topic either for teachers to teach (Chick & Baker, 2005) or for children to understand (Benko & Maher, 2006). The following extract comes from a staff meeting in which the Year 8/9 teacher described her frustration with students' writing about probability.

> In the end I thought I failed that activity because being in the 8/9 area maybe I needed more focus on probability which is one of my weaknesses whereas every other strand we've done seems to be more input, more guiding, more modelling, more blah, blah, blah. With probability you need them to come out and strategise. I guess it is quite hard in this case. Probability isn't a good subject. I just don't like it. (Y8 Teacher, Staff meeting, 5 November, 2007)

At the same time, probability has a reputation for being more language dependent than other topics (Watson, 2006) as the next extracts highlight.

Y12 Teacher: I was just looking at the level two [internal NCEA assessments] at the moment and the first internal [assessment] was sampling, the second was normal bell curves...

Tamsin: Yeah,

Y12T: ...and probability, and the third one was practical trig, and practical trig was just an equation that they basically had to do. The sampling one was just about all writing so in a way they had to learn heaps of words, heaps of grammar, heaps of words. At that stage of the year, right at the start because of random number, *tau matapōkere,* just heaps. (Interview, September, 2008)

T. Meaney et al., *Collaborating to Meet Language Challenges in Indigenous Mathematics Classrooms*, Mathematics Education Library 52,
DOI 10.1007/978-94-007-1994-1_7,

Uenuku: I think the kids sometimes feel that lack of language muscle in that area too. The other thing is when you think about probability even if you try to sit down and, say, explain what chance is. What's the probability that something? You try to explain, what you mean by the probability, it does get pretty complicated. It is one of the more convoluted levels of language in English. That's what I've noticed too in Māori. So when they get happy, is when you start falling back on those rules. Because it is a pattern of language they can use again rather than trying to pluck these ideas out of the air.

Tamsin: So as a whole school, where you have probability being taught all the way up is there a strategy that you can?

Y4 Teacher: We have talked about this before

Tamsin: Mmm

Y4T: Maybe we can sit down and say these are the things we can use. For *whakatauira* [demonstrating] we had a *e wha o ngā mea rima* [four out of five things], you know. We were going to sit down and put together a whole lot of things we can be saying because we will be modelling them now even though they might not be able to read them but we model. They shouldn't change throughout. (Staff meeting, November 5, 2007)

Amir and Williams (1999) stated, "[t]he question of language arises both in the refinement of technical language on which probability draws, and in the contrasting languages used by children in their home and as the medium of instruction" (p. 105). Interference from connotations from a student's first language is known to affect his or her acquisition of second-language probability terms (Kazima, 2006). Without explicit teaching, students may often be unaware that the same word has different meanings in the everyday context and the mathematical context. Yet, at Te Koutu we did not discuss the possibility of such complications in describing probability events, even though we were aware that the language the children spoke at home, usually English, was not the same as the language they used at school, *te reo Māori*. This remains a challenge that has not been recognised, or at least not acknowledged, at Te Koutu. One of the tensions in working in a school where there is such a strong emphasis on *te reo Māori* is that it becomes difficult to discuss issues that stem from students (or teachers) having English as their first language. Instead, these issues are dealt with indirectly as the school community works through broader challenges, such as how to use *te reo Māori* to think probabilistically.

The challenges associated with probability meant that it was an ideal topic for considering how speaking and writing supported students' mathematical thinking. Over the years, but particularly in 2007, we video recorded a number of probability lessons as well as collected students' writings about this topic. In interviews and at meetings, teachers also talked about their experiences of teaching probability.

Students Learning About Probability

Probability is a relatively new topic in curricula around the world, and there is limited research on how students learn it (Chick & Baker, 2005). Many of the teachers at Te Koutu would not have learnt about probability when they were at school, or if they did, it would have been in the final years of secondary school. Consequently, teachers have not had much opportunity to develop the relevant pedagogical content knowledge (Chick & Baker, 2005).

Research has suggested that the different facets of probability are difficult for students to grasp, and so the topic and its components need to develop over a number of years (Nickson, 2000). As a result, it is not a topic that one teacher can cover, but must be part of a whole-school approach. In a longitudinal study of how junior high school students developed ideas about probability, Green (1983) found that:

> 1. The concept of ratio is vital to children's understanding of probability
> 2. The level of understanding of the language of probability is poor (e.g. words such as 'certain' and 'least'
> 3. A systematic approach to the teaching of probability and statistics in schools is necessary to overcome children's misconceptions in connection with the subject. (cited in Nickson, 2000, p. 94)

Since then, probability has come to be seen as even more complicated. In 2005, Chick and Baker suggested the following list of conceptual issues that are needed early in probabilistic thinking:

- ideas of long-term probability (i.e., probability as a long-term phenomenon),
- variation and random behaviour,
- sample space,
- likelihood, and
- fairness.

These issues are expanded further in Fig. 7.1, which is taken from Watson (2006). In this diagram, Watson highlighted the main ideas about chance, the precursor to probability, and the relationships between them. In the diagram, the ideas of both Green (1983) and Chick and Baker (2005) can be seen.

For Watson, the lack of clarity in regard to specific terms such as "random" was a major inhibitor in students' understandings about chance. She suggested, "the complexity of the concept makes it important to begin discussions with students early and continue them, hopefully with increased sophistication, throughout the school years. Exploring contexts where students believe random happenings take place is an important foundation for later work" (p. 131). From the work of others, Chick and Baker (2005) suggested that there is a need for teachers to provide appropriate intuitions in primary schools in order to stop a possible decline in students' understandings about probability because of "the complex interplay between intuitive understanding and educational experience" (p. 233).

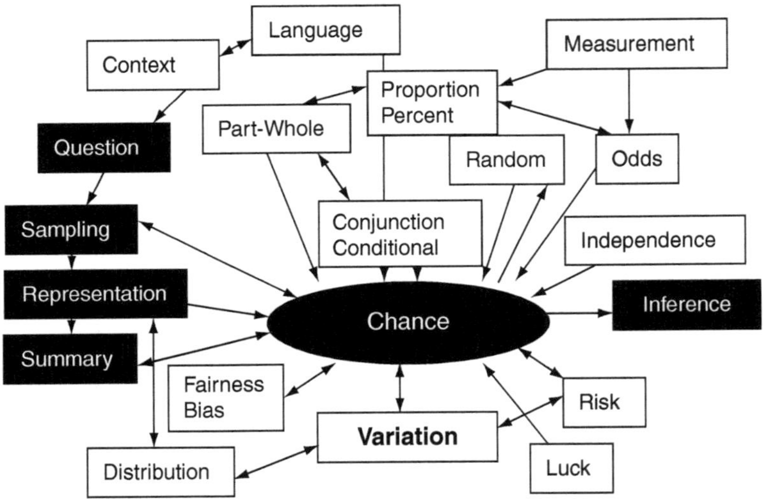

Fig. 7.1 Links between ideas and statistical elements related to chance understanding (from Watson, 2006, p. 130)

Over the years, researchers have suggested different developmental pathways that students follow when learning about probability. Jones, Langrall, Thornton, and Mogill (1997) emphasised the importance of looking first at concepts to work out theoretical probability, which is when "symmetry, number, or simple geometrical measures can be used as the basis for determining probabilities" (p. 104). They outlined four main constructs for theoretical probability understandings: sample space, probability of an event, probability comparisons, and conditional probability. They felt that concentrating on understanding students' thinking about theoretical probability enabled connections to be made more easily to their intuitive thinking. Only once these connections had been made should students' thinking about experimental probability, or probability based on relative frequency, be explored. On the other hand, work by Amit and Jan (2006) has shown that students can build understandings about theoretical and experimental probability simultaneously whilst engaging in activities that culminated in the presentation of their reasoning about the likelihood of different events occurring. Commenting on changes in curricula in regard to probability, Watson (2006) noted that there is a move to having primary students work with an empirical frequency-based approach to probability, before moving towards concerns to do with theoretical probability. This turns on its head; Jones et al.'s (1997) views that an initial concern with theoretical probability is more appropriate.

Nonetheless, Jones et al.'s (1997) work remains useful because it illustrates the types of answers that students provide when explaining probability events. Table 7.1 outlines the four levels of answers that are connected with each of the four constructs.

Table 7.1 Framework for assessing probabilistic thinking

Construct	Level 1 Subjective	Level 2 Transitional	Level 3 Informal quantitative	Level 4 Numerical
Sample space	• Lists an incomplete set of outcomes for a one-stage experiment	• List a complete set of outcomes for a one-stage experiment, *and* • Sometimes lists a complete set of outcomes for a two-stage experiment using limited and unsystematic strategies	• Consistently lists the outcomes of a two-stage experiment using a partially generative strategy	• Adopts and applies a generative strategy which enables a complete listing of the outcomes for a two- and three-stage case
Probability of an event	• Predicts most/least likely event based on subjective judgments • Recognizes certain and impossible events	• Predicts most/least likely event based on quantitative judgments but may revert to subjective judgments	• Predicts most/least likely events based on quantitative judgments including situations involving non-contiguous outcomes) • Uses numbers informally to compare probabilities • Distinguishes "certain", "impossible", and "possible" events, and justifies choice quantitatively	• Predicts most/likely events for single-stage experiments • Assigns a numerical probability to an event (it may be a real probability of a form of odds)

Table 7.1 (continued)

Construct	Level 1 Subjective	Level 2 Transitional	Level 3 Informal quantitative	Level 4 Numerical
Probability comparisons	• Compares the probability of an event in two different sample spaces, usually based on various subjective or numeric judgments • Cannot distinguish "fair" probability situations from "unfair" ones	• Makes probability comparisons based on quantitative judgments (may not quantify correctly and may have limitations where non-contiguous events are involved) • Begins to distinguish "fair" probability situations from "unfair" ones	• Makes probability comparisons based on consistent quantitative judgments • Justifies with valid quantitative reasoning, but may have limitations where non-contiguous events are involved • Distinguishes "fair" and "unfair" probability generators based on valid numerical reasoning	• Assigns a numerical probability measure and compares • Incorporates non-contiguous and contiguous outcomes in determining probabilities • Assigns equal numerical probabilities to equally likely events
Conditional probability	• Following one trial of a one-stage experiment, does not give a complete list of outcomes even though a complete list was given prior to the first trail • Recognizes when certain and impossible events arise in non-replacement situations	• Recognizes that the probability of *some* events changes in a non-replacement situation, however, recognition is incomplete and is usually restricted only to events that have previously occurred	• Can determine changing probability measures in a non-replacement situation • Recognizes that the probability of all events change in a non-replacement situation	• Assigns numerical probabilities in replacement and non-replacement situations • Distinguishes dependent and independent events

From Jones et al. (1997, p. 111)

The four levels show how students' responses move from idiosyncratic to those that link related ideas about sample size to determining a theoretical probability of an event occurring. An example of the sorts of typical responses for the four levels in regard to the probability of an event were provided by Johnson, Jones, Thornton, Langrall, and Rous (1998, p. 204) and are reproduced in the following section:

Levels	Example: What gum ball is most likely to come out of a machine that has 6 red and 3 yellow gum balls?	Thinking
Level 1:	Subjective	"Yellow, it's my favorite color".
Level 2:	Transitional	"Red because there's more red and I like red best".
Level 3:	Quantitative	"Red, because there are 6 red and only 3 yellow".
Level 4:	Numerical	"Red, because the chance of red is 6 out of 9, and yellow is only 3 out of 9".

Jones et al. (1997) suggested that students may not be at the same level for all constructs simultaneously, nor will progression through the levels always be in a forward direction. In the eight case studies that they used to refine their framework, they found several instances where students' answers indicated that they had reverted to an earlier level in their understanding. As Chick and Baker (2005) noted, declines in probabilistic thinking have been recorded by several researchers. Jones et al. (1997) argued that the instruction programme, which included both theoretical as well as experimental probability activities, might have contributed to students reverting to subjective reasoning. However, the work of Amit and Jan (2006) discredits this as a reason for the decline. Based on their background experiences, intuition seemed to make a significant contribution to students' thinking.

The influence of students' background experiences on their probabilistic thinking has also been explored. Amir and Williams (1999) investigated the role of culture and language on first year, high school students' understandings about probability. They found that large numbers of students held non-stochastic understandings.

> Devices that are normally considered as random, such as dice and coins, are not seen so by all children. There is a significant proportion that sees the behavior of these random devices as not equiprobable, dependent on some superstition, or on how you operate with them (i.e., a causal, rather than a chance, explanation). This should be clarified before using them in teaching or research. (Amir & Williams, 1999, p. 104)

Students need to use and discuss these devices before they simply become background information in written questions about probability. Watson (2006) had also noted the importance of students developing understandings about randomness. Amir and Williams (1999) suggested that without appropriate informal probabilistic knowledge, it is unlikely that students will develop formal probabilistic knowledge. Figure 7.2 illustrates how they see the relationship between children's background and the development of formal probabilistic knowledge.

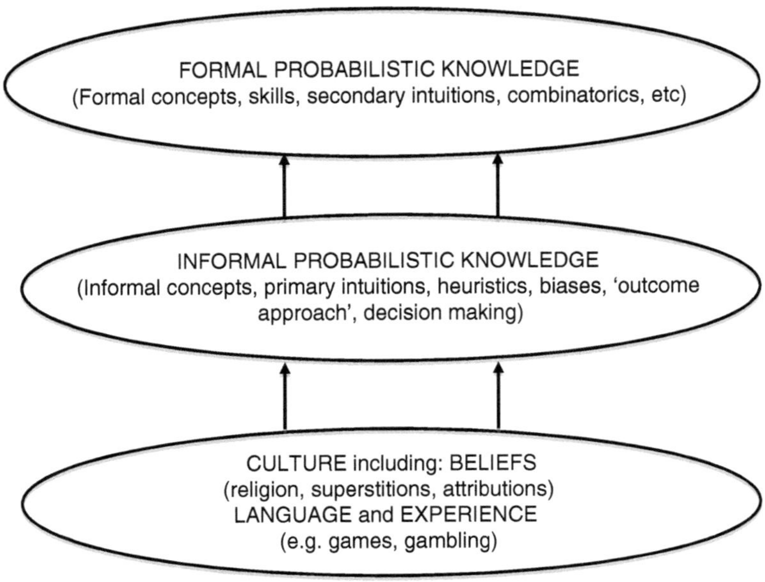

Fig. 7.2 Culture and probabilistic thinking: theoretical model (from Amir & Williams, 1999, p. 88)

In order to support students to develop the necessary understandings, research was also conducted on the impact of students' writing about probability. Johnson et al. (1998) investigated the writing done in solving probability problems by fifth-grade students. They found that students relied both on mathematical symbols and words to describe their understandings. "In fact, the evidence documented in this study suggests that students whose writing moved to the higher generalizing and relating levels developed a more complementary relationship between their writing symbols and their attendant mathematical symbols" (p. 218). This corresponds with our findings in the previous chapter, which was that when students provided explanations and generalisations, they were more likely to include different modes to express their ideas. Johnson et al. (1998) suggested that it was the teacher's requests of the students, which resulted in the use of two or more modes. They also indicated that probably "rich written interchanges between student and teacher enabled learning to be maximized" (p. 220).

In much of the research on probability, there has been a concentration on comparing students across a single year level. Although Watson (2006) discussed understandings about probability held by students from different age groups, there is no clear indication of how probabilistic thinking develops over time. Even Jones et al.'s (1997) work on levels of probabilistic understanding was developed from working with Year 5 students only. In this chapter, we explore how the teachers at Te Koutu in different year levels worked to develop students' probabilistic thinking, especially in relationship to supporting their oral and written explanations and justifications.

Learning to Think About Probability

Across the school, there was a recognition that students needed to learn specific terms and expressions in order to discuss probability. This is not surprising given that most of the students were second-language learners of *te reo Māori* and that the terms were newly developed and had not yet stabilised in the wider community. While Māori had words to express quantity, space, shape, and measurement, there were very few words to express probability. Traditionally, events were influenced by the *mana* (power) of certain individuals and not seen as random events that were more or less likely to occur. In this section, we examine data that we have collected from the different classrooms to show how probabilistic thinking was developed across the school. Although we do not have a complete set of activities, it is possible to give an indication of how the ideas were developed.

Developing the Idea of Likelihood in the Beginning School Years

The teachers in years 0–2 spent much of their probability teaching time discussing ideas about likelihood. This was in contrast to what research suggested. For example, Chick and Baker (2005) suggested, "sample space is a central component of quantifying likelihood, and is often a focus of early probability activities" (p. 234). Yet at Te Koutu, the children engaged in few activities where sample space was the focus. One of these was a problem-solving task that was done by students at different year levels across the school (Fig. 6.4 was an example from a Year 11 student from this task). The problem was to find the number of combinations that could be made from different amounts of coloured blocks. To begin with, the students had to find the number of possibilities when they made columns from two blocks of two different colours. They then moved on to finding the number of possibilities with three blocks and two colours and then three blocks and three colours, and so on. Figure 7.3 shows children working on this problem, putting the blocks together in the different combinations and then recording those combinations.

Fig. 7.3 Children using blocks to find combinations

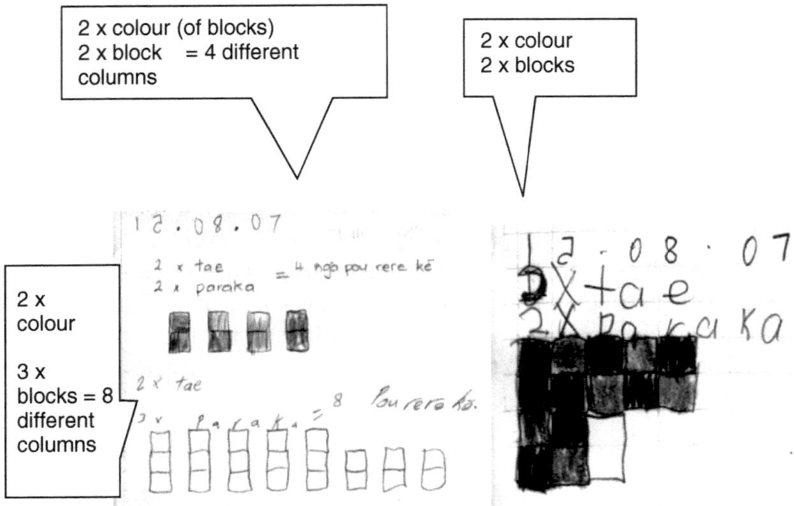

Fig. 7.4 Two examples of Year 2 students' written work

Figure 7.4 shows two examples of Year 2 students' work. In the first one, the initial part of the recording is in the teacher's handwriting indicating that she had shown this student how to set out her results. The second example is in the child's handwriting, but follows the teacher's example. Note that only the first four combinations are drawn for the second example in the first child's writing.

Although the children had the skills to draw the blocks and add words and symbols to describe the combinations, the teacher's intervention was needed to help them structure their recordings systematically. Systematic recording supports students being able to identify all the possibilities in a situation, but often this needs to be explicitly taught for students to understand its value.

Whilst this activity was about finding the number of possibilities and, therefore, was related to sample space (Watson, 2006), the activity did not require the children (or the teacher) to explicitly discuss this connection. Nevertheless, given that Jones (1974) found "that significant numbers of grades one through three children were not able to list the outcomes of a one-stage experiment" (cited in Jones et al., 1997, p. 104), supporting students to record outcomes systematically may well be beneficial for later lessons on probability.

Not all activities resulted in students developing their probabilistic thinking. Figure 7.5 shows work by a Year 1 student. The task was designed to build up students' intuitions about randomness by identifying what the outcomes could be. The student had to predict the outcome of whether the coins would come up by circling one of the options on the worksheet: *māhunga*, *māhunga* (head/head) or *māhunga*, *whiore* (head/tail) or *whiore*, *whiore* (tail/tail). The children then tossed the coins to see what the results were. Figure 7.5 indicates that the child became confused and circled the actual result as well as their predictions. The need for children to get the "right" answer may have circumvented the teacher's aim to have the child recognise that outcomes of specific events are independent of previous events.

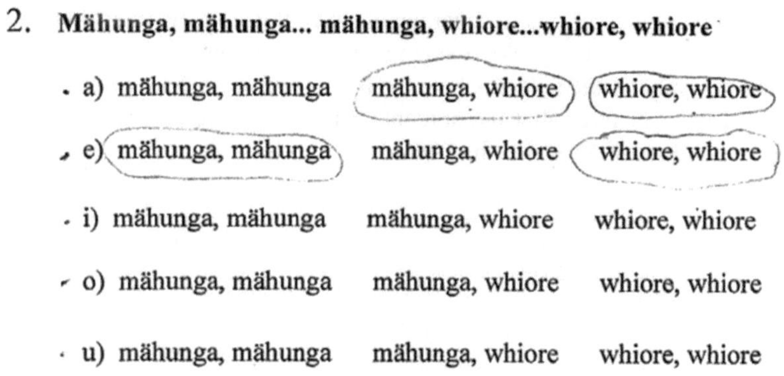
2. Māhunga, māhunga... māhunga, whiore...whiore, whiore

a) māhunga, māhunga   māhunga, whiore   whiore, whiore

e) māhunga, māhunga   māhunga, whiore   whiore, whiore

i) māhunga, māhunga   māhunga, whiore   whiore, whiore

o) māhunga, māhunga   māhunga, whiore   whiore, whiore

u) māhunga, māhunga   māhunga, whiore   whiore, whiore

Fig. 7.5 A Year 1 student's worksheet on probability

Sample space and randomness were not the main focus in the early years at Te Koutu. Instead, the teachers built up students' probabilistic thinking by developing their understanding of different terms about likelihood. The teachers did this by modelling the terms and providing opportunities for the students to use the new terms. It has been suggested that language learners need to hear a word seven times, at spaced intervals, to acquire it (Thornbury, 2002 cited in McNaughton, MacDonald, Barber, Farry, & Woodard, 2006). At this early stage of acquiring new terms, the teacher takes on most of the responsibility for developing the conversation, as students are not expected to know or use the terms by themselves (see Chapter 10 for more information about the acquisition of the mathematics register).

In 2005, the Year 2 teacher used the story of Little Red Riding Hood to introduce the term *tērā pea* (perhaps) as part of a series of lessons on probability.

Y2 Teacher:	He aha kei roto i te ngahere?	Year2 Teacher:	What is in the forest?
Student1:	He kau	Student1:	A cow.
Y2T:	He kau i roto i te ngāhere [boy laughs], tērā pea. Student2, i kite ia i te aha?	Y2T:	A cow in the forest, perhaps. S2, what will she see?
Student2:	He wūruhi	S2:	A wolf.
Y2T:	He wūruhi, āe. Kua kite kē ia i te wūruhi. [nods to another child] Kua wareware, he aha ngā momo mea ka kite a Pōtae Whero kei roto i te ngāhere?	Y2T:	A wolf, yes. She'll see a wolf instead. [I've] forgotten, what sort of things will Little Red Riding Hood see in the forest?
Student3:	He kiwi	S3:	A kiwi.
Y2T:	He kiwi, āe, tērā pea	Y2T:	A kiwi, yes, perhaps.
...		...	
Student4:	He manu	S4:	A bird.
Y2T:	He manu, āe. He manu kei roto i te ngāhere. He whakaaro anō?	Y2T:	A bird, yes. A bird in the forest. Any other ideas?
Student5:	Ka kite i te whare	S5:	[She] will see a house.
Y2T:	He whare, āe, tērā pea, ka kite ia i tētahi whare. He whakaaro anō?	Y2T:	A house, yes, perhaps she'll see a house. Any other ideas?

In the following lesson, the teacher again asked the students to evaluate the likelihood of Little Red Riding Hood seeing different things in the forest that were suggested not only by the students but also by herself. By offering her own suggestions, the teacher ensured that the students practised using all the terms based on their understanding of what they had heard.

Student:	I kite ia i te kuia.	Student:	She saw the grandmother.
Y2 Teacher:	I kite ia i tēnā kuia i te ngāhere?	Y2T:	Did she see that grandmother in the forest?
Students:	Kāo	SS:	No.
Y2T:	Student1, hōmai tāu, Student2?	Y2T:	S1, give me your [suggestion], S2?
Student:	Rau	S:	Leaves.
Y2T:	Rau, he rau i te ngāhere?	Y2T:	Leaves, were there leaves in the forest?
Student:	Āe	S:	Yes.
Y2T:	Āe, he rau. He mea anō Student3?	Y2T:	Yes, leaves. Anything else, S3?
Student:	Kāo	S:	No.
Y2T:	Kāre, i kite he mea he aha ngā mea roa ngā mea teitei nei he parauri te tinana?	Y2T:	No. She saw something [else]. What are the long things, tall things that have a brown "body"?
Student:	Kakī roa	S:	A giraffe?
Y2T:	Nā he aha ngā mea kara kākāriki?	Y2T:	Now what are the green-coloured things?
Student:	Rākau	S:	Trees.
Y2T:	Rākau. I kite ia i ētahi rākau Student4?	Y2T:	Trees. Did she see some trees, S4?
Student:	Āe	S:	Yes.
Y2T:	Hipo	Y2T:	A hippopotamus?
Students:	[laugh]	SS:	[Laugh]
Y2T:	Student2, i kite ia i ētahi hipohipo i te ngāhere?	Y2T:	S2, did she see some hippopotami in the forest?
Student2:	Kāo	S5:	No.
Y2T:	Student 5 i kite ia i ētahi hēpara i te ngāhere?	Y2T:	Some lions?
Students:	Kāo	SS:	No.
Y2T:	He raiona?	Y2T:	Yes, you're quite right. Did she see, S2, some cars?
Students:	Kāo	S2:	Perhaps.
Y2T:	Āe, tika tāu. Student2, i ētahi motuka?		
Student2:	Tērā pea		

The familiar story meant that students already had ideas about the certainty of what Little Red Riding Hood was likely to see. With the teacher's support, they connected their intuitions to their probability vocabulary and understandings. In using *tērā pea*, the teacher highlighted both the concept of uncertainty and the word that described it.

Watson (2006) advised that it was important to ask students to provide examples of events that have different possibilities of occurring as "[t]his provides the teacher with information on the contexts that are available to students in imagining chance happenings and on the appropriateness of initial understandings" (p. 134). In so doing, some of the difficulties identified by Amir and Williams (1999), with regard to students' previous experiences interfering with their informal probability understandings, could be counteracted. Figure 7.6 shows representations by two Year 1 students of probability phrases, which were part of a lesson in 2007. This activity developed understandings about different possibilities and also highlighted the value of combining a written description with a diagram. At a later stage these could be developed into explanations or justifications. Beneath the diagram is an extract of an interview with the two students about this work.

In this activity, the work of the students did not represent real events, as had been the case in the work presented in Fig. 7.5. Instead, they imagined "real" events, which would have the appropriate likelihood of occurring. On the whole,

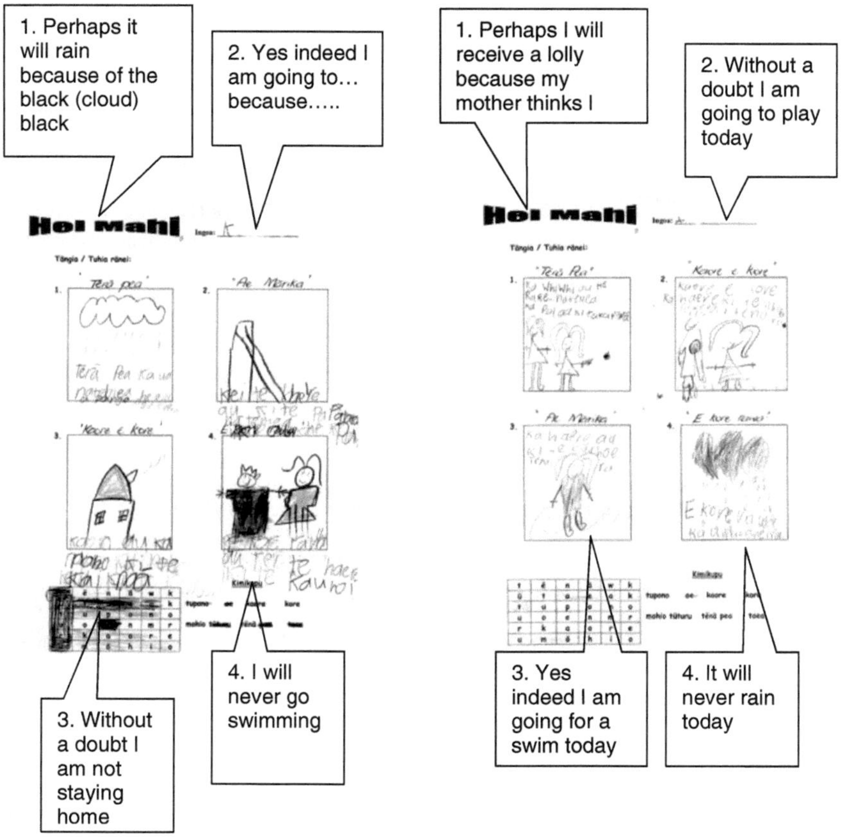

Fig. 7.6 Students' definitions for specific probability terms

Interviewer:	Nā tērā, he aha te tikanga o te pikitia tuatahi i tō pepa? Te mea e kī ana "Te Rā"	Interviewer:	Now, that, what is the purpose of the first picture on your paper? The thing that says, "The Sun".
Student1:	"Tērā pea".	Student1:	[No,] "perhaps".
I:	Āe. He aha te tikanga o ngā pikitia kei tō pepa? E whakaatu aua pikitia i te aha?	I:	Yes. What's the purpose of the picture on your paper? Those pictures show what?
Student1:	I tā au i "tērā pea" ka ua, na te mea ka pango ngā kapua.	S1:	I drew " 'perhaps' it would rain because the clouds were black".
I:	Ka pai. Papatākaro, i te mea.	I:	Good. The park because...
Student1:	(reads her paper) I kï ahau [i] te mea tuarua "Āe marika!" na te mea kei te kī kei te haere pea ki tētahi wāhi – kei te pono koe? [pause] Te mea tuatoru ...	S1:	[Reads her paper] I said in the second one "For sure!" because it said that I may go somewhere – do you believe me? [Pause] The third one ...
I:	Mmm	I:	Mmm.
Student1:	[reads her paper] "Kāore e kore" i te mea tuatoru, kore au e noho i te kāinga i te wā me haere au ki te kura.	S1:	[Reads her paper] "Without a doubt" in the third one, [as] I wouldn't stay at home when I should go to school.
I:	Nē. Pai tērā. Tuawhā ...	I:	Is that so? That's good. [The] fourth ...
Student1:	I runga i te mea tuawhā ka kī "kore" au e haere ki te kauhoe.	S1:	On the fourth one I said "I'll 'never' go swimming".
I:	Na te mea...	I:	Because...
Student1:	Na te mea, i wareware au aku pūeru kauhoe.	S1:	Because I forgot my swimming costume.
I:	Student2, ka tae te whakamārama mai i tāu i mahi ai?	I:	S2, could you explain what you did?
Student2:	Ko te mea tuatahi ko "tērā pea". Ka whiwhi au he rare na te mea ka pai au ki taku māmā.	S2:	The first one is "Perhaps". I'd get a lolly because I am good to my Mum.
I:	Tuarua...	I:	[The] second ...
Student2:	"Kāore e kore" [ka] haere ki te whīra, tākaro i reira.	S2:	"Without doubt" I'd go to the field to play there.
I:	Ka pai	I:	Good.
Student2:	Ka haere au ki te kauhoe [i] tēnei ra. "E kore rawa" ka ua [i] tēnei ra.	S2:	I'll go swimming today. It "certainly won't" rain today.
I:	Ka pai. He aha i tuhi ai koe i ērā?	I:	Good. Why did you write that?
Student2:	I te mea i te hiahia ahau.	S2:	Because that's what I want.

the students showed a good understanding of the terms, but there were exceptions. Student 1 stated that she would never go swimming because she had forgotten her swimming costume. However, the use of "never" in this situation is not appropriate as she would be able to go swimming the following day if she bought her swimming costume. Student 2's reasoning for writing that "it certainly won't rain today" was not based on logic, but instead seemed to be something the she wanted to happen, or subjective reasoning (Jones et al., 1997). This discussion provided valuable insights into the students' understanding of probability terms.

For young children, the discussions about the different probability terms occurred around the connection between actual and imagined "real" events. The justifications that they gave helped them to resolve uncertainties about the meaning of different terms. The following extract is from a meeting between the teachers of Year 0–2 and Tamsin, in 2009, just after probability had been taught again.

Y2 Teacher: We talked about a tiger. Do you think you might see a tiger? Well there was a big argument on the floor. One said "No way" and the other one said "*Ae*" [yes]. *Pehea koe* – and how will you? And young Student1 says – on my *pouaka whakaata* [television]. On the TV. I thought – oh *hika* [dear]!

Y0 Teacher: When I was with [Year1] class – Student2 was doing "you can climb a tree" – "you're great at climbing a tree" – "you're alright at climbing a tree" – and "you can't"... And he goes – *Ae*. And I sat there – and go "oh yeah" so I went to go and help someone else, and then I said "where is Student2?" And I went – *Whaea* [teacher], Student2's out there climbing the tree – because he said "*Ka tino taea*", I can climb that tree. And I was sitting there helping this child, looked out my window and he was trying to climb that skinny tree out there. Student2 – *hoki mai ki roto* – come back in here. And he's going like this with his hands... and he looks back at his paper when he walks back in here – "*ka taea e koe*? [are you able to?] *Ka titiro mai ki au* [Look at me]", but he still – *ka tino taea* [very able], although he couldn't climb that tree at that time – he can climb. It was just like – oh, he's right. It's amazing what little kids do. (Interview, November, 2009)

Thus, the students in the early years at Te Koutu were being introduced to the language identified by Green (1983) as being problematic for high school students. This language was being built on informal experiences with probability and so would not directly lead the children to developing quantitative understandings about the likelihood of events, because of the lack of connection to sample space. According to Jones et al.'s (1997) framework this would ensure that the students remained somewhere between level 1 – subjective – and level 2 – transitional. For example, the child who needed to climb the tree in order to justify to himself that he could actually found that he could not climb the tree outside the classroom. His response that he was "a very good climber" was subjective. His intuitive understanding of the situation was not overcome by participating in one actual event. However, the need to prove and justify responses is a foundational requirement in probabilistic thinking – how do you know that this event is a certainty unless you have evidence to prove it? It is anticipated that as students progress through school, what counts as evidence would become logical, rather than based just on a physical experience. The discussion, which was occurring in the Year 1 and 2 classes of five to seven year olds, pushed children to describe their answers in ways that would convince others. Yet there was a need to further push the children to think carefully about the logic of their responses in order to make the link between intuition and logical reasoning explicit. This would support these young students to develop more advanced skills in thinking probabilistically.

Developing Ideas About the Probability of Events at the End of Primary School

At the end of primary school, the teachers continued with developing students' probability language. At times, this was done in similar ways to those experienced by the children in Year 1 class. At other times, the connection to sample space and randomness became more explicit when games were played with a spinner. This began a movement from physical evidence – that which could be seen, touched, and so on – towards logical evidence.

In Year 4, students, like those in Year 1, were channelled into using the appropriate probability language. At the introduction of this topic, the students had to place events on a continuum from "definitely not going to happen" to "definitely going to happen" (*Kāore e taea*; *ka taea*; *ka tino taea*). An example of this is shown in Fig. 7.7.

Placing events along a continuum from "never going to happen" to "certainly will happen" is a beginning stage of thinking about providing a quantitative judgement about the likelihood of an event occurring. Certain events can be given an outcome

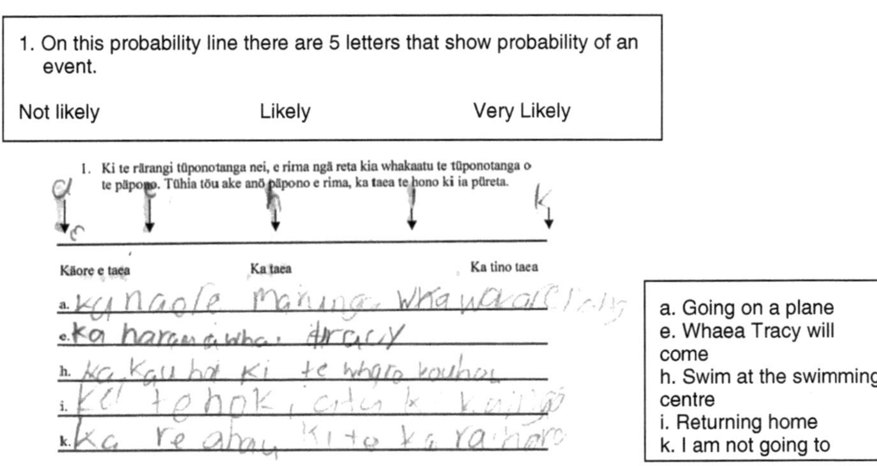

Fig. 7.7 Using probability expressions in Year 4

of 1 and events, which will never happen, are given an outcome of zero. A fifty percent change of something occurring can then be seen as something that has as much likelihood of occurring as not occurring.

Introducing students to quantifiable probabilities was done through using spinners. Watson (2006) stated that "[h]ands-on simulations as well as software simulation packages now available, make it imperative to explore the nature of probability distributions, especially in light of the opportunity to refine understanding of variation" (p. 180). The transcript from five minutes of a Year 5 lesson is provided with stills from the video recording of the lesson. The first photo shows the spinner, which was divided into eight sectors, with four shapes appearing on two of the sectors. In the video, the teacher's voice was quite clear, but the voices of the children were often difficult to hear. An example of students' writings on this topic is provided after the transcript.

Y5 Teacher:	He aha te tūponotanga ka toa, kōtiro? He aha te hautau, he aha te hauraro?	Y5 Teacher:	What is the probability of winning, girl? What is the fraction, the denominator?
Student:	Waru. (unclear)	Student:	Eight
Y5T:	Āe, ko te hauraro ko te waru, nā reira e hia – he aha te hautau mo te . . . ka toa koe?	Y5T:	Yes, the denominator is eight, therefore how many- what is the fraction for.. you winning?
Student:	Me tuhi māua ināianei	Student:	We will write it down now.
Y5T:	E hia ngā.	Y5T:	How many?
Student:	Tekau mā ono? (answer unclear)	Student:	Sixteen?

It is unfortunate that the students' discussion was too difficult to transcribe to see whether they made the connection between the areas of the individual sectors compared to the total area, fractions, and theoretical probability. These probability ideas are connected to the idea of ratio. Green (1983) had identified ratio as being an

Y5T:	He aha? Kāo. E hia ngā mea pai?	Y5T:	What? No. How many good ones?
Student:	Rua	Student:	Two
Y5T:	Āe. Nā reira, rua hauwaru. Rua hauwaru nē? Koira tō tūponotanga ka toa. ia wā, rua hauwaru nē?	Y5T:	Yes, therefore two eighths. Two eighths, yes. That is the winning probability. Each time two eighths, yes?
Student:	Ko ērā mea?	Students:	Those things?
Y5T:	Āe. (mic covered) Ko tēnā te tūponotanga ka tau te pine i runga i ēnei, nē? E rua hauwaru. Nā reira, he aha te tūponotanga ka tau te pine i runga i tētahi atu?	Y5T:	Yes . . . . . That is the probability the pin will land on these, yes? Two eighths. Therefore, what is the probability that the pin will land on another one?
Student:	Kāre anō . . . . . . (unclear)	Student:	Not yet
Y5 Teacher:	Āe. Engari, he aha te tūponotanga ka tau te pine i runga i tētahi atu āhua?	Y5T:	Yes, but what is the probability the pin will land on another figure?
S:	Waru	Student:	Eight
Y5T:	Engari, he aha . . . . . . Ka pai, ono hauwaru. Nā reira, mehemea e rua hauwaru ka tau te pine i runga i te mea pai, anei te tūponotanga ka tau te pine i runga i ētahi atu āhua, he pai ake te tūponotanga ka tau i runga i te porohita, e tau i runga i tētahi atu āhua? He aha ōu whakaaro?	Y5T:	But, what is . . . . That's good, six eighths. Therefore if the pin lands correctly on two eighths, here is the probability the pin will land on another shape. Is the probability better if it lands on a circle or land on some other shape? What do you think?
Student:	(unclear)	Student:	(?)
Y5T:	Nē? Pai ake tō tūponotanga ka tau te pine i runga i tētahi porohita? Ko tēhea hautau he nui ake?	Y5T:	Is that correct? Your probability is that the pin is more likely to land on a circle? Which fraction is the bigger?

Student:	He nui ake	Student:	Bigger
Y5T:	Āe. Tika koe, nā reira, he aha te tūponotanga ka tau te pine i runga i tētahi atu āhua?	Y5T:	yes, you are correct, therefore, what is the probability that the pin will land on another shape?
Student:	I tīmata	Student:	Started . . .
Y5T:	Āe, nā reira, he nui ake te tūponotanga ka tau te pine i runga i tētahi atu, nē. Nā te mea he iti ake tēnei ki tēnei. Āe. He tika tēnā?	Y5T:	Yes, therefore, the probability that the pin will land on another shape is greater? Because this is fewer than this? Yes, is that correct?
Student:	Āe.	Student:	Yes
Y5T:	Kei te mārama? Ā, tākaro, a te wā ka kite, nē?	Y5T:	Is it clear? Play the game and [you] will see, yes.
Student:	Āe	Student:	Yes.

issue for junior high school students learning about probability. Ratio is a complex idea in relationship to which area is larger – the area with the circle shapes or the area with other shapes? In order to support the students to determine this, the teacher channelled them into using fractions to write about the probability of the pin landing on the different parts of the spinner. In the end, she suggested that the students use the spinners to see if the game was fair if two people played it.

Figure 7.8 provides an example of a Year 5 student's writing about probability over the course of one week in August, 2007. It is possible to see how the child's thinking about probability fluctuated between levels 1, 2, 3, and 4 of Jones' et al.'s (1997) framework for assessing probabilistic thinking. Each day the students worked with spinners, although the activities and reflection questions were different. At the end of the week, the child showed some ability to describe the probability of an event by using fractions. However, although he used the fractions appropriately, there was still an element of subjectivity in his reasoning when he stated that he would win because the probability of the pin landing on his circle shape was 2/8. He later changed this to state that there was a greater probability of the pin landing on the other shapes. It may be that the number of games played were too few for the child to realise that the theoretical probability of the circle shape winning, represented by the fractional amount, was less than the other shapes' chances of winning. The teacher would need to ensure that the games were played enough times for the students to see a stable relationship between the area of the spinner and the likelihood of an event occurring. Even so, the child's belief in their ability to win still had to be disrupted if understandings about theoretical probabilities were to develop. Therefore, given the complexity of the interactions between the different influences on the child's thinking, it is not surprising that this is something that takes a while for children to develop.

In having the students reflect on the activities, the teacher was asking them to make predictions and to justify them. This required a higher level of writing than just

6/8/07
1. I think this game is about probability.
2. I think this is a probability game because you think about a colour, is it correct, is it wrong?

8/8/07
1. If the pin falls on a smaller number you know it will land on a bigger one [next]. If it lands on a bigger one it will land on a smaller one [next].

9/8/07
1. I think the odd number will win because the next number above it is even.
2. I think odd numbers will win because there are more odd numbers than even numbers.

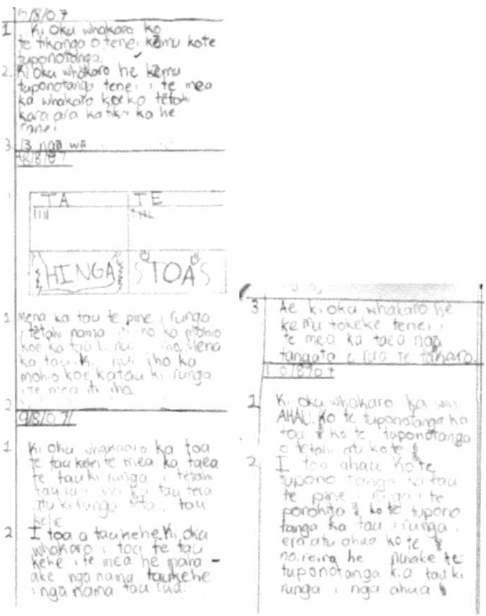

3. Yes, I think this is an unbiased game because 2 people can play.

10/8/07
1. I think I will win. The probability is 2/8. The probability of some one else winning is 6/8.
2. I won because the probability of the pin (dial) landing on the circle is 2/8. The probability of landing on the other shape is 6/8. Therefore the probability is greater to land on the 6/8 shape.

Fig. 7.8 A Year 5 student's description of a probability experiment

describing events. It was also clear that students were still grappling with expressing their ideas about what made a fair game, maybe because they were unsure what constituted a fair game – if two people can play the game does not make it unbiased. By the end of the topic, the teacher felt that some students had grasped they had less chance of winning because there were fewer sectors on their spinner, even though the reality had been that they had won a particular game. An understanding of randomness is needed to make sense of how the results from a spinner are distributed. Other students had not understood this idea. Given that Jones et al.'s (1997) work with children who were a year older had also found significant variation in students' understanding about the probability of an event occurring, it is not surprising that this was also the case for students at Te Koutu.

In 2007, the Year 6 students also worked with spinners and wrote about their experiences. As can be seen in Fig. 7.9, the children worked on the probability of the pin landing on various parts of three different spinners. This student is able to predict which number has the most likelihood of having the pin land on it by comparing the different size of the sectors. Jones et al. (1997) saw this as being at an earlier level than the situation for the Year 5 students, who had to include fractions in their justifications. Nevertheless, like the Year 5 students, some students

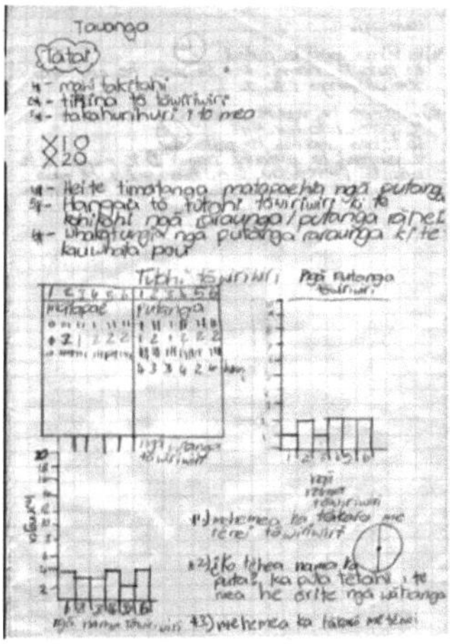

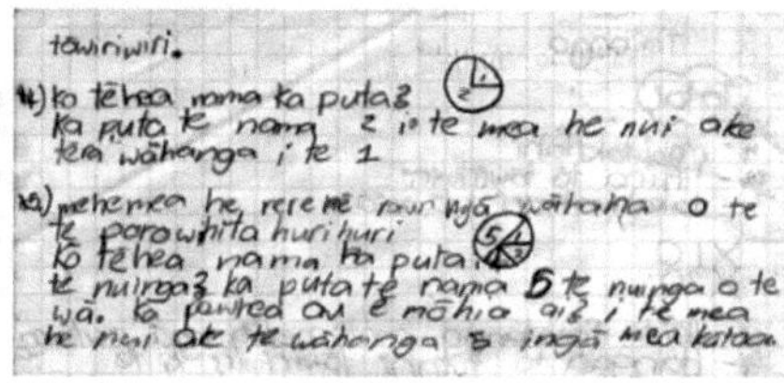

1. Work alone
2. Fetch you *tōwiri* [spinner]
3. Rotate the thing x10, x20
4. At the beginning, predict outcomes
5. Construct a frequency chart to collect the data that comes up
6. Show the data outcomes on a bar graph

1. If played this is the spinner
2. What number comes up? One will come up because the parts are the same

4. Which number will come up? The number 2 because that part is bigger than the 1
5. If the parts of the spinner are very different, what number will come up most? The number 5 will appear mostly. How do I know? Because the 5 part is bigger than all the others.

Fig. 7.9 A Year 6 student's explanation about the connection between the spinner's sectors and the chances of winning

struggled with being able to write appropriate explanations and justifications. At the end of the unit, the Year 6 teacher told how one child had not yet recognised that if the portions were the same size, they had the same chance of winning.

As had been the case with writing about probability in Year 5, Year 6 students also were expected to use a range of different modes. They had to keep tables of results, draw graphs, and explain what had occurred using diagrams of the spinners. However, it is only the diagrams of the spinners and the written explanations and justifications that could be said to be integrated.

The teacher, like many of the others, provided students with a vocabulary list on the wall so that students could use the most appropriate terms in their explanations and justifications. The vocabulary list, along with the one for measurement, *ine*, can be seen in Fig. 7.10.

Given that some students struggled in Year 6 with making judgements about the likelihood of the pin landing on certain numbers, it may seem unnecessary to have

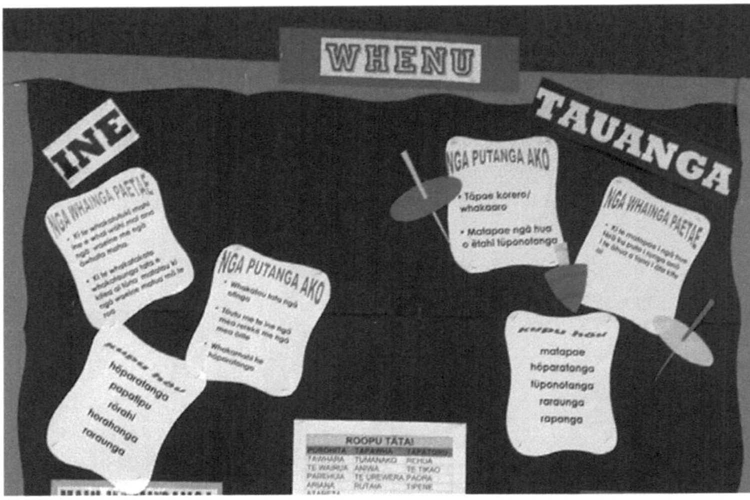

Fig. 7.10 Learning outcomes, learning intentions, and vocabulary lists for measurement, *ine*, and probability, *tauanga*

introduced fractions as had been done in Year 5. However, some students were able to use fractions to describe their understandings in Year 5, and so fractions might have been useful as well to some of the Year 6 students. Students at the end of primary school work at different levels, and they need to be provided with a variety of experiences that can be described in a range of ways in order, for their thinking to develop. By having students provide explanations and justifications related to their understandings, teachers can better tailor their teaching to meet the needs of the students.

Developing Ideas About the Probability of Events in Intermediate and High School

Once students begin high school, many of the ideas that they have been working on in primary school are dealt with more explicitly. Yet, some of the same issues regarding how to support students to make connections between theoretical and experimental probabilities remain. The traditional approach of concentrating on theoretical probability in the New Zealand curriculum has meant that the contribution of experimental probability to these understandings is often not made clear to students, possibly because the teachers themselves are struggling with understanding these connections.

The students in Years 8 and 9, combined in the one class, were also involved in playing a game. In this case, they had to choose a number between 1 and 100. They then had to throw two dice ten times to try for a total that came close to their chosen number. They could use a calculator to keep track of the cumulative total. After each game, they could revise their chosen number. It was expected that if after the first game their chosen number was much smaller than the total, in the next game their chosen number would be bigger. At the end of five games, they had to write about whether their final chosen number was a "good" number for the total, whether they had a strategy for choosing the possible total, and what they could see in the data that may have given them some idea about why the chosen number was a "good" number. It was hoped that in playing the game, the students would use the fact that seven, with six and eight, was the most likely outcome from throwing two dice and so a good chosen number should be around 70 (7×10).

The following is a reflection about the activity from the teacher's notes provided in September 2007. She found the students' responses on the whole to be limited but used them to reflect on her use of this game.

Findings

The actual game overtook the time required, limiting the writing time.

Meaningless thoughts – for example – "I wanted to/or not wanted to reach the end.", "I liked this number".

All not supported.

Some students calculated at the end of the activity [not after each game], making the activity invalid.

Good Findings

"Chose a big number, the first number was high, other numbers that followed were high. Will make it to the end". Year9 Student

"The possibility of the number 7 plus being thrown is high. Therefore reaching the end at 70 plus". Year8 Student

Probability

Obviously a weak strand.

Writing is atrocious both legibility and grammatically.

Explanations are weak, very rushed thoughts.

The activity was complicated and needed students to make several connections in order to produce an appropriate response. This also had been the case with some of the spinner activities that were done at the end of primary school. Students needed to combine their understandings about sample space with understandings of the likelihood of different events occurring – totals formed by different combinations of the dice. The students also needed to see how the randomness from throwing dice ten times, per game, may not always give them a result close to the likeliest outcome. In order for students to bring all these understandings together, they would have needed to have been exposed to all these different understandings, including knowing how to determine the likelihood of different amounts coming up from throwing two dice. As observed by Amir and Williams (1999), it may be that some students still had superstitious beliefs about the likelihood of different outcomes from throwing the dice. Without appropriate prior experiences, students had to identify the commonalities between the different totals from each of the five games and to see how adjustments of their predicted totals were connected to these commonalities. These skills of noticing and attaching appropriate significance to what they noticed also required a high level of thinking from the students. However, answers, such as

"I liked this number", suggest that some students were still at the subjective level of probabilistic thinking. To expect them to be able to bring all the appropriate knowledge and skills together may have been unrealistic.

Nevertheless, to find out what the students are capable of, it is useful to provide them with an activity, which may be beyond what they had been shown previously. As a consequence of evaluating the students' responses, the teacher was determined to work with the students on their writing of explanations and justifications. She also saw the need to provide adequate time within a lesson for the writing. Figure 7.11 is the response of the Year 8 student who the teacher felt showed the most understanding of probability.

This student was able to discuss how seven was a likely total from throwing two dice and therefore a total of around ten lots of seven was a good total to aim for. This certainly indicates a sophisticated understanding about probability. However, the way it is expressed is not clear. The reader needs to read "between the lines", through adding extra information in order for it to make sense. Although the student used a table to tally his dice totals, his justification of his strategy for throwing the dice did not mention it. This is one of the few examples of writing about probability that only used one mode.

In Year 11, students were expected to work with problems that used each of the four constructs described by Jones et al. (1997). Figure 6.4 is an example of a Year 11 student's response to the sample space problem of the number of outcomes from combining different amounts of coloured blocks. Figure 7.12 provides an example of a student detailing the different outcomes from throwing two dice. The student became confused about whether (2, 1) was a different combination from (1, 2) when the two dice were thrown. It is only in the second diagram where the individual outcomes are represented by tally marks that the spread of possibilities becomes clearer.

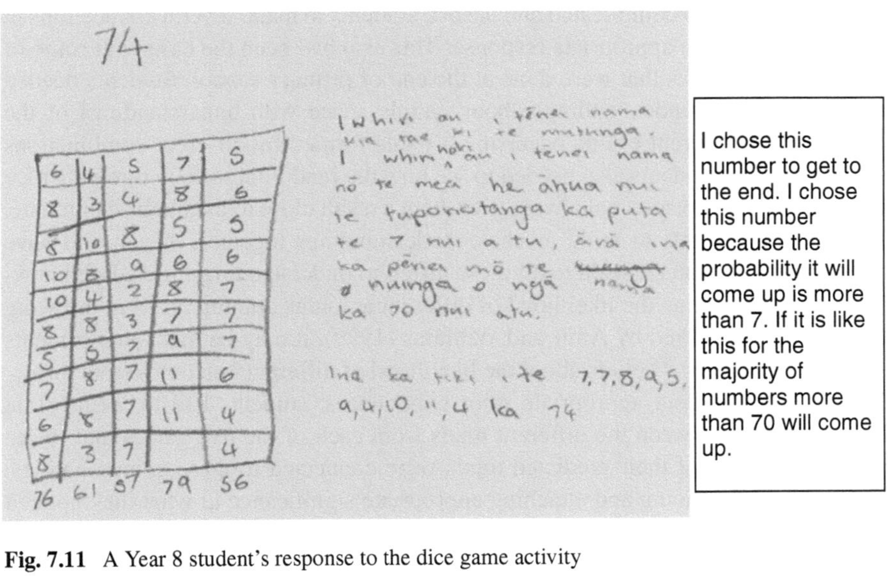

Fig. 7.11 A Year 8 student's response to the dice game activity

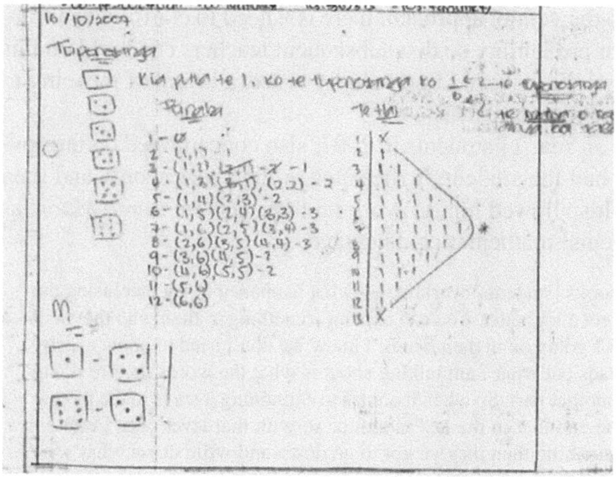

16/10/07
Probability

For 1 to come up, the probability is 1/6

1 the probability
/6 the number of outcomes

Fig. 7.12 A Year 11 student's description of the results from tossing two dice

An example of using conditional probability is provided in Fig. 7.13 and shows the depth of ideas that students need to gain in order to pass the relevant NCEA assessment in Year 11. It discusses the likelihood of different outcomes when one event is dependent on another.

By the end of high school, it is expected that all students have had sufficient experiences with probability activities to develop at least informal probabilistic understandings. Some of these students would also have been able to use information about using fractions in an earlier year level to describe their understandings.

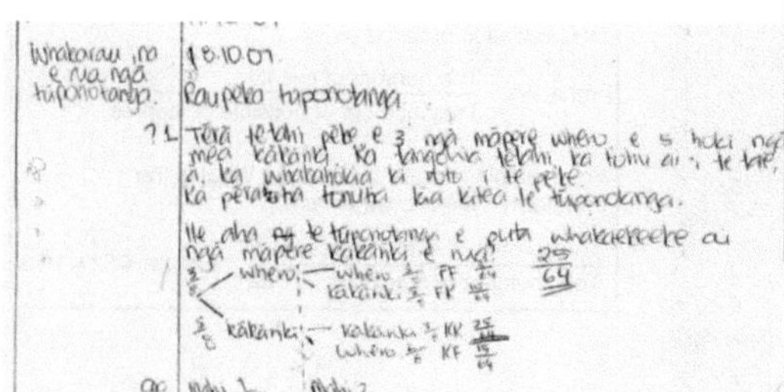

Multiply if probability of two
1.There is a bag with 3 red marbles and 5 green also. If one is taken indicate the colour and return to the bag.
Carry on like that to see the probability.
What is the probability of getting 2 green marbles?

$\frac{3}{8}$ red
- red $\frac{3}{8}$ FF $\frac{9}{64}$
- green $\frac{5}{8}$ FK $\frac{15}{64}$

$\frac{5}{8}$ green
- green $\frac{5}{8}$ KK $\frac{25}{64}$
- red $\frac{3}{8}$ KF $\frac{15}{64}$

$\frac{25}{64}$

Fig. 7.13 A Year 11 student's writing about conditional probability

It may be that in an across-the-school approach, there is a need to clarify when fractions will be introduced for probability so that subsequent teachers can build on this knowledge. Further discussions in Year 11 add clarity or extra layers of meaning to students' understandings about these ideas.

Uenuku, as the teacher of Year 11 students in 2007, also concentrated on improving students' writing. He had the students write rules in their own words and then discuss these with him. This allowed him to work on their general *te reo Māori* as well as improved the students' mathematical language.

> You will see in those notebooks [students' workbooks] what happened is, by increasing the writing, their vocal ability got a lot better. You can explain something to them and they will do it but I am not sure what's going on in their heads. I know we don't need to know exactly what's going on in their heads but what I am talking about is what the words they're doing from one part to add on to another part. So when it comes to explaining it stays in the picture ... so you'll notice near the end that in the last month or so with that level ones [Year 11 students], I'll give them something then they've got to go down and write down what's the rule. If it has been a show and tell sort of thing go down write the rule, bring it back, let's discuss it and I'll say something, such as "that means to me that I start up here" and they will say "No, no you are meant to start down here". So, [I ask them] "why use 'Kei' you should be using 'E' so I know to go from here to there. Little things like that". (Uenuku Fairhall, Meeting, November 2007)

Figure 7.14 provides an example of a student's work that described the students' understandings of the rules in probability. As can be seen, a combination of different modes is used to support the explanation.

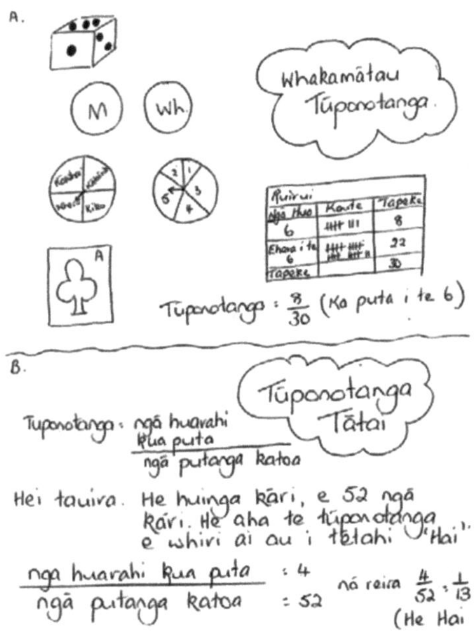

A. Probability test

Heads tails

Shakes		
Outcomes	Frequency	Total
6		8
Not 6		22
Total		30

Probability = 8/30 (the number of times 6 appeared)

B. Calculating probability

$$\text{Probability} = \frac{\text{The number of events}}{\text{Total number of possible outcomes}}$$

For example: a pack of 52 cards. What is the probability I will choose an ace?

$$\frac{\text{The number of events (aces)}}{\text{Total number of outcomes}} = \frac{4}{52} \text{ therefore } 4/52 = 1/13.$$

Fig. 7.14 A Year 11 student's rules about probability

The improvements in the students' use of vocabulary and grammar reinforced Uenuku's continued use of a conferencing approach to their writing explanations and justifications. Although the school continues to work on how to coordinate their approach to teaching probability, there was an increased expectation across all the classes that students should justify their understandings.

Meeting the Challenge of Using Language for Thinking Probabilistically

The challenge of supporting students to use language to think probabilistically comes from all stages of the meeting challenges cycle. By not acknowledging the interference from a child's first language on his or her learning of probabilistic terms in the second language, the teachers failed to recognise an important aspect of this challenge. On the other hand, by concentrating, across the school, on having students explain and justify their understandings, the teachers have normalised new, more appropriate ways to teach probability. These new practices may overcome the possible interference of the first language. Yet, the teachers also recognised that the teaching of probability at Te Koutu needed further improvement. Therefore, there were other parts to this challenge, which fell between the two extremes of the cycle.

From the analysis that we have undertaken for this chapter, it would seem that the school is on its way to try to determine an across-the-school approach to the teaching of probability. However, work needs to be done to maintain this emphasis. By 2009, only one of the teachers highlighted in this chapter taught the same year level as the teachers had done in 2007. Some teachers had changed class levels and so were still teaching probability, but other teachers had moved on and been replaced. This changeover in staff is not unusual in *kura kaupapa Māori*, and it means that an emphasis on explaining and justifying needs to be reiterated as new staff take up position at the *kura*. New teachers could resist adopting the practices of having students explain and justify their answers if they do not understand its purpose.

The sorts of activities being done at the different year levels also need to be more clearly linked to the progressions that can be expected of students in learning about probability, as well as to the different constructs mentioned by Jones et al. (1997). It was interesting to find that Year 6 students had engaged in spinner activities that were likely to result in them working at lower levels of Jones et al.'s (1997) framework than that of those done by Year 5 students. A relationship to sample space and randomness also seemed not to be made explicit to the students until they were in Year 11, although many of the activities done in earlier years required students to have some understanding about them. There is also a need for them to experience how the different constructs, such as sample space and randomness, connect to each other in the different activities.

The teachers were unsatisfied with the progression of ideas across the school, but were unsure about how it could be improved. Professional development on teaching probability in *te reo Māori* is one approach that is likely to help. The fact that the teachers were talking about how to change what was being done indicated that they were finding activity spaces for acting differently.

Given the newness of probability within curricula, it would seem that teachers need professional development, which is based on the language of instruction being *te reo Māori*. It is only through doing an analysis of the students' learning across the school that we ourselves have become aware of the need to be more explicit about ideas on sample space as one of the areas that bridge between theoretical and experimental probability. Unless there is an across-the-school approach to the teaching of probability, it is unlikely that teachers will see how the ideas build together over the years to form students' understandings.

Part III
Meeting Community Challenges

Māui

Taranga's tears were already falling as she swept Māui into her arms, lamenting her long lost son and relinquishing a long-held fear.

Once she was able, Taranga gently led her son back into the meeting house. A sentry at the door had already informed the people of their return, and they waited silently. She led Māui to the centre of the house and turned to face her other children and the rest of the tribe.

"This is my youngest son. When he was born prematurely I thought that he was dead and so put him in the sea to be carried away by the tide". Taranga continued with her son's story, and finished by saying, "And so I name this son Māui-tikitiki-a-Taranga, Taranga's Māui of the topknot!"

Some rushed forward to welcome the new member of the tribe. Others held back incredulous, including Taranga's other children, a matter sourly noted by the fifth Māui.

Upon returning to their house the other children had to face even more disappointment. Taranga motioned to her new found son to sleep beside her. The older brothers said nothing but gave each other looks that expressed their resentment as they readied their mats for the night. The sister, Hinauri, also looked troubled as she lay down upon her own mat.

The expectation of many who were part of the revitalisation movement in the 1980s was that *te reo Māori* would become a language of communication in situations where at that time English dominated. The use of *te reo Māori* as a language of instruction in schools would support this wider aim by providing young people with language they could use regularly once they left school. Thus, expectations about how the mathematics register would be used by others, apart from students in schools, formed the background to the development of the mathematics register. However, like Māui's birth, this wider aim has almost been lost in the broader community, when the focus, almost by default, stayed within the education sector.

In this part, we examine the use of the mathematics register in radio and television broadcasts in *te reo Māori*, as well as how students who leave Te Koutu use the mathematics register in their adult lives. The results suggest that there are few who use the mathematics register extensively outside of Year 1–13 schooling. We describe the opportunities for learning the mathematics register for teachers, as one group of adults who have an essential need to learn the language. Their opportunities for learning the mathematics register are limited, both through formal teacher education programmes and informal exposure to the language. Meeting the challenge of making the mathematics register in *te reo Māori* more widely available is the responsibility of the broader community, under the terms of the Treaty of Waitangi. The challenges in this part concern moving from resistance to an understanding about what needs to be done to making progress on this issue.

In the heyday of the push for self-determination by Māori, the Treaty of Waitangi was invoked in order to gain some redress for past policies that had worked against Māori. The Treaty written in both *te reo Māori* and English was signed in 1840 by Māori chiefs and the British Crown (Earp, 2004). In summary, the Treaty gave sovereignty to the British crown in exchange for the "ownership" (protection) to Māori of certain resources and treasures. After much agitation, in 1976, the New Zealand government established the Waitangi Tribunal to investigate claims raised about breaches of the Treaty (Earp, 2004).

In the mid-1980s a group of Māori, Ngā Kaiwhakapūmau i te Reo Māori, brought a claim before the Waitangi Tribunal that the Crown had failed to protect the Māori language and that this failure was a breach of the Treaty of Waitangi. The Waitangi Tribunal found in favour of the claimants stating that, under Article Two of the Treaty of Waitangi, the Māori language was a *taonga* [treasure]:

> The Crown did promise to recognise and protect the language and [. . .] that promise has not been kept. The 'guarantee' in the Treaty requires affirmative action to protect and sustain the language, not a passive obligation to tolerate its existence and certainly not the right to deny its use in any place. (Waitangi Tribunal, 1986, p. 1)

It is fortunate for *te reo Māori* that the Tribunal found in support of the claimants, and its subsequent recommendations to the government of the day provided the impetus leading to significant changes in language and education (Waitangi Tribunal, 1986). The Tribunal recommended that Māori language be used in the courts and all dealings with government, a body be established to foster the use of the Māori language, an inquiry be instituted to ascertain better ways of ensuring that Māori students could learn Māori at school, more be done in regard to broadcasting in Māori, and Māori-English bilingualism in the public service be fostered (Waitangi Tribunal, 1986). Planning initiatives of this kind are known as acquisition planning (Cooper, 1989).

However, not all these suggestions have been implemented. After nearly thirty years of language revitalisation efforts, the survey of the health of Māori language carried out by Te Puni Kokiri (2007) suggested that the decline of Māori language has been arrested, but it is still in a precarious state. It requires continued strategic planning and policy to ensure its survival as a modern, vibrant language appreciated by Māori and non-Māori alike. Peddie (2005) argued that New Zealand has never had a comprehensive language policy at higher levels of government with planning and policy tending to proceed on an *ad hoc* basis. Harlow (2003) expressed a similar sentiment in regard to Māori language planning, also arguing the *ad hoc* nature of development. While Starks and Barkhuizen (2003) agreed that New Zealand may not have a "national" languages policy, they argued this has not prevented a range of language-planning initiatives from taking place in New Zealand.

The provision of opportunities for *te reo Māori* to be used in a range of circumstances was not endorsed by some. Karetu (1995) highlighted that since the passing of the Māori Language Act there has been opposition to its active promotion, to its use in the media and its use in public places. As Māui was resented by his siblings after his mother's recognition, *te reo Māori*'s recognition was resented by some European New Zealanders who were uncomfortable with another language being used. Karetu (1995) provided a number of examples of his experiences as a Māori language speaker and former commissioner of Te Taura Whiri (The Māori Language Commission) including open hostility to the promotion of Māori. He considered this to be symptomatic of the thinking of the majority of New Zealand (Karetu, 1995). Anecdotally, there were stories about students being actively discouraged from learning *te reo Māori* as a second language at school, because it was not seen as a language that could take them overseas, like French and German. Such stories illustrated the low status that *te reo Māori* had in the wider community.

In order to have *te reo Māori* used widely, including its mathematics register, there is a need for planning policies which instigate opportunities for its use. This planning needs to include overcoming the resistance of New Zealanders who speak only English, so that programmes to increase the use of *te reo Māori* are more accepted, and thus more likely to be successful. The Treaty of Waitangi is based on a partnership whereby non-Māori are equally responsible for the achievement of its aims, including the support for the language as a treasure. Having non-Māori accept this responsibility will affect the likelihood of increasing the use of *te reo Māori*. The cultural-discursive orders and arrangements (Kemmis & Grootenboer, 2008) form the backdrop for practices to change. Without a change in the cultural-discursive orders and arrangements around attitudes to *te reo Māori* in the wider society, it is unlikely that its use will increase.

Chapter 8
Using the Mathematics Register Outside the Classroom

Te reo Māori was the language of communication when the first European settlers arrived in Aotearoa/New Zealand. Since then it has been used continuously as a first language in some parts of New Zealand. More recently, it has become the second language of many others. This increase in second-language speakers has come about as a result of the revitalisation movement. An aim of this movement in the latter part of the twentieth century was to have *te reo Māori* serve as many functions as possible, and not be relegated to talking only about traditional matters. However, this remains a challenge with much uncertainty about how to ensure that the best opportunities are made available for *te reo Māori* to be used in a range of domains. An ongoing tension is the need to increase opportunities for the language to be used, whilst at the same time ensuring the integrity of the language. Given that the mathematics register can be used in solving real-life problems that need a mathematical solution, the focus for this chapter is how the specialised terminology of the mathematics register is being used in the wider community. We also look at how ex–Te Koutu students transition from school to work and further study and how they use and learn more mathematics.

Before the 1980s, many prominent Māori advocated the use of English, especially by the rapidly growing, post-war, urban population. English was seen as the language of social improvement from the beginning of the twentieth century, and many Māori stopped using *te reo Māori*, except on the *marae* or in religious services (Te Puni Kōkiri, 2004). Consequently, *te reo Māori* was used for a limited number of functions. Since the 1980s, with the revitalisation of the language, there have been hopes that once again *te reo Māori* would be used to fulfil a similar range of functions to that of English. However, for these hopes to be realised a number of factors must come into play. The first and foremost of these is having a vibrant community of speakers. Figure 8.1 shows that in 2006 the number of speakers who felt that they spoke *te reo Māori* very well, or well, had since 2001 increased in the four youngest age groups.

Given that this age group would mostly be in the workforce, with some in the youngest group coming from Māori-immersion schooling, there is potential for these speakers to use the mathematics register in *te reo Māori* in their everyday

T. Meaney et al., *Collaborating to Meet Language Challenges in Indigenous Mathematics Classrooms*, Mathematics Education Library 52,
DOI 10.1007/978-94-007-1994-1_8,

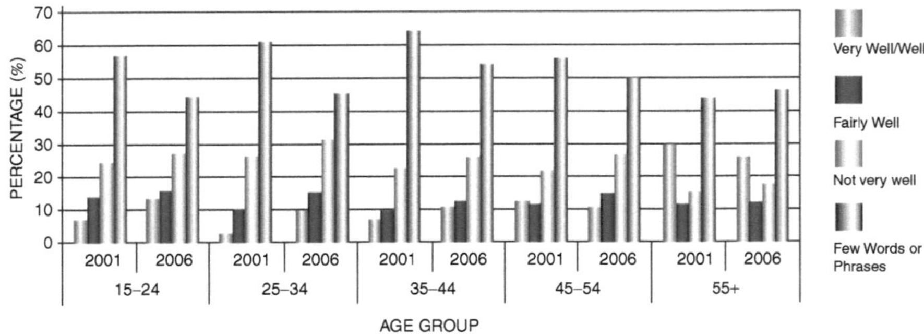

Fig. 8.1 Māori-language-speaking proficiency by age (from Te Puni Kōkiri, 2007a, 2007b, p. 5)

Table 8.1 Percentage of Māori adults who speak Māori outside the home, and whom they speak to

Person spoken to	All Māori/Māori equally	Mostly Māori with English	Some Māori no Māori (i.e. less than/50 percent Māori)
Visiting relatives/friends	9	10	82
Working	7	7	85
Sport	3	7	90
Helping out at school	20	12	68
Shopping	1	4	94
Religious activities	21	11	68
Club or interest group	20	16	63
Hui	23	19	58
Marae activities	32	17	51
Socialising	3	8	89

From Statistics New Zealand (2002)

conversations. Yet in 2001, information gathered from nearly 5000 Māori respondents showed that only 14 percent used *te reo Māori* at least 50 percent of the time (Statistics New Zealand, 2002). In social situations, where the mathematics register might also be used, such as with sport or shopping, the use of *te reo Māori* was even less. This information is shown in Table 8.1. By 2006, there were more Māori using *te reo Māori* "some" or "most of the time" in all domains. However:

> In ... 2006 fewer people spoke Māori for half or more of the time while shopping (7%), at sports (9%), while socialising (10%), at work (15%), and while visiting relatives, friends or neighbours (20%). This information suggests that the most use of Māori language in community settings is in cultural practices and formal occasions. More instances of Māori language use outside of these settings is needed until normalisation of the language is achieved. (Te Puni Kōkiri, 2008a, p. 32)

An increase in the use of *te reo Māori* is not simply a matter of more people being proficient. People need to see the possibilities for using the language and then want to take advantage of them. In relation to this issue, the attitudes of non-Indigenous people to the use of Indigenous languages has been a concern (May, 2000). In a 2006 survey of attitudes to *te reo Māori*, 80 percent of non-Māori respondents and almost a hundred percent of Māori respondents thought that it was a good thing for Māori to speak Māori in public (Te Puni Kōkiri, 2006). To temper this positive response, Boyce (2005) found, in regard to attitudes to *te reo Māori*, that "the wider community will express more positive responses to generic questions, but will indicate more negativity as questions become more specific" (p. 94). In a report by Te Puni Kōkiri (2001) on students' learning *te reo Māori* as a second language in high school, one of the reasons given by Māori students for not learning it was that there was "little real-life application for *te reo*" (p. 8). In the same survey, parents suggested that having the language might contribute to their children gaining jobs in tourism or with some government departments. However, a report on Māori tourism in Rotorua commented that the use of *te reo Māori* would not necessarily have positive outcomes for the language.

> Participants generally felt that tourism could be used to promote Te Reo in a far more positive light. One of the other impacts on Te Reo was the tendency to use transliterations because the depth of reo Māori "proper" would require more explanation and may be too complex for tourists to comprehend.
>
> Thus, tourism is seen as a mixed blessing for Te Reo, with positive outcomes requiring a more concerted effort overall. (Tahana, Grant, Simmons, & Fairweather, 2000, p. 53)

So the likelihood of *te reo Māori* being used in workplaces depends on a number of aspects aligning, even if there is a sufficient number of language speakers. For instance, the communication partners need to be at a similar level of proficiency if the language is to be used in more than a superficial way and in a way that is not detrimental to the language. Using the mathematics register to solve problems requires a higher level of proficiency than would be necessary to have a discussion about the weather, which also requires the use of aspects of the mathematics register.

Although there have been many surveys about Māori people's use of *te reo Māori,* there has been little research about whether the new registers created for schools have transferred into the wider community and incorporated into everyday speech. Of the small amount which has been done, there is no research that investigated the features of the mathematics register that appear in *te reo Māori* use in the wider community.

In order to determine where the mathematics register was used, it was necessary to look at situations in the wider community where aspects of the register were likely to appear. One of the domains where *te reo Māori* is spoken outside the classroom is the media, particularly radio and television, and it also has an increasing presence in e-media mostly driven by young Māori (Te Puni Kōkiri, 2010). Currently, there is minimal language print media in the medium of Māori outside a few localised newspapers. In the next section, we briefly describe the history of Māori broadcasting before presenting the results from a research project that monitored the different *te reo Māori* mathematical terms used on television over a two-week period in 2009.

Te Reo Māori and Broadcasting

Internationally and nationally it is recognised that the media has a role to play in the revitalisation of an Indigenous language (Hinton & Hale, 2001). Broadcasting in *te reo Māori* in Aotearoa/New Zealand has a long and complex history, with much discussion over whether the funds expended on it contribute significantly to the revitalisation.

Up until recently, there was very little broadcasting in *te reo Māori*. It was not until the Second World War that radio airtime was dedicated to discussing in *te reo Māori* the exploits of the NZ Māori (28th) Battalion. *Kaumātua* or Māori elders had petitioned parliamentarians to gain this time slot (Te Ua, n.d.). From then on, a 15 min news bulletin was broadcast once a week, written and read by Wiremu Parker. Originally all material was censored by the Prime Minister's Office. It is reported that Wiremu Parker did not use English words and by implication transliterations in his broadcasts (Te Ua, n.d.). However, this was not the norm either in later *te reo Māori* broadcasts, or in the earlier *te reo Māori* newspapers. These newspapers had flourished in the early part of the twentieth century, but most had ceased publication in the 1930s.

In the 1930s and 1940s when *te reo Māori* was still the first language of many Māori, the main employing industry, agriculture, would have been discussed extensively in *te reo Māori*. It is quite likely that discussions of selling and buying produce through markets would have required the use of the mathematics register. However, as mentioned in Chapter 2, this was in the era when transliterations were prevalent in the language, and the mathematics register probably did not develop much from the level it had reached in the mid-1800s.

The use of technology of any kind in recording *te reo Māori* is likely to have implications for the language. Kamira (2003) warned that technology, even the simplest form such as pencil and paper, has had a significant impact on Indigenous knowledge. She felt that Pākeha written accounts of Māori in the early 1800 transformed knowledge by first separating it from spirituality and then by "a systematic process by which Māori knowledge was discarded, modified or validated to suit an English international strategy for colonization" (p. 466). It is possible that newspapers and early broadcasting contributed to the proliferation of transliterations in the mathematics register, thus contributing to the distortion of Māori knowledge. In the first half of the twentieth century, when *te reo Māori* was still spoken by the majority of Māori, language change occurred slowly and Kamira's concerns perhaps are not relevant when Māori themselves chose to use their language in the media. Yet the impact of new technologies on *te reo Māori* cannot be understated, as is illustrated in the way that English has had to grapple with the new dialect of text language.

While there was more regular use of *te reo Māori* on the radio after the war (Te Ua, n.d.) and intermittent use on the new medium of television (Middleton, 2010), it was not until the 1980s that significant progress was made towards a Māori broadcasting policy. Along with the recognition by the Waitangi Tribunal of *te reo Māori* as a treasure came the requirement for joint responsibility to maintain it by the government and Māori. The role of broadcasting, much of which, at that time, was owned by the government, was criticised for its contribution to the decline of the language (Hollings, 2005). This resulted in the government being forced to support a number of legislative and policy remedies including funding Māori language radio and television. A large number of *iwi* (tribes) set up their own local radio stations as a direct consequence of being given radio frequencies to transmit on. A pilot television station was also operated for a twelve-month period between 1996 and 1997 in Auckland (Te Puni Kōkiri, 2008b). However, Middleton (2010) stated:

> The Government's early efforts to respond – such as a Māori-Crown bid for a third channel and the pilot for a Māori channel – were isolated and under-funded measures rather than the expression of a coherent, long-term strategy underpinned by solid foundations. (p. 167)

In considering how to respond to the Waitangi Tribunal's findings, the role of mainstream television and radio came under the spotlight (Hollings, 2005). By having *te reo Māori* as a language for mainstream broadcasting, it was believed that its profile would be raised. On the other hand, there was also concern that non-Māori would react negatively to being force-fed *te reo Māori*, which they could not understand.

Nevertheless, there was still a lack of evidence to show that "broadcasting has an impact on language revitalisation" (Hollings, 2005, p. 115). Successive governments remained unsure how best to meet their Treaty of Waitangi obligations with the funding that they have to allocate. To some degree this uncertainty was resolved in the publication of a report to the New Zealand Treasury by Grin and Vaillancourt (1998), who put a strong case for broadcasting being one of the most efficient ways of achieving language use and revitalisation. Consequently, the government of the time put money aside for a dedicated Māori television station. This station was opened in March, 2004. The Māori Television Act (2003) detailed that the station had to:

> ensure that during prime time it broadcasts mainly in te reo Māori; and ensure that at other times it broadcasts a substantial proportion of its programmes in te reo Māori; and ensure that, in its programming, the Service has regard to the needs and preferences of children participating in te reo Māori immersion education; and all persons learning te reo Māori; and provide broadcast services that are technically available throughout New Zealand and practicably accessible to as many people as is reasonably possible. (cited in Middleton, 2010, p. 162)

Currently, a few mainstream broadcasters also include some *te reo Māori*. For example, news on Radio New Zealand National is broadcast in *te reo Māori* at 6.27 am, 8.45 am, 5.45 pm, and 6.45 pm (www.radionz/national/programmes/waatea). There is little or no *te reo Māori* heard on commercial radio. Although there has been an increase in the use of *te reo Māori*, particularly in salutations on mainstream television programmes, the perceived disregard by many broadcasters to the correct pronunciation of Māori words raises the ire of many Māori. In an opinion piece, Willie Jackson (2009), a Māori activist and broadcaster, wrote:

> Māori language is mispronounced daily not just on small radio or TV channels but on networks like Radio NZ, TVNZ and TV3. And while top broadcasters like Simon Dallow, John Campbell and National Radio's Geoff Robinson make a big effort, their attempts are negated by fools like Leighton Smith, Paul Henry and Michael Laws, who don't give a damm about pronouncing Māori correctly.

The resistance of some non-Māori to listening to *te reo Māori* on the television may be lifting with the advent of Māori television and its popular programming choices. The 2006 ANZAC day dawn service beamed from Gallipoli received enormous ratings, which included many non-Māori. Anecdotally, it also seems that programmes provided with English subtitles allow non-speakers not only to hear the language, but also to understand what is occurring. For example, the 1960s American children's comedy *Mr Ed* was revamped into *te reo Māori* and is broadcast at 8 pm on Wednesday evenings in prime time. Replacing the language relocated *Mr Ed* to New Zealand. On the whole, the programme has been well received by both Māori and non-Māori. The following reaction is typical of those found on blogs, twitter, and facebook.

> A horse is a horse, of course, of course ... But is it a horse ... if it's in Māori??? The nation of New Zealand does an excellent job of making the most of its bilingual status of making sure that both English and Māori languages are fairly evenly presented in most mediums. Nowhere did this seem more obvious lately when we discovered the 60's sitcom, Mr Ed was being broadcast on the Māori television channel. (Spelling corrected) (from http: regator.com/p/153060911/mr_ed_maori/)

In deciding how broadcasting can be used for the revitalisation of the language another tension concerns meeting the different needs of learners as well as proficient speakers. For a revitalisation effort to be successful, television and radio broadcasting must support both groups to gain meaning from *te reo Māori* programmes. In 2006, a survey was undertaken to see how often radio and television in *te reo Māori* were utilised (Te Puni Kōkiri, 2008b). It found that proficient users of *te reo Māori* were the most frequent listeners. When users who had limited proficiency were asked why they did not listen, they responded that they could not understand what was being said.

Table 8.2 shows the number of hours that Māori accessed *iwi* (tribal) radio in the week before the survey was taken.

By 2006, Māori television was able to reach homes in 90 percent of the country. In another survey, 71 percent of Māori adults responded that they had watched at least one television programme in *te reo Māori* in the preceding four weeks (Te Puni Kōkiri, 2008b). Of these viewers, 28 percent watched Māori language programmes

Table 8.2 Duration of listening to *iwi* radio during the week

	Percentage of listeners	
Listening duration	2001 (percent)	2006 (percent)
Less than an hour	19	24
1–5 h	45	47
6–10 h	14	13
11–20 h	9	6
21–30 h	6	5
31–40 h	3	2
41 h or more	3	3

everyday. These results combined with the results from Table 8.2 suggest that there is an audience for programmes broadcast in *te reo Māori*.

The Use of the Mathematic Register on Māori Television

What remains unclear is whether programmes make use of the mathematics register, either in the form of transliterations or the new terms promulgated by the Māori Language Commission in the 1990s. Hollings (2005) lamented the lack of a methodology to determine the impact of broadcasting on the revitalisation of a language. As a first step, we have monitored *te reo Māori* programmes to determine whether the mathematics register was used. If the mathematics register was not used extensively, then the aim of revitalisation to support the increase in the number of functions served by *te reo Māori* is not going to be realised through the use of broadcasting.

In 2009, a range of radio and TV broadcasts in Māori was listened to over a two-week period. Instances of the mathematics register were recorded, and the results presented in Tables 8.4, 8.5, 8.6, and 8.7. The television items included a mixture of news and current affairs and programmes for children and youth. It was anticipated that children's programmes designed for those attending Māori-immersion schools were the most likely to include the newer terms in the mathematics register. More Māori youth had been documented as watching Māori television programmes for entertainment than any other age group (Fryer, Kalafatelis, & Palmer, 2009). Therefore, entertainment programmes designed for this age group were monitored. The youth age group included recent graduates of Māori-immersion schooling, and thus they would be able to interpret the mathematics register if it was found in these programmes. Table 8.3 sets out the programmes that were monitored for the use of mathematics register.

The frequency of different terms varied from just a couple of instances to over a hundred. In Tables 8.4, 8.5, 8.6, and 8.7, the frequency of terms used in different mathematical topics is shown in different tables beginning with number terms.

Generally, whole-number terms were by far the most common mathematical terms used across the programmes, particularly numbers between zero and 100. These terms were coined during the early 1800s with the advent of trade between Māori and European. There were also many references to numbers between 1000 and 10,000. Most of these numbers referred to years, for example, *I haere ia ki Taupo i te tau 1965* (he/she went to Taupo in 1965) and quantifying measurement terms – *6500 heketea o te whenua e watea ana* (6500 ha of land is available). Very large numbers such as a million or billion almost always were used in reference to money.

There were only two references to fractional numbers, one the recently coined term for a half (*haurua*) and the other to the transliterated version – *hawhe* (half). For a long time, Māori used the words *hawhe* and *koata* (quarter), which were borrowed from English. With the development of the teaching of mathematics in the medium

Table 8.3 Programmes, amount of times monitored, and brief description of programme type

Programme	Length	Times monitored	Description
Miharo	1 h	6	Designed for five- to eight-year-olds, it is specifically aligned with school curriculum areas.
Te Kaea (Māori news)	30 min	13	Nightly news programme.
Tau Ke	30 min	6	Screening between 4 pm and 5.30 pm, it is aimed at children attending *kōhanga reo* and *kura kaupapa Māori.*
Te Tēpu	30 min	2	Current affairs with some of the country's best practitioners of *te reo Māori* sharing their views on local, national, and international issues with presenter Wai horoi Shortland.
Whare Puoro	30 min	3	New Zealand musicians performing entirely in *te reo Māori.*
Pukoro	30 min	3	Educational show for children.
Kupuhuna	1 h	3	A language-based game/quiz show.
Haa	30 min	7	Daily info-tainment magazine programme for young teenagers.
Kaimanga	30 min	3	Music show.

of Māori, *hau*, an old word that meant fraction in the sense of over and above a complete number, is now used to express fractions in the mathematical sense. The use of fraction terms came first and then time terms were based on these.

The time terms were popular in news items about different events. Common time terms were not new. Although Keegan (2005) suggested that the new terms for the days of the week and the months of the year had gained some acceptance in Māori language broadcasting, it seems that it very much depended on the individual broadcaster which versions of the terms were used. New terms were created for the days of the week and the months of the year by the Māori Language Commission. However, they have not been well received by older speakers of *te reo Māori* (Harlow, 2003). Keegan (2005) related the story of a *kaumatua* (elder) who, when talking to students in the local *kura kaupapa Māori,* would chide any student using new terms for the days of the week and months of the year, because he did not agree with them. There is some rationale for the months of the year, in that they originated from a specific dialect of Māori and therefore can be considered traditional. Nevertheless, Harlow (2003) made some valid points in his scathing commentary on the abandonment of the borrowed terms, which have been in long use, for the recently invented pseudo-traditional days of the week, such as *rāhina* (moon-day) for Monday. As noted by Keegan (2005), these days are based on the seven-day Judeo-Christian tradition and are calques of French words that would have originally come from Latin and transferred into English after the Norman invasion of 1066. Given that traditionally Māori

Table 8.4 Number terms and their frequency in Māori television programmes

Term	Frequency	Examples	Translations
Ordinal numbers	37	*I tae tuatoru mai*	She/he came in third
		Ko te mea tuarua...	The second thing.
		Ko au te tuatahi	I'm the first.
		Tokorima...	Five (people – toko indicates that it is people who are being counted).
0–100	110	*62 (t)ōku pakeke ināianei * (not good Māori)*	My age is 62 now.
		Kerēwa te tokorua rā	That couple (two people) over there are clever.
		Kua mau kāpehu tatou, hei ārahi i a tātou ki ngā tai e whā.	We've bought a compass to guide us to the four sea coasts.
100–1000	18	*102 ngā tauira ki te kura*	There are 102 students in the school.
		200 m te roa o te kaukau	The swim is 100m long.
1000–10,000	65	*8 850 m te teitei o te maunga*	The height of the mountain is 8850 m.
		1 450 taara te utu hei...	In order to .... the cost is $1450.
		6 500 heketea o te whenua e wātea ana	There are 6500 ha of land available.
		3 754 mita te teitei o tērā maunga	The height of that mountain over there is 3754 m.
		I tēnei tau 2009	In this year, 2009.
		I toa ia i ngā whakataetae i te tau 2007	She/he won the competition in the year 2007.
		I te tau 2006 ka whakaputa i tōna kōpae Pākehā	In the year 2006, she/he put out her English-version disc.
		5000 taara ka tukua ki te kura ka toa.	$5000 will be given to the school that wins.
		Mō ngā whakataetae 2010	For the 2010 competitions
		I haere ia ki Taupō i te tau 1965.	He went to Taupo in 1965.
10,000–million	10	*2.5 miriona taara te utu hei...*	The cost to ... is 2.5 million dollars.
		10,000 mano taara i tohaina ki...	Ten thousand dollars was distributed to ....
		Kotahi miriona taara	One million dollars
		Piriona	Billion
Fractions	2	*Haurua i te hāora ka huakina te "Kete Panui"*	Half past the hour, "Kete Pānui" will be opened/started.
		Hāwhe	Half
Operations	2	*Whakarau 24*	Multiply (by) 24.
		I tātaitia ngā tāngata...	The (number of) people were calculated.

Table 8.5 Measurement terms and their frequency in Māori television programmes

Term	Frequency	Examples	Translations
General terms	1	*Te teitei o Maungawhau*	The height of Mt Eden
Time terms	75	*Ia pō, ia wiki ka whakaatuhia tētahi mea papai o tēnei tau*	Every night, every week something good from this year was shown.
		I tēnei pō	Tonight
		He wiki kei te toe!	There is (only) a week left.
		Hoki mai āpōpō.	Come back tomorrow.
		Kua tae ki waenganui o te wiki	(We've) arrived in the middle of the week.
		Hei te mutunga o tēnei marama	At the end of this month
		Inanahi rā i tū te . . .	Yesterday the . . . . was held.
		Transliterations	
		Paraire	Friday
		Wenerei	Wednesday
		Māori Language Commission terms	
		I tēnei Rātū, 1 o Hakihea	This Tuesday, the 1st of December
		I te marama o Hongongoi. . .	In the month of July
		Tēnei Rāpare ataahua rawa atu	This incredibly beautiful Thursday
		Kohitātea	January
		Rāmere	Monday
Metric terms	2	*170 kirokaramu te taumaha o taua manu*	The weight of that bird is 170 kg.
		74 kiromita te tere o te rere i ia haora tā tēnei manu.	The speed of the flight on this bird per hour is 74 km.

Table 8.6 Space and geometry terms and their frequency in Māori television programmes

Term	Frequency	Examples	Translations
Shape	11	*He whīra ōrite ki tētahi tapawhā.*	A field (shaped) like a square
		He ōrite ki tētahi tapawhā.	Just like a square
		Anei he tapatoru rite.	Here's an equilateral triangle
		Anei he tapatoru hangai.	Here's a right angle triangle
Location	10	*Huri noa i te motu*	All around the country
		He tata rawa koe ki a ia.	You are really close to her/him
Compass points	35	*Nō ngā topito katoa o te motu*	From all the compass points [corners] of the country
		He haumuri	A north wind
		Mai i te tonga	From the south
		Mai i te raki	From the north
		Kua tau ki te Tai Raki	Has descended on the North Coast
		Kua haere ki runga	He/she has gone south [down]

Table 8.7 Probability terms and their frequency in Māori television programmes

Term	Frequency	Examples	Translations
Probability	20	*Tērā pea ka toa koe i tēnei whakataetae.*	Perhaps you'll win this competition
		Ākene pea ka tīnihia taua āhua	Maybe that situation will change
		A tērā tau pea ka toa tōku kura	Next year, perhaps my school will win
		Mahi matapōkere noa	Doing it blindly/randomly

did not use "weeks" as a measurement term from time, there was no equivalent for the days of the weeks.

Metric measurement terms were used sparingly, and the two examples came from the programme *Miharo*. Metric terms are transliterations. As they are labels, it would make little sense to try to find a more Māori set of terms that shows the same relationship between prefixes.

The use of shape terms came from the children's television programmes, but even so there are very few of them, given that there were over ten hours of viewing these programmes.

In broadcast about the weather, the new directional terms were used but were highly localised. Compass points were related to the speakers' locations. The major compass points were based around the new terms of East–West. When speakers came from the north, *muri* (behind) was used. Up meant to go south as in going to the head of the fish and relates to the fact that the north island of New Zealand was deemed to be a fish, hooked and brought to the surface by Māui. The canoe that he and his brothers were fishing in was the south island.

Many of the terms are used in everyday language. For example, there were many examples of the use of *tērā pea* (perhaps). Yet, their inclusion in general conversations suggests that these terms represented only a loose relationship with the probability understandings taught in schools.

Over the two-week period, 27 and a half hours of television were monitored. It is disappointing to see that the variety of mathematical terms used is quite limited. If a comparison was to be made with the mathematics register terms used in the Māori newspapers of the 1930s, the main differences probably would be in the terms used in relationship to the days of the week and months of the year, as well as the compass directions. Nevertheless, the variety of terms is likely to be of a similar number. Although the use of the older terms for the days of the week and months illustrates the resistance to the use of the new terms, it is interesting to see other transliterations such as *hāwhe* still being used almost two decades after the Māori Language Commission recommended that these terms be replaced. The use of different terms with the same meaning can be confusing for language learners. It also illustrates the tension around the use of technology and ensuring the integrity of the language. It may be fortunate for the language that only two examples of fractions occurred over the monitoring time, of which one was a transliteration.

The lack of variety in the range of terms used is perhaps not as surprising as it first seems. In-depth analysis of statistical information on modern broadcasting is often relegated to being too hard for listeners or viewers to understand, and presenters will often stop these types of conversations before interviewees can explain their point. It would be interesting to do a similar analysis of English-medium broadcasting to see how often the mathematics register occurs across a similar range of programmes. If the mathematics register is to be used in fulfilling a set of non-school functions, then arguably the wider community actually needs to value such discussions, not just in regard to the media. Until this happens, it is unlikely that there will be an increase in the use of mathematics register in the wider community.

The Use of *Te Reo Māori* by Students Once They Finish Their Māori-Medium Schooling

When students finish their schooling in *te reo Māori*, there are possibilities for them to continue to use it as a main language in social, work, and study situations. As discussed in the first part of the chapter, interactions in the medium of Māori can only occur if a number of factors come into alignment, including having conversational partners and opportunities for discussions (Te Puni Kōkiri, 2004). The information that we received from two ex–Te Koutu students indicates that *te reo Māori* still remains a significant language of communication for them, even though they also use English to fulfil some language functions. Nonetheless, it seems that they do not engage in much mathematics that draws on their knowledge of the mathematics register.

Using Te Reo Māori for Further Study

In relationship to the possibilities for further study in New Zealand in *te reo Māori*, teacher education is the main opportunity for completing a university degree in *te reo Māori*. These programmes are discussed in more detail in the next chapter. There is very little research into students' experiences of doing university content courses in *te reo Māori*.

The exception is a paper by Keegan (1998) on the teaching of computing through the medium of Māori. In 1993, as part of the challenge presented by the co-ordinators of the BA in Māori Studies to other departments at Waikato University, the computer science department began offering an introductory course in *te reo Māori*. The average mark received by students completing this course was similar to that gained by students completing a complementary paper in the medium of English. Keegan (1998) commented on how he felt that the nature of the class was more collaborative than the English-medium equivalent. Reminiscent of the development of the mathematics register, in an earlier article, Barbour and Keegan (1996) discussed the difficulties in translating many of the computing terms into *te reo Māori*. Many of the students, especially in the first year, commented on the difficulties of learning a large amount of new *te reo Māori* terms in one semester. Although the paper was originally run over the same time period as the English-medium one, in 1996 the paper was spread over a whole year to enable students who were both learners of *te reo Māori* and computing to get longer time to adjust to the material. This resulted in a significant increase in the average mark gained by the 1996 cohort of students (Keegan, 1998). In 1996, these students were briefly questioned about their opinions of the paper. They felt that:

- There was a strong support for the course due to the fact that it was in Māori, thus was supporting the Māori language.
- The course was beneficial to potential teachers, who will need this information in Māori when arriving in total immersion situations.
- It also was needed by the younger generation who would arrive with a stronger Māori language base.
- It proves that the Māori language can survive in new environments.
- Students who were learning the language had another environment where they could consolidate what they have learnt in a language course. (Barbour & Keegan, 1996)

Of the two ex–Te Koutu students, only one had gone on to do further study. This young man had completed his secondary schooling in Canada, attended university in the United States and then had done further study in New Zealand. All of this study was done in English. Nevertheless, even five years after he had left Te Koutu, he drew on his Māori to think mathematically: "I was studying a degree in Māori Development which involved Accounting and Finance and I must admit I would do the equations speaking in Māori" (ex-Student1, Survey, 2011).

In 2006, after he had spent one year in Canada, he explained in an interview his strategy for completing word problems:

> If I'm thinking right, then you have a word equation and then I translate it into Māori and then of course you are going to get the numbers right. Like, you are going to say two has to be taken away from this number and this number and this number. So then once I read that part in Māori I sort of just write the numbers and if it's minus or multiply or such and such, and then I mean because numbers you can read in any sort of language, right? (ex-Student1, Interview, 2006)

Although he was aware that his translating could introduce errors through mistranslation, he was able to use the numbers, which were the same in both languages, as a bridge for his thinking between English and *te reo Māori*.

Even if students from Māori-immersion schooling do not continue their education in *te reo Māori*, some of them are likely to continue to think in Māori. Yet, the transfer to a new language of instruction was not simple. In 2009, when the previously mentioned student was asked about the circumstances in which he would talk about mathematics with friends, he said that they did not talk about mathematics as such in *te reo Māori*. However they did spend a lot of time talking about the difficulties of moving from learning mathematics in *te reo Māori* to learning mathematics in English. In the 2006 interview, he gave an example of this difficulty:

Tamsin: And is it harder to learn something new through English than if you were learning it in Māori?

Ex-Student1: Māori being my first language I would say yes to that. I'm not saying I'm not used to learning anything in English but it's having sort of 11 years of experience in maths [learning in *te reo Māori*] and one thing that's different is there are words in English that I cannot identify you know if I think the words in Māori. And when I think of words in Māori that may mean that, for that reason, sometimes it just doesn't fit and looking in the dictionary doesn't help either because words in English aren't sometimes in the Māori dictionary.

Tamsin: So give me an example.

Ex-Student1: Okay, this may sound funny, but hey as I told you earlier that I've been 11 years here right at the *kura* and suddenly going over there. For the first maths class I did not know what a square root was because in Māori we've always learnt it as *pūtakerua* right so that's one of the examples of the difference between the English.

Tamsin: How did you overcome that problem with the square root?

Ex-Student1: I'm trying to remember. Well at first I sort of thought to myself you know I can do this without asking anybody, so you know then the teacher goes "Okay you'll have to square root this number and square root this", so I sort of got the idea that "Okay, well if you're going to do this to this and this to that, square root this is the equivalent of square root that then I was thinking well maybe it's *pūtakerua* so I used that for the beginning and then when we finally got our test results back it worked out that I was right so that's how I sort of found out".

Given that most students from Māori-medium schools do not have much opportunity to continue their education in *te reo Māori*, some language support is required by these students when they begin to learn in English. As these students often have good conversational skills in English, they risk having their language-learning needs going unrecognised. Further education facilities, such as universities, are responding to this need.

Using Te Reo Māori at Work

The possibilities for using language at work depended very much on what the work was. If there were others who had good *te reo Māori* skills and the work being done required *te reo Māori*, then students regularly used this language to communicate.

When the context was a predominantly English one, *te reo Māori* might be used for thinking but not for communication with others. In the 2001 survey of the health of *te reo Māori*, Te Puni Kōkiri (2004) found that just over 50 percent of proficient speakers were likely to use Māori most of the time in the workplace, whereas less than 15 percent of non-proficient speakers used it most of the time. The numbers had increased slightly when the 2006 survey was undertaken (Te Puni Kōkiri, 2008a).

The strong *te reo Māori* language skills of Te Koutu students meant that several of them ended up working in Māori television. In the 2011 survey, one ex-student wrote: "At the moment I'm working on a children's program called PUKORO on Māori Television, and on the program we sometimes teach preschoolers how to count and measure" (ex-Student2, Survey, 2011). This workplace actively supported the use of *te reo Māori* and only employed presenters who were proficient in the language.

To use the mathematics register requires a purpose. For this young man, his role as a presenter meant he was aware that he was using aspects of the register in teaching preschoolers. Unless the mathematics that young adults are doing matches what they know mathematics to be from their experiences at school, then perhaps they do not always recognise the sorts of mathematics that they do in their work.

When these young adults were not working in a Māori industry, they often used different languages for thinking mathematically. For example, the first ex-student wrote about the mathematics he used in his work, "I was a Cashier performing customer services in the tourism industry so handling the money and doing money transactions I used English and sometimes Māori and sometimes Spanish" (ex-Student1, Survey, 2011). By this stage, he was fluent in all three languages and used whatever language came to mind when he was completing the calculations. This is not dissimilar to what others have reported about choosing a language for doing mathematics. In research on multilinguals' choice of language for doing calculations, Dewaele (2007) found the following:

> Frequency of general use of a language appeared to be a very strong predictor of use of a language for mental calculation across all 5 languages. Clearly, a constant use of a language can make that language become the inner language used for cognitive operations. Two participants, Meral and Elisabeth, who used their L1 and an LX with equal frequency in their daily lives, reported that they performed arithmetic operations with ease in either language, and that the language choice tended to be dictated by the situation. (p. 368)

The environment, in which students who had completed Māori-immersion education worked, influenced whether *te reo Māori* was a language of communication or a language for thinking. The sort of work that they were required to do also influenced their awareness of using the mathematics register.

Using Te Reo Māori for Socialising

Young people in the age group 15–24 years are documented as spending more of their time socialising than any other age group (Te Puni Kōkiri, 2004). Therefore, understanding the languages used in this socialisation is an indicator of how successful *te reo Māori* is being used outside of school settings.

For both our participants, *te reo Māori* was the language they used when communicating with their immediate family as well as with their extended family on their mother's side. These family members were all fluent in *te reo Māori,* and so it was the natural language of choice. Te Puni Kōkiri (2008a) reported that 28 percent of children or dependents lived in households where there was at least one Māori speaker, and therefore opportunities existed for intergenerational language transmission. This is a very small proportion of the population, given that there is also a need for the adult with *te reo Māori* proficiency to want to pass these skills on to the younger generation. It would seem that these two young people were brought up in a situation where they had opportunities to learn *te reo Māori*, not just at school but also from their extended family. This is likely to have had an impact on their proficiency levels. "[O]ver 80 percent of the people who could speak Māori 'well' or 'very well' said that adults had spoken to them in Māori during their childhood" (Te Puni Kōkiri, 2004, p. 21). At the time the survey was taken, 2001, this did not include adults who had attended Māori-immersion schooling.

As well in Rotorua, many of the adults that these young people socialised with used *te reo Māori* as their main language for communication. This is probably one of the few geographical areas in New Zealand where this is the case (Te Puni Kōkiri, 2004). The proportion of Māori in the population is high in this city, and connection to the tourism industry means that *te reo Māori* skills are valued. When asked in the survey about his use of *te reo Māori* for thinking, ex-Student2 wrote "All the time really!! *He ngāwari atu māku te whakaaro i roto i te reo Māori, i te mea kua tupu ake au i roto i te reo Māori*". (It's easier for me to think in Māori, because I grew up in an immersion situation). However, ex-Student 1 is now working overseas and so has few people in his immediate circle of acquaintances with whom he can converse in Māori. It was only through contact with his family that he used his Māori regularly.

From the very small sample of ex–Te Koutu students, it would seem that *te reo Māori* remains a viable language for them to use, even if it has mostly become an inner language for thinking with the limited number of conversational partners in the different fields of their lives. The ex-students felt that because they had leant mathematics in *te reo Māori* for a long time, it came more naturally when they had to think mathematically. However, the more engagement in mathematics in an English-speaking environment, the more they were able to use their mathematical understandings to switch between languages and gain fluency in the English mathematics register.

Meeting the Challenge of Having *Te Reo Māori* Spoken in the Community

How *te reo Māori* can become a language of choice in the wider society in New Zealand is not a simple challenge. There are a large number of factors that need to interact to support the use of this Indigenous language within a predominantly

English-speaking general population. The situation is summarised well in the following quote:

> It is likely that there will be a range of conflicting forces within individual speech networks, some of which support the threatened language and some which support the majority language. The challenge for language revitalisation is to ensure that the forces that favour the use of the threatened language are able to predominate in a significant number of speech networks, to the point where there is a critical mass of people who *can and do* use the threatened language. (Te Puni Kōkiri, 2004, p. 22)

Broadcasting in *te reo Māori* provides a variety of opportunities for the language to be used. Yet it seems that only limited aspects of the mathematics register are being used. If the aim of revitalisation is for the language to fulfil a variety of functions, then opportunities may need to be provided which specifically support such a specialised register. Although *Miharo* was advertised as a programme specifically aligned to the school curriculum areas, very little of the mathematics register was used in this programme. However, it did provide the only two instances of the use of the metric system. The lack of high-level discussions about mathematics in the media is not just an issue in Māori broadcasting, but is a situation that also occurs in the English media. Listeners in any language are not expected to understand much mathematics or statistics. The only way that this situation is likely to change is if attitudes to mathematical discussions in general are changed.

On the other hand, it seems that graduates of Māori-immersion schooling would continue to use *te reo Māori* in a variety of ways once they are operating as young adults in the wider world. Therefore, there is hope that when more of these young adults graduate, the critical mass of *te reo Māori* speakers will see it become the language of choice, enabling more conversations about a range of topics to occur.

The challenge posed by the desire for everyday use of *te reo Māori* is being acted upon. It is recognised as a challenge, but there is no clear path yet available for resolving it. Increasing numbers of Māori-immersion graduates may have a greater impact than any other outcome.

Chapter 9
Teachers as Learners of the Mathematics Register

Communities connected to *kura kaupapa Māori* face an ongoing challenge of ensuring that teachers have sufficient knowledge of the mathematics register so that they can work with students. Many Māori-immersion schools struggle to employ and retain registered teachers (Education Review Office, 2002). Ensuring that all teachers have sufficient mathematics knowledge is part of this wider challenge. Although the Māori mathematics register is used in some radio and television programmes, there are limited opportunities for the mathematics register to be acquired outside formal education. Consequently, the learning needs of teachers in this area tend to be met by initial teacher education, professional development programmes, and more often than not on-the-job experiences. We consider the issue of providing the most appropriate support to the teachers as learners of the mathematics register to be a challenge that is the responsibility of the school and education community.

This chapter describes the contexts in which teachers learn the mathematics register and the strategies that they use for learning new terms. Children learn the mathematics register, so that they can understand and use mathematics, whilst teachers learn it so that they can teach children. The different purposes influence how and what is learnt. The linguistic challenges that learning the mathematics register in *te reo Māori* pose for teachers is not unique to mathematics, and there are similar challenges for other subjects such as science and technology (Stewart, 2007). For the teachers in the primary section of Te Koutu, mathematics was one of seven different content areas with specialised language that they had to learn before they could teach it (Trinick, 2005). Having a better understanding of how teachers learnt the terms and expressions of mathematics whilst on the job is helpful as it is informative for those who provide initial and ongoing teacher education for immersion schools.

Given that around the world many teachers teach in languages, which are not their first languages and in which they may not be fluent, this issue is significant internationally (Cleghorn, 1992; Bakalevu, 1999). For example, Bakalevu (1999) expressed how as a mathematics teacher who taught in her second language, her lack of facility in English resulted in her teaching in a manner that emphasised procedures.

T. Meaney et al., *Collaborating to Meet Language Challenges in Indigenous Mathematics Classrooms*, Mathematics Education Library 52,
DOI 10.1007/978-94-007-1994-1_9, © Springer Science+Business Media B.V. 2012

> Looking back on my experience as a mathematics teacher, I remember well the difficulty of helping students make meaning of mathematical problems, particularly at the upper secondary level, where mathematics is more abstract. Some concepts just never registered. My colleagues and I followed the text closely for fear of teaching the wrong thing. Fear of misrepresentation often restrained us from straying away from the given statements and exercises (foreign as they were). When students found it difficult to grasp a point, we faced the problem of finding alternative ways of expressing it while keeping the meaning as close as possible to the original. In the end, against our best intentions, many resorted to drill and practice methods. (p. 64)

As is the case at Te Koutu, the option of teaching mathematics in the teachers' and the students' main language was not possible. For Bakalevu, Fijian, her first language, did not have political support to be the language of instruction in schools, and she was not herself taught mathematics in Fijian. At Te Koutu, the school's policy is that all subjects must be taught in *te reo Māori*. This policy also reflects a political decision, but one made at the school rather than government level. There is an emphasis on the students being able to think mathematically and use the mathematics register of *te reo Māori* fluently (see Chapter 10). Teachers, as the main agents for ensuring that students achieve both aims, need to have sufficient background both in mathematics and in *te reo Māori*.

Nevertheless, many teachers at Te Koutu did not begin their teaching with fluent knowledge of the mathematics register in *te reo Māori*. The following two extracts from different meetings provide examples of how teachers felt about their unpreparedness in this regard, and they mirror the comments made by Bakalevu (1999) mentioned earlier.

Year 4 Teacher: For my own planning I need to be aware. You plan your unit but with all the *reo* that's involved you can't just go and copy everything you have to go. I chased my tail for my *āhuahanga* (shape) unit. Because I came to you [Uenuku] asking how to do the rotation when I came upon it I didn't know how to say to them clockwise, anti-clockwise and all that. So I should have had that at the beginning. If I had that structure instead of running out of the classroom when I saw you coming past. So my classroom practice would mean me being a bit more onto it and going through things and knowing how to say this and this and being ready and since we are all doing the same *kaupapa* [knowledge] we should all be using them. Perhaps we could do it together a bit more team planning. (Meeting, November 2007)

Year 0 Teacher: The language is a barrier for me, the maths language itself, and new Māori language that's out now. I was never exposed to that at school.

Year 1 Teacher: And it's different in science

YOT: And when you come in here, it is a language barrier. I, just kind of, go with the language that I'm used to. (Meeting, October, 2009)

Many of the teachers at some point during their time at Te Koutu commented on having to learn different aspects of the mathematics register. Christensen (2003) also found that the majority of the 20 teachers in his research about the *Poutama Tau* professional development programme "knew only 'some' of the pāngarau [mathematics] words that were relevant to the junior level of school" (p. 35). Consequently, developing teachers' skills and knowledge in this area needs to be a focus both in initial teacher education and in professional development, at the primary and secondary level. In the next section, we examine how the knowledge of the mathematics register could be conceptualised as part of the pedagogical content knowledge that teachers need for teaching.

Language Knowledge as Part of Pedagogical Content Knowledge

Often in discussions about the knowledge that teachers need in mathematics classrooms, Shulman's (1986) description of pedagogical content knowledge is raised. Ball (2000) elaborated the meaning of this in relationship to mathematics:

> Included here is knowledge of what is typically difficult for students, of representations that are most useful for teaching a specific idea or procedure, and of ways to develop a particular idea, for example, the ordering of decimals or interpreting poetry. What are the advantages and disadvantages of particular metaphors or analogies? Where might they distort the subject matter? For example, both "take away" and "borrowing" create problems for students' understanding of subtraction. These problems cannot be discerned generically because they require a careful mapping of the metaphor against core aspects of the concept being learned and against how learners interpret the metaphor. Knowing that subtraction is a particularly difficult idea for students to master is not something that can be seen from knowing the "big ideas" of the discipline. This kind of knowledge is not something a mathematician would necessarily have, but neither would it be familiar to a high school social studies teacher. It is quite clearly mathematical, yet formulated around the need to make ideas accessible to others. Pedagogical content knowledge highlights the interplay of mathematics and pedagogy in teaching. (p. 245)

Having teachers with strong mathematical pedagogical content is essential if students are to become mathematical thinkers (Kennedy, 1997). Nonetheless, as can be seen in the previous quote, the use of terms or metaphors even in English-medium situations can cause confusion unless there is awareness of how the interaction between the terms and mathematics interferes with students' learning. Shulman (1986, 1987) saw the use by teachers of metaphors that aided students' learning as a vital component of pedagogical content knowledge. In Māori-immersion classrooms, where the majority of students are second-language learners, teachers' understanding about how the language impacts on the learning is essential. In the next chapter, we deal with the pedagogical knowledge about teaching the mathematics register, but in this chapter our focus is on the content knowledge of the register that teachers need for teaching.

In many ways, the knowledge of the terms and grammatical expressions needed by teachers is similar to that needed by students. However, as the role models and facilitators of students' mathematical language, teachers need a meta level of understanding about the contribution of language to students' learning. This meta language is made available when a subject and its associated language is intellectualised. This intellectualisation involves the development of new linguistic resources for discussing and disseminating conceptual material at high levels of abstraction (Liddicoat & Bryant, 2002). A key component of intellectualisation is the development of academic discourse in the language at various levels of education, including initial teacher education. This is a characteristic of most languages where there is an expansion of functions that are being discussed (Finlayson & Madiba, 2002). In expanding languages, intellectualisation is a way of providing "more accurate and detailed" means of expression, especially in the domains of modern life, for example, science and technology (Garvin, 1973, p. 43). Common mathematics terms like half (*hāwhe*) and quarter (*koata*) have been available as transliterations for some time in general daily Māori conversations. However, it is the ability to articulate, discuss, and represent the underlying mathematical concepts in *te reo Māori* that is important for the teaching and learning of mathematics.

In Fig. 9.1, Murphy, McKinley, and Bright (2008) outlined a number of reasons for why teachers need this broader understanding. When mathematics is being discussed, knowledge of the technical vocabulary and grammatical expressions, as well as the cultural background to these, is vital. Knowledge can only be discussed and

Teachers	May lead to	**Students**
Poor teacher language proficiency	→	Poor student language proficiency
No understanding of technical language and concepts	→	Students will be hampered in their learning - and may only learn through English
Teachers are often the only language role models for their students	→	Students have no option but to learn from teachers, whether good or bad
If there is an absence of personal iwi and hapū experience, teachers struggle to put language into a cultural context for students	→	Students learn language without any of the richness and experience associated with culture
Planning, implementing and evaluating unit plans encourages reflectivity, critical thinking, and evaluating through practice	→	Greater chance students will learn well, and have a thorough understanding of what is being taught

Fig. 9.1 Possible links between teachers' language ability and student achievement (from Murphy et al., 2008, p. 36)

understood through making connections using language (Meaney, 2005a, 2005b). The option outlined in Fig. 9.1 of teaching mathematics in English is not a possibility for Te Koutu. As a political decision (see Chapter 3), the sole use of *te reo Māori* for the language of instruction needs to be respected, and support should be provided to teachers so that their understanding of the mathematics register is improved.

In the following extract, the teachers at Te Koutu indicated that they were aware of the need to provide appropriate models of the mathematics register in *te reo Māori*, even for students in the first years of school. They saw it as their responsibility to ensure that they were providing "proper" language to their students. This was an example of the critical thinking, reflectivity, and evaluation advocated by Murphy et al. (2008).

Tamsin: Okay, if you wanted to tell student teachers what were the important things about teaching in *kura kaupapa Māori* – what would you say? In teaching maths, in particular, but maths as part of the whole education?

Year 1 Teacher: *Ehara i te huarahi ngawari*, not an easy path. But you know for me, the hardest thing about it is creating those resources. Making sure it's in *te reo Māori*, and it's proper *reo*, good *reo*. You know that's why it's important to give it to someone – like Uenuku. To peruse and make sure that it's right. Because you can put out stuff that's not right. But for me for my babies, it's making sure, yeah, use those proper words, but also break it down so that they understand what that word means. And to enjoy teaching maths. I hate maths, I've never liked maths. But I've learnt to enjoy it and make it fun and happy for the kids. Some of those are hard to teach. Some of the *whenu* [mathematical topic strands in the curriculum] are quite easy. Probabilities – oh my god. *E taea ana – e kore* (Able to teach it- no). (Meeting, October, 2009)

Consequently, knowledge of language and the contribution of lamguage to the teaching of mathematics need to be seen as a component in pedagogical content knowledge. The need to recognise the importance of language knowledge is even more important in situations, such as Te Koutu, where the language of instruction is an Indigenous language and/or the second language of teachers and students.

If teachers have a strong language and mathematical knowledge, then they can be flexible in how they support students in developing mathematical concepts. This flexibility is valuable in providing a bridging language that links the conversational language with the formal language of the mathematics register (Herbel-Eisenmann, 2002). From research with two Year 8 classes, Herbel-Eisenmann (2002) classified the language that the teachers and students used into the three main categories as shown in Table 9.1. Contextualised language is the language used frequently when particular types of problem are discussed. She provided an example of the term "per" which occurred repeatedly in discussing problems related to "rate".

Table 9.1 Ways of talking about mathematical ideas

		Bridging languages (BL)		
	Contextual language (CL)	Classroom-generated language (CGL)	Transitional mathematical language (TML)	Official mathematical language (OML)
Definition	Language that is dependent on specific, re-occurring contexts or situations.	Language that is student- or teacher- generated. It pertains to the mathematical object, but is particular to the classroom in which it is generated (idiosyncratic).	Language that describes a location or process that is associated with a particular representation (i.e., the graph, equation, or table), which includes certain set phrases that are often repeated in the classrooms, but do not include a contextual reference.	Language that is part of the mathematical register and would be recognized by anyone in the mathematical community.

From Herbel-Eisenmann (2002, p. 102)

Classroom-generated language includes non-standard terms that generally arise out of a specific shared experience. These terms and expressions can be idiosyncratic and thus others, who were not present in the class when they were coined, may not gain any meaning from them. On the other hand, transitional language often provides an analogy that is accessible to others, even though it may not be part of the standardised mathematics register.

Herbel-Eisenmann's research was done in an English-medium situation where the students and the teachers were native speakers. There is an even greater need for teachers to provide bridging language for second-language speakers, because of the need to increase conversational language fluency at the same time as fluency in the mathematics register.

At Te Koutu, the use of alternative terms was sometimes considered an appropriate way to develop students' understanding of mathematical concepts. The newness of the mathematics register may have contributed to the students' and the teachers' feeling that they had more freedom to play with some of the terms, than if they had been working with more established terms.

Like the native English speakers in Herbel-Eisenmann's (2002) study, in the following extract from Uenuku's Year 10 class in 2005, these students invented mathematical terms.

Y10 Teacher:	He aha te wāhi e rite ana tētahi whāritenga ki tētahi atu?	Y10T:	Where is the place where the equation is the same as another?
Student:	Te "t"? Āe.	Student:	At the "t"? Yes.
Y10T:	He aha ai?	Y10T:	Why?
Student1:	( )	Student1:	( )
Y10T:	Koinā tētahi kōrero anō,	Y10T:	That's another way of putting it,
Student1:	He mea nui tērā.	Student1:	That's important
Student:	( )	Student:	( )
Y10T:	Whakamāramahia anō tō kōrero.	Y10T:	Explain what you said again.
Student1:	Ka ōrite ngā whāritenga ki te tūtakitanga o ngā rārangi, ā, pūwāhi ōrite.	Student1:	The equations are the same at the intersection of the lines, that is, the same point.
Y10T:	Pūwāhi ōrite. Mehemea ka kōrero mō ngā pūwāhi, ahakoa kei kōnei, ahakoa kei kōnei, ahakoa kei kōnei, ngā mea katoa ka tika mō tēnei nē? Na, i tēnei rārangi i kōnei, kōnei, kōnei, kōnei, ngā mea katoa ka tika mō tēnei. Nē, kōtiro? Nē?	Y10T:	The same point. If you talk about the points (along the line), whether here, whether here, whether here? It is true for this [equation], isn't it? Now, with this line here, here, here, and here, everything is true for this one. Isn't it, my girl?

In explaining the term *tūtakitanga* (point of intersection of two lines), the student used *pūwāhi ōrite* (the same point). The teacher, Uenuku, repeated this term to remind students about what they already knew about the points on a line and how this related to the point of intersection. The teacher reused this invented term because he saw it as "enabling" rather than "distracting from" the students' understanding of the concept. Making such a judgement required teachers to have pedagogical content knowledge that included an understanding of how classroom-generated language supported students learning mathematics.

In another example, students in Uenuku's senior class in 2006 coined the term, *whakawhānau*, meaning "making families" for identifying "like terms" in algebraic expressions, such as $-4x - 3x + 3x^2 - 5x^2$. In *Te Reo Pāngarau* (Christensen, 2003), the dictionary of *te reo Māori* mathematical terms, *rōpū* is the official term for "grouping" like terms. It was through discussion of what was happening when "like terms" were gathered together that the students felt that *whakawhānau* was more appropriate. For them, the connotations that this term invoked more readily fitted their understanding of what was happening. Discussion of Western mathematics in the mathematics register needs to be done in a way that makes explicit use of the cultural connotations of the Māori words, as was suggested in Fig. 9.1.

At the same time, inventing words does have implications for maintaining the integrity of the language. Traditionally, *whānau* (family) terms were rarely used to describe inanimate objects, unless there was a special relationship with the object. Using it in this way in the mathematics discussion moves it away from its traditional connotations. Therefore, the term *whakawhānau* may need to be considered as a step to helping students understand the mathematical idea, and would be replaced later with the standardised term *rōpū*. If invented words are to act as a bridge into formal mathematical language and not have an adverse effect on *te reo Māori*, then the teachers need a strong understanding of mathematics and *te reo Māori* so that they can balance the conflicting demands within the situation. However, this may be a battle that the teachers lose because young people are generally the main instigators of adding new terms to a language that contribute to language change (Kerswill, 1996).

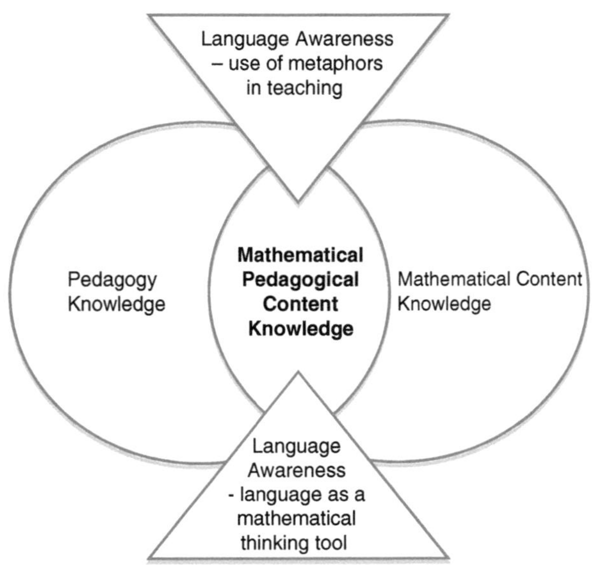

Fig. 9.2 The contribution of language awareness to mathematics pedagogical content knowledge

Teachers need to have a strong language knowledge that is integrated into their knowledge of mathematics and pedagogy if they are to support students' learning. Awareness of language, therefore, needs to be considered as part of their pedagogical content knowledge. This relationship is illustrated in Fig. 9.2. However, unlike other aspects of their mathematics pedagogical content knowledge, knowledge of the mathematics register and how to make the best use of metaphors in *te reo Māori* will not be provided within English-medium school education or teacher education. It is most likely only to be found in teacher education programmes designed specifically for Māori-immersion education.

Initial Teacher Education for Māori-Medium Teachers

To understand how teachers of mathematics in the medium of Māori come to be teaching in a register that they are still learning requires an understanding of the historical and contemporary contexts, in which teachers from schools like Te Koutu find themselves in (see also Chapters 2 and 3). The majority of teachers at Te Koutu had their own schooling in the medium of English. As the first generation of *kura kaupapa Māori* graduates enter the teaching profession, an increasing number of teachers will be conversant with the mathematics register in *te reo Māori*. However, the current situation at Te Koutu and elsewhere is that many teachers continue to learn the language as they teach mathematics.

Although future teachers may want to do their teacher education in Māori-medium programmes, there are only limited opportunities to do this. Murphy et al. (2008) identified eight teacher education programmes that provided immersion education, which delivered 80–100 percent of the content in *te reo Māori*. However, some of these programmes were for early childhood centres rather than for primary schools. As well, Murphy et al. (2008) stated:

> Participants were also asked about the connection between language proficiency and the ability to deliver the marautanga [curriculum]. They noted that some teacher educators have problems teaching the marautanga in Māori because they themselves do not have a good grasp of the Māori vocabulary and concepts associated with each curriculum area. This raises concerns about whether the programme graduates will in turn be able to teach the marautanga in schools. (p. 25)

In reality, it is more likely that only four tertiary institutions currently provide comprehensive Māori-medium teacher education for primary and secondary schools. These institutions tend to be in large urban areas. In recent years, the satellite programmes that had flourished in smaller towns such as Rotorua have been withdrawn as universities are put under pressure to remain financially viable and have had the numbers of students that they can enrol capped. These considerations have had a greater impact on Māori-medium programmes than English-medium programmes because of limited access to Māori-medium teacher education programmes, often in the areas where there is a high concentration of *kura*. Consequently some *kura kaupapa Māori* teachers start their careers as unqualified teachers (Rutene, Candler, & Watson, 2003) and then do their qualifications whilst still based in their school.

For student teachers, who live outside the main urban areas, the availability of Māori-medium, initial teacher education is restricted. Hence, many Māori who are interested in a career in Māori-medium education have little choice, but to participate in English-medium or bilingual programmes to gain a teaching qualification. As well, graduates of Māori-medium education also may decide to do their teacher education in English so as to increase their career opportunities, or because they want to support Māori students in mainstream education. Some of the reasons for choosing different teacher education options can be seen in the following discussion between three teachers of the young children at Te Koutu.

Tamsin: And when you did your training, you did it to come into Māori immersion?

Year 0 Teacher: No, I didn't.

Tamsin: Ah.

Y0T: I already had the *reo* [language], so I did it just so I can go either way. Most of my training was in mainstream, like, my [teaching] practicums, not most of it, all of it. So coming in here, and then being faced with *Poutama Tau* [numeracy professional development programme], it's like a new thing. It's like a culture shock for me. So there is a lot I have to learn.

...

Y0T: I've been through *kura kaupapa*, right up till 7th form (end of high school), but we never had the mathematics language.

...

Year 2 Teacher: Yeah, whereas with mine, when I went home and trained, because I went back into my people, so I could learn in my *reo*, you know? Train in my *reo*. But what I did was, I already had a bank of words, standardised words, or pretty much, like the *reo* and maths and social studies, all those things. There's a general word bank there. But, my people have their own words as well. I just used, because I was at home, I could just *tāpiri* (add) those words at the time, but always being aware that there are others. There's a standardised, so that, if I'm at home and talking to my people, then they know what I'm talking about. But if I come out of there, I just change over. For me that was quite easy to adapt. Because I had to go from home there, and did my practicums all over the place. And I chose mainstream and *kura kaupapa* [Māori], mainly for the same reasons that Y0T has just said, so that I could adjust to either, without any problems. And of course, the majority of my training was mainstream. Even though I had minimal, like I did *Kaiarahi Reo* [Language Support] for seven years, in *kura kaupapa Māori*, and then I decided to go training. So, I had a good background of *te reo*, and *kura kaupapa Māori*, and just went training. My practicums were in mainstream. *Pehea koe Year 1 Teacher*? (What about you, Year 1 Teacher?)

Year 1 Teacher: All ours was in Māori. We were trained only to go to *kura kaupapa Māori. Kura Pouako*, [a satellite initial teacher education programme] through Auckland College [of Education]. And Tony [Trinick] was one of our tutors. And I've been in *kura kaupapa Māori* for 20 years, so.

Y2T: But the reason I did mainstream was to target Māori kids that didn't have *kura kaupapa Māori*, but you could establish relationships and *tikanga Māori* [Māori cultural values] that you knew they weren't getting. So it was all about passing that to them and making them aware that it was there for them.

Y0T: And the roll is 90 percent Māori. Practically, in a lot of mainstream schools.

Y1T: We were told if we were planning on going to mainstream, to leave, because we were basically only for *kura kaupapa Māori*. For our training. That was [principal of another kura], she was the one that set it up. She said, she wanted just, *kura kaupapa Māori* teachers that hadn't been exposed to mainstream because of the poisoning [of the mind].

From this conversation, it can be seen that there were a variety of reasons why teacher education students chose particular programmes. However, having only a limited number of teacher education programmes available in *te reo Māori* restricts the opportunities for an "intellectualisation" of the teaching and learning of mathematics in the medium of Māori. In the present system, Murphy et al. (2008) lamented the current state of affairs where initial teacher education programmes struggle to deliver material in *te reo Māori*:

> Technical language of the various subject areas within the Marautanga [curriculum] is identified as a barrier to delivering curriculum papers fully in the Māori language, even when both the tutor and student are proficient Māori speakers because of the huge number of neologisms. The findings suggest that teachers and students lack sufficient knowledge about the marautanga and marautanga-specific language. This is compounded by a lack of time within programmes to give more than a basic grounding in each subject area. (pp. 44–45)

In teacher education, it is important to introduce an intellectualised variety at a high academic level to develop professional competencies of teachers, who will then pass on this intellectualised variety in primary and secondary schools. It would also be likely to lead to the enhancement of the school-based, Māori-medium, mathematics register. Data from Te Koutu, and anecdotal feedback from teachers in other schools, suggest that many graduates of teacher education programmes who now teach in Māori-medium education lament the lack of opportunity to develop their academic mathematics discourse.

Learning on the Job: The Situation at Te Koutu

Most teachers at Te Koutu had both their own education as well as their teacher education in the medium of English. In 2010, teachers of mathematics were surveyed on their learning of the mathematics register in *te reo Māori*. For the majority of teachers, the first time they encountered the Māori mathematics terms was when they started teaching in Māori-medium education (see Fig. 9.3).

According to the teachers, they have learnt (and are still learning) the specialised *te reo Māori* mathematics terms as teachers. In the next section, we describe some of the strategies that the teachers used for learning the mathematics register on the job.

The need to learn on the job affected how teachers rated their own knowledge of the mathematics vocabulary. As can be seen in Fig. 9.4, generally the teachers at Te Koutu felt that they had either basic or good knowledge of the mathematics vocabulary, with the teachers working in the secondary section of the school rating their knowledge as extensive. Given that these teachers teach more abstract mathematics through language, they would need to have an extensive vocabulary. Christensen (2003) noted the following:

> there are 43 different types of number listed in the curriculum document that use the base word "tau" (for example, taukehe *odd number*, tau tōraro *negative number*, taurahi *scale factor*, and so on). If students are introduced to the specialised vocabulary relevant to their level, they will experience less difficulty when further terms are added as they move to higher levels. (p. 35)

In the survey, the teachers were asked, "[A]re there any mathematics terminology you have found challenging and why?" As adult learners, the teachers found two groups of specialised terms challenging to learn and remember. These included terms from the statistics (refer to Chapter 7 for discussion on probability) and algebra strands, which rarely were heard and used.

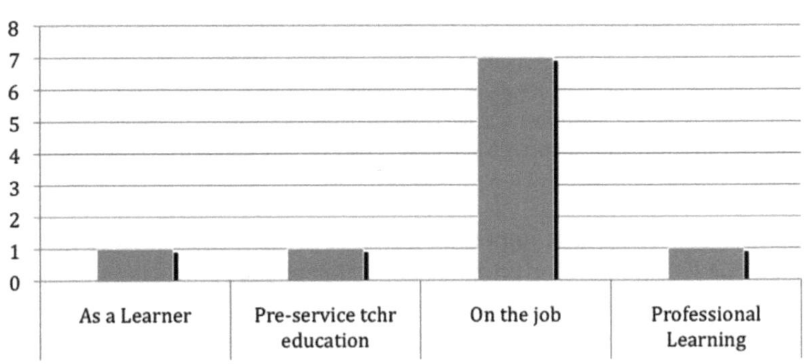

Fig. 9.3 Ways that the mathematics register was learnt by teachers at Te Koutu

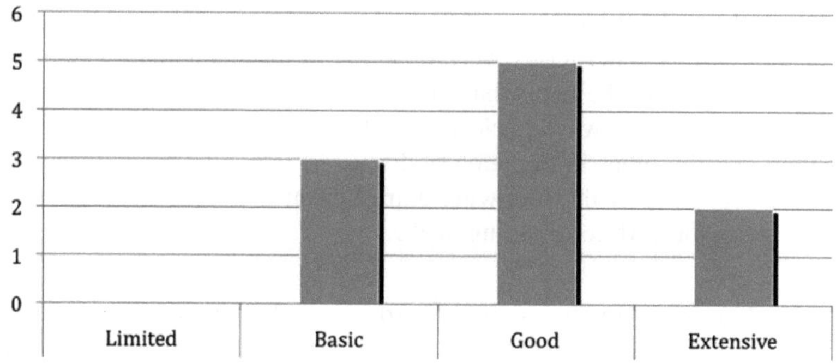

Fig. 9.4 Teachers' self-rating of their knowledge of mathematics vocabulary

The other group of terms that the teachers found difficult were labels for concepts, particular techniques, and problem-solving strategies. In some cases, this is the academic discourse that teachers may use but in discussion with other mathematics educators and not with children. For example, some teachers noted they had difficulty understanding and remembering terms that denote the stages of the number framework (Ministry of Education, 2008a, 2008b), currently used by many schools in New Zealand. These included terms like *tikanga paremata* (compensation), *te wāwāhi whakarearea* (multiplicative partitioning), and so on.

Part of the challenge facing teachers is their own resistance to using the standardised terms (see Chapter 2). Often transliterated terms, or terms used by older generations of speakers, are the identity markers of their generation of speakers. The older generation of speakers are often equated with more authentic speakers of Māori. Using the newly created terms could indicate that they were a recent and/or second-language learner *of te reo Māori*. However, teachers commented that using the traditional transliterated terms often lightened the linguistic load on them and their students. In describing her own choice of terms, one teacher wrote the following:

> I tend to use the ones [terms] I have learnt from previous teaching or the words commonly use by the school.

Teachers at Te Koutu were asked whether there were any standardised mathematics terms that they had easily accepted and why this was the case. The reasons for acceptance varied, but generally can be categorised by the fact that there was a linguistic clue in the word, which helped them remember it.

> If I understand the origins of the word and why it was chosen I tend to accept words more. When the words are closely related.

The second group of standardised terms accepted readily by the teachers were words that are used less frequently, such as *whenu* (cosine), and which did not exist as mathematical terms prior to the 1980s. Traditionally, the word *whenu* was one of the strands or warps used in weaving that ran horizontally.

Teachers who had not had their own education in *te reo Māori* were not always aware of the reasons some terms were included in the mathematics register. The transparency of some of the terms that had been clear to the developers of the mathematics register was not always so clear to the teachers. In 2008, two staff meetings, one in June and the other in September, discussed issues to do with the vocabulary. At the beginning of the following extract from the June meeting, there was discussion about the verb for dividing, *wehe*.

Y6 Teacher: Are there certain words that help the kids learn what you are trying to teach them? For some words there's not because you want them to have that wide use of vocab and they are not using that wide use, they are still going back to what they know, so they are not giving you a clear idea of

Y5 Teacher: What they really know

Y6T: of the concept and how deep that knowledge is.

Y5T: Mmm, so you are saying that's not a very good word then?

Y6T: Not if you are trying to teach it as division properly to them because *ko wehe* – to make

Y5T: *Ko wehe*, ah yeah, *ko wehe kia kitea e hia nga huanga* (divided to see how many elements)

Y6T: They've got a group and they are trying to make other groups of

Y5T: Because another way is to share. *Ae* [yes], and that's not, *whakawehe* [to divide].

...

NCEA Teacher[1]: I just think that when you look at the Māori ones there's a lot more if you just look at the word and you don't know what they are talking about you could just about work out what it was, you know. The English ones you wouldn't know what the hell half of them were about. So it's quite choice when you see things, even like *tua ōrite* (equal number), you are allowed to be same, you know? That sort of stuff. *Tau whānui* (range). All those sorts of things.

...

Y7 Teacher: In some cases, *āe* (yes). It is like NCEA Teacher said you can sort of see the meaning. Ah, yeah that's a good word for that. Another instance is measurement *horahanga* [area] and what perimeter, *āwhiotanga*, although there's another word that's sort of slipped in *takanga*, the

[1]This teacher taught mathematics to students in the final years of school. He had only begun at Te Koutu 6 months previously.

perimeter. You're describing to the kids around the outside of the shape for that. And what's the other one? *Horahanga* is spreading across the surface of the table, for the area. So whoever invented these words that's good for us. I think that in this case for teachers who teach maths we're just grateful that we have these words otherwise I would never have known what words to use in this case.

NCEA Teacher: I suppose we know what the words are and we know that there's a thing, but we don't stop to think what to actually teach the kids, you know what I mean? Even though we probably think, of myself I don't stop and tell the kids that. That's how that word came about because I am just sort of trying to get that maths across and you are getting us to use the language more and that's probably the first time I've stopped and thought about it.

Y1 Teacher: Yeah for that language thing to happen you've got the resources and use the *rauemi* [resource], so they can actually see the *kupu* [vocabulary], even like with *mua* [before] and *muri* [after] always get them to stand in front of another child, so they get *mua*, *muri*. (Meeting, June, 2008)

In the second meeting in September 2008, Tony Trinick talked with the teachers about the principles used in developing the mathematics register. These meetings helped the teachers to think about how they were introducing new terms to the children and whether providing an insight into where the terms came from might help the students. Christensen's (2003) research also suggested that "where teachers understood the Māori origin and mathematical context of the word, acquisition was greatly assisted" (p. 35). In the next section, we describe some of the strategies that the teachers used to acquire the mathematics register.

Strategies for Learning the Mathematics Register Whilst at Work

In the 2010 survey, teachers were asked about how they learnt new mathematics language. Then, they were asked to explain the strategies they used when they required a term, or they came across one which they did not know in *te reo Māori*.

The most common strategy was asking a colleague for help when they identified the key terms in the planning stage but realised that they were unsure of the meaning of some of them. Uenuku Fairhall was often the person that teachers checked the terms with before teaching a new unit of work.

Year 4 Teacher: I had to ask Uenuku what *tūpono tauanga* means, what is the word for probability. When he explained it to me I thought oh well to get it across to them [the children] would be to play that. They got it out of year 1 but not in depth.

Uenuku: What I like about *tūpono tauanga* is it stops kids saying "āe tua..." [yes, next].

Y4T: You are using *tūpono tauanga* now, so am I.

Uenuku: It means an accident with a bad outcome and so they were always using that because it reminded them of their little dictionaries but next to accident they put āe. The thing about *tūpono*, it just happened by chance it wasn't meant to happen. So that idea of when probability comes up that whilst you might have only a selection of things that can possibly happen you can't state for certain what is going to happen this next time but you are able to state what generally might happen over a long period of time. It's got good wash back on their other language. (Meeting, November, 2007)

As well, teachers frequently used the Māori-medium mathematics dictionary (Christensen, 2003) to look up terms. The dictionary provides the origin of the mathematics term, including its general meaning(s). However, when local, school terms had been introduced to students in previous years, it became difficult for a new teacher to insist on the dictionary term becoming accepted.

NCEA Teacher: I use that book quite a bit, this one here [*Te Reo Pāngarau*].

Tamsin: Yep.

NCEA Teacher: It's really my best mate. This book's really good because it's got some examples and that sort of stuff and the only problem we talked about on Monday is some of the students, all of them, they have different words from here so when I explain something then I am using this because obviously I don't know anything because this is my first year teaching in Māori so I use the points out of here but they've already ....

Tamsin: Uenuku has invented his own ones.

NCEA Teacher: He's got some other words and that sort of stuff. If I get something like that I just try to write it down then I make sure that I use it. Even some things like, square root, it's just the way that, *pūtakerua*, now these are different, all the words are different so when it's like, *pūtakerua* and they'll say "hey, what are you talking about? It's not that, it's ... de, de, de, de". So I write it down, I try and write down what those are.

Tamsin: So are you writing it in the book or.

NCEA Teacher: No, I'm writing it in here. I should actually [write it in the book]. I've got a couple of these. I write it all over the place. I should just use one of them.

...

NCEA Teacher: Yeah, because just like the other day when we were doing volume, which is *rōrahi*. When you split those words up, *rō* is short for *roto* that says insides and *rahi* being size, so the size inside. (Interview, June, 2008)

It had been suggested by Christensen, Trinick, and Keegan (2003) that recently trained teachers, like the NCEA teacher mentioned earlier, would be more willing to accept the terms in *Te Reo Pāngarau* (Christensen, 2003) – the mathematics dictionary – than older speakers. Nevertheless, the established use of non-standard terms at Te Koutu resulted in the teacher learning and then using these terms rather than the ones in the dictionary.

On a few occasions, teachers admitted to "making the word up", generally by transliterating the term, but this was the exception rather than the rule. Another strategy was to look for linguistic clues in the word:

> I try and look for linguistic clues. It is also a good strategy to teach children – particularly the numbers.

However, teachers noted that some words provided little if any linguistic clue to its meaning:

> Some words provide no linguistic clues just like the English word rectangle.

To be able to decipher meanings using the linguistic clues requires a high level of fluency in *te reo Māori*. For example, most teachers who are reasonably fluent when confronted with words like *tau whakanui* will know it has something to do with the "number that enlarges". On the other hand, there are words like *kauwhata* (graph) that present minimal linguistic clues as to its mathematical meaning. *Kauwhata* is not a word commonly used today in modern everyday Māori discourse, but belongs to the era when Māori was very much an agrarian, rural society. A *kauwhata* or *whata*, as it was more commonly known, was a framework erected to hang and dry food such as corn or eels. It is difficult to see the association between a framework to store and dry food and a framework to illustrate information, unless you were involved in the original development.

Some terms, although based on Māori cultural understanding, were affected by interference from English. Compass directions in *te reo Māori* were orientated to east–west rather than north–south. Teachers struggled with this because they were so used to Western conceptions of compass points being based on a north–south orientation. The teachers had to understand the cultural background to the terms as being related to where the sun was in the sky during the day. Unless the background for how different terms were chosen during the development of the mathematics register in *te reo Māori* is made apparent to teachers, they will not be able to pass this on to their students. As the original developers of the mathematics register head towards retirement, there is a need not only to provide a dictionary such as Ministry of Education (2004) but to also include information on why terms were chosen. It may be that some terms need to be reconceptualised as is currently happening with science vocabulary in *te reo Māori* (Ministry of Education, 2009a, 2009b). This should not be seen as a failure of the original choices, but as a reappraising of the conflicting interests that contributed to those choices being made.

Professional Development for Teachers of Mathematics in *Te Reo Māori*

The lack of Māori-medium initial teacher education is compounded by the lack of opportunity for teachers to take part in professional learning opportunities in the medium of Māori (Meaney, Fairhall, & Trinick, 2008). Professional learning opportunities often concentrate on teachers' perceived lack of content or pedagogical knowledge (see Chapter 12) and rarely have the flexibility to work on the issues that teachers perceive as being what they need. From time to time, there are workshops, but they tend to be about the latest Ministry of Education initiative, based on the needs of the English-medium education community. Often the Māori-medium initiative is a translation of its English-medium counterpart, rather than being about the ongoing needs of Māori-medium teachers, including their need to continue to develop their linguistic proficiency.

Nevertheless, the Ministry of Education initiatives still provide possibilities for language learning. Trinick (2005) stated that involvement in *Poutama Tau* supported teachers to develop language models for areas such as proportion and ratio, which previously had not had an established way of being described in *te reo Māori*. Christensen (2003) reported that the teachers who attended *Poutama Tau* professional development sessions felt that their *te reo Māori* had improved in relationship to mathematics teaching. Professional development such as *Poutama Tau* often resulted in teachers having to learn a large amount of new vocabulary, but when such programmes provided opportunities to discuss the language, they were very beneficial for teachers.

The mathematics register is more than a set of vocabulary as it includes the grammatical expressions that illustrate how ideas are connected. The following extract comes from an interview with a teacher new to Te Koutu and to teaching mathematics in *te reo Māori*. It shows how he grappled with using the traditional passive voice structures in *te reo Māori* so that he could support the authenticity of the language.

NCEA Teacher: The rest of the class, there is quite a big range in this class of ability and then we went over and just went through a little bit of grammar. If I said in English three plus four equals seven and if I said that in Māori, I could say that in Māori, but I would be using the English sentence structure, so I would go "*toru tāpiri whā ka whitu*" [three plus four

> equals seven], but what I should be saying using Māori sentence structures and Māori words "*ka tāpirihia te toru me te whā, ka whitu*" [three is added to five which equals eight]. I just went over that trying to have a go correcting those grammatical things because it's probably one of the things that someone talked about it at a *hui* [meeting] I was at. You are just saying an English sentence with Māori words. I try and have a crack at those every now and then. It's hard to get kids out of just seeing numbers and just going "*toru me te whā, ne*" [three and four, yes]? So we had a little bit of a dig at that (Interview, September, 2008)

From a discussion at a professional development meeting, he understood why it was important to have students use appropriate Māori sentence structure. He then worked with the students in his class to change their current ways of talking, so that the students used better sentence structures. However, the interference from English was strong for him and for his students, even though many of them would have not have been taught arithmetic in English.

Another advantage of the *Poutama Tau* programme was that it came with a set of resources including books, which set out questions to ask students. Originally there had been material only in English so that individual teachers had to do the translation into *te reo Māori*, which did not always result in appropriate language being used (Christensen, 2003). A bilingual presentation of material was well received by the teachers in Christensen's research. Having the material in both English and *te reo Māori* is a similar approach to that taken by some initial teacher education programmes, who front-ended their teaching by providing material in lectures in English but having tutorials for teacher education students in *te reo Māori* (Murphy et al., 2008). However, even with the materials from *Poutama Tau* being available in *te reo Māori*, the teachers felt the need for more resources.

> Year 1 Teacher: But it also gives them meaning, why they're doing it. That's why I like those *Poutama Tau* books, because they've got some really fun activities in there. Utilising the resources and stuff. But it doesn't have many worksheets, *āe* [yes]?
>
> Year 0 Teacher: Yeah it's got the masters for the actual games and things, but not the worksheets.
>
> Year 1 Teacher: Little worksheets that can go with it. You've got to still make your own. That's the frustrating thing. Making your own. (Meeting, September, 2009)

As has been documented in other chapters in this book, Te Koutu teachers continue to be active in participating in learning opportunities to develop their language fluency. This has included weekly afternoon sessions or week-long sessions with other general subject teachers, or with teachers from other institutions who teach *te reo Māori*. These are helpful and can provide some opportunities for the teachers to practise and develop the specialised mathematics register. The following extract comes from an interview, which was held the day after one of the in-house professional development sessions on mathematics language had taken place. At this session the teachers were given a high school mathematics problem in English. They had to first determine the answer and then rewrite the question in *te reo Māori*. The extract begins with a short discussion about the teachers' upcoming attendance at a week-long Māori language professional development workshop in Christchurch.

Year 2 Teacher: Christchurch will be interesting. I think we must be the only school that goes off for PD [professional development] on our holidays.

Tamsin: I think as a group, on mass, yeah.

Y2T: I think it's really good, I think it's really good. It's what I enjoy working here. But that's one week of my holidays gone. But never mind, it is for the benefit of the kids and oneself really. When you look at it in the positive like that, it's suppose to be . . .

Tamsin: It was interesting in that meeting yesterday, you know. Just watching people deal with the language and . . .

Y2T: Oh, yeah. Because I looked at that sentence and I went "oh, my god". I wouldn't even bother going there with my ones. They'd be lost. I thought that there were three major ideas. There were more, quite a few. Just to get this one answer.

Tamsin: Yeah, but I've done a lot of work with students who have English as a second language and it's sort of like if I was going to make it meaningful for students who didn't have English as a native language, what would I do to deal with it? And I'd actually break it down into simpler sentences and put a diagram with it.

Y2T: Yeah you would, would you? Yeah, that's right because I thought about doing the diagram with the clock, just to show the, but NCEA Teacher, because we didn't, Year 6 Teacher and I said forty minutes because we don't know. So we asked him [NCEA Teacher], knowing that he'd know the answer. So he just showed us on a diagram. Just common sense really.

Tamsin: Well, part of it, that's what we want kids to think about it and these are the sort of things that don't just sit and reside in here where special rules apply but actually if you just stop and break it down and think about it, what do you know, what do you need to know and how are you going to get there? Rather than going

Y2T: Well see, Uenuku changed the whole, because the actual sentence began with "how long will it take", in translation you would say "*e hia te roa*" or "*he aha te roa*". Well he

> changed it to make it simpler and I though geez, that's so right, and Year 6 Teacher and me, we did it and we actually got it down in about two sentences. I think, two and a bit sentences. Whereas before we'd had all this and lost it really, which is really what you don't want to be doing. (Interview, April, 2008)

Gaining fluency in the mathematics register is closely tied to having good mathematical understandings. Professional development in-school or through outside agencies needs to consider how to build on teachers' mathematics knowledge at the same time as their knowledge of the mathematics register. Having a school where the teachers teach students of all ages means that the language and mathematical strengths of different teachers can be combined. For the Year 2 teacher in the extract mentioned earlier, once she understood the mathematics of the question, she was then able to develop the Māori version of the question. Both parts of this learning were done with others. The learning of the mathematical register, like other teacher learning assignments at Te Koutu, was a collaborative activity.

Meeting the Challenges of Teachers Learning the Mathematics Register

In this chapter, we have concentrated on teachers learning the mathematics register. Given the historical development of Māori-immersion education, it is not surprising that many teachers do not have fluency in the mathematics register. As the rejection of the use of *te reo Māori* as a language of instruction had been a political decision made at the government level (May & Hill, 2005), it is the responsibility of the wider community to provide appropriate resources and opportunities for teachers to gain this knowledge. It should not be left up to the Māori community alone to have to provide in-school support. Given that there are only a limited number of teachers in Māori-medium education, it is a wider community responsibility to provide these opportunities. The acceptance of the importance of the Treaty of Waitangi in the 1980s after a long struggle by Māori resulted in the government being forced to accept responsibility for joint maintenance of *te reo Māori*.

> In 1986, the Tribunal's landmark inquiry into the te reo Māori claim (Wai 11) concluded that te reo Māori was a taonga [treasure] guaranteed under the Treaty, and that the Crown had significant responsibilities for its protection. (Waitangi Tribunal, 2010, p. 1)

It is not just Māori and governments who have obligations to maintain the language, but it also is the responsibility of all New Zealanders under the Treaty of Waitangi (Meaney, Trinick, & Fairhall, 2009a).

Over the last twenty years, initial teacher development programmes as well as professional development opportunities have been provided for teachers in Māori-medium schools. To some degree, these initiatives are alleviating some of the teachers' needs, and so it could be said that this challenge is in the process of being met. Certainly, this is an improvement as there were no such programmes when *kura kaupapa Māori* were first set up. However, given the many factors that contribute to teachers being learners of mathematics, we suggest that the challenge itself is still evolving, and consequently the resolutions are also evolving.

For example, although there is teacher education in *te reo Māori*, not all potential *kura kaupapa Māori* teachers will be able or willing to enrol in these programmes. For some of these teachers, they had fluency in *te reo Māori* but were not aware that the mathematics, or other technical registers, may have changed since they had been at school. Whatever the reasons, it is likely that there will be continuing numbers of teachers entering *kura kaupapa Māori* who will need to learn the mathematics register before they can teach it to their children.

Each new professional development project often produces another set of terms and sentence structures for the teaching of specific content. With new words constantly being introduced into the mathematics register of *te reo Māori*, language issues need to remain as a core focus of any professional development programme.

Teachers at Te Koutu have taken on the responsibility for learning the new register themselves. They have made use of resources such as the mathematics dictionary, *Te Reo Pāngarau* (Christensen, 2003). However, the individual words that have become embedded into the mathematical discourse at Te Koutu mean that there are often differences to those used in the dictionary. Consequently, the teachers talk and work together to determine what language they should be introducing to the children. Uenuku, as the designated local authority, was often the one called upon to determine how different ideas would be discussed. At times, it was through discussions in planning meetings that teachers as a group decided what vocabulary and sentence structures would be introduced to the children.

It is likely that the use of teacher education programmes and resources such as the *Te Reo Pāngarau* will be the main support for teachers to become fluent in the mathematics register. The discussions and collaborative planning done by teachers at Te Koutu also provide models for sustaining continual growth in the mathematics register. These models would be useful in other situations where teachers are having to learn the register of a subject in a non-native language.

Part IV
Meeting Pedagogical Challenges

Māui

Later on in the night, it was Māui-pōtiki's turn to be surprised. Just before the dawning of the new day, Taranga rose from their mat and groped around in the dissipating darkness for her clothing. Once dressed, she quickly left and was not seen again till darkness. His mother's manner made it clear that he was not to ask about where she had gone, or to even comment about her absence during the day.

And so it was, day after day. Māui-pōtiki's brothers and sister were not able to explain where their mother went each day, nor did they seem to find it unusual. Taranga had done this as long as they could remember. Their newfound brother, however, was not at all ready to accept the situation so unquestionably. His siblings' nonchalance both baffled and frustrated him. The only relief for Māui's annoyance was his mother's continued invitation to share her sleeping mat.

Māui-pōtiki's curiosity soon got the better of him and he devised a plan whereby he could discover his mother's daily destination. One night, after every one was sound asleep, Māui rose from their sleeping mat and proceeded to fill in every chink in the walls of the house with odds and ends. He made sure the window and door were similarly sealed. He even covered over the smoke-hole. It was so dark it was difficult to find his way back to the mat.

Taranga woke at the usual time, but seeing that the house was so dark she went back to sleep. She woke several more times, and with growing unrest lay back down to sleep. Finally her unease became unbearable, and she groped her way to the window. Pulling the shutter away she was horrified to see the beams of light that streamed onto the floor. Stifling a cry Taranga grabbed the closest piece of material at hand and ran in panic from the house, barely covering her nakedness.

Māui, who had not slept at all, quickly sprang up and followed his mother, making sure that he was not seen. His caution was unnecessary as his mother ran as fast as she could out of the village and into the wilderness. Eventually she reached a large clump of reeds, which she yanked out of the ground. Then, to Māui's surprise, she jumped into the gaping hole, pulling the reeds back in place above her.

Māui gingerly approached the reeds and hesitatingly tugged at them and they easily came away. He was amazed when he looked into the hole. It was actually a long twisting tunnel from which light shone. It was the entrance to another world!

Māui returned home and tried to convince his brothers to join him and go down to that other world. None of them would agree, as much from fear as from their disbelief and their resentment. Finally the eldest, Māui-mua, agreed to accompany him to the clump of reeds. He was even more amazed than Māui-pōtiki when he saw the tunnel and the light that emanated from below.

He looked back at his brother and asked, "How are you going to get to the other end?"

"Like this!" replied his brother. Māui-mua nearly fainted as he saw Māui-pōtiki slowly change shape until before him, perched on a nearby branch, was a beautiful, proud *kererū*, or pigeon.

Once his brother had retaken his human form, Māui-mua asked him if he would teach him how to achieve such a transformation. Māui-pōtiki promised to do so upon his return, one of the few promises he was to keep.

This final part concerns the challenges that the teachers faced when teaching mathematics in *te reo Māori* at Te Koutu. In earlier chapters, there has been much discussion about different linguistic aspects of teaching mathematics in *kura kaupapa Māori*. Pedagogical challenges related to teaching mathematics have been acknowledged for many years. However, the focus of research in this area has tended to be on cognitive issues dealing with specific topics such as the research on probability mentioned in Chapter 7. What has tended to remain undiscussed, even in non-Indigenous situations, is non-cognitive issues around the teaching of mathematics (Valero, 2007). Like the continuing absence of Taranga, there is a need to explore what these issues are and the likely impact of them on students' learning. In this part, we explore, like Māui, the often undiscussed norms that for our project concerns practices associated with the teaching of mathematics in *te reo Māori*.

As shown in earlier parts, the situation for Māori students at *kura kaupapa Māori* is more complicated and thus more challenging than it would be if students were learning through English, their first language. Pedagogical issues are very much identity issues for both students and teachers – how can students/teachers be Māori whilst learning/teaching mathematics? Consequently, the challenge is not just about how to support students to mentally construct appropriate mathematical knowledge. It is through discussing these challenges that there are opportunities for "normal" teaching practices to be reformed and alternative approaches adopted. In some respects, some alternative teaching practices are already the norm at Te Koutu.

However, identification of why these strategies are appropriate is as important as acknowledging them as being appropriate.

Each of the chapters in this part deals with pedagogical issues that have been the focus of investigation at Te Koutu. Language learning has been a reoccurring theme throughout the book. This is not surprising given that *kura kaupapa Māori* require that *te reo Māori* be a central concern for teaching. Chapter 10 is different because it looks at how the teachers supported the students to gain the terms and expression of the mathematics register. The Mathematics Register Acquisition model described in this chapter was elaborated and refined during our research and provides a different set of lens to those described in other research.

For *te reo Māori* to be considered a language that is functional in all aspects of life, it is essential that more is known about the process of register acquisition. As Khisty and Chevl (2002) stated, "[i]n essence, those with power are literate or in control of a discourse" (p. 167). In order for students to gain this power, the role of the teacher is crucial in providing the environment in which learning can occur (Anghileri, 2002). This environment includes expectations about the interpretation and production of mathematical language (Khisty & Chevl, 2002). For example, Khisty and Chevl (2002) showed that when teachers did not use mathematical language fluently, their students were unable to describe ideas mathematically, but instead did so in ways that highlighted non-mathematical aspects. However, apart from a few exceptions, little research about this process has been undertaken for any language.

> Nowhere is the focus on the teacher more important than in study of the acquisition of cultural tools and, in particular, language. How students learn to speak, read, and write science and mathematics, and what is taking place in the classroom, laboratory, or informal learning context are critical areas for research. (Lerman, 2007, p. 756)

The lack of research in this area has been noted for some time. In investigating students' writing about mathematics, Morgan (1998) concluded that there was a general lack of knowledge about language and language teaching. Consequently, she was unsure that students could adequately express themselves mathematically. This is supported in research by Bicknell (1999) in which New Zealand secondary teachers voiced their belief that the process of writing explanations and justifications should be explicitly taught to students. Without research on how this could occur, teachers are left on their own to work out what is the best way to continue.

In Chapter 10, we describe what we have learnt about students' acquisition process of the mathematics register. This process is often chaotic because some terms and expressions are being introduced, whilst others are being used by students with more or less fluency. It is the interaction between the students and the teachers that supports the acquisition process. However, the point in time in the topic development also affects what sorts of interactions are most likely to occur. Although we found some differences between individual teachers in the approaches that they used, there were many similarities in the strategies. When a topic was introduced, the strategies were aimed at making the students notice the terms, whilst at the end of the topic, the teacher facilitated students' use of the terms by providing opportunities where it would be natural for the students to use those terms. In Chapter 12, we look in more detail at how teachers made use of this information to change their practices.

Another undiscussed phenomenon is the issue around how being Māori affects the teaching practices used at Te Koutu. Cross-cultural research on mother–child interactions suggests that the ways scaffolding is undertaken are culturally determined (Kermani & Brenner, 1996). As well, research in reading classrooms for Hawaiian students showed that reading achievement increased when discourse interaction patterns more closely matched those of a traditional Hawaiian cultural activity, such as talk story (Au, 1980). This is supported in mathematics learning by the work of Lipka and colleagues (Lipka, Mohatt, & Cisulstet Group, 1998; Lipka et al., 2005) Therefore, Māori teachers teaching Māori children in *te reo Māori* may not use the same scaffolding strategies as those identified by Chapman (1997) in English-medium classrooms. Nelson-Barber and Estrin (1995) commented that "[u]nfortunately much of the knowledge on culturally influenced notions of good teaching remains unrecorded and unformalized because, as a whole, educators (researchers and practitioners alike) have made little effort to elicit the perspectives and experiences, or study the classrooms, of teachers who are highly effective with non-mainstream students" (p. 5).

Chapter 11 investigates Māori pedagogical practices – whether there are such practices and what benefits do the teachers at Te Koutu see in employing them. Discussions by Māori academics acknowledge the use of Māori pedagogies in *kura kaupapa Māori,* but in general do not detail what these pedagogies are in practice. In Chapter 11, we discuss the difficulties of being silent about these pedagogies before describing practices from Te Koutu that could be considered as supporting students to be Māori learners of mathematics. We did not want to be like Māui's siblings and accept the situation whereby Māori pedagogies are used in *kura kaupapa Māori* without a discussion about what they really are. There has been a great need to identify *kura kaupapa Māori* as being different to mainstream schools, and this leads to a presumption that different teaching approaches would be used. However, this presumption ignores the fact that the sort of knowledge, such as mathematics, that is taught within *kura kaupapa Māori* is the same as that in English-medium schools in New Zealand. Unfortunately, the possibility that comments that there are no differences in teaching approaches could be made seems to us to have paralysed any likelihood of a discussion about this issue. The consequence is that discussions about what could be pedagogical approaches to support students to be Māori mathematics learners are also lost. In Chapter 11, we challenge the status quo by beginning such a discussion.

Chapter 12 looks at how and why teachers changed their practices whilst they were engaged in the different research projects that happened at the school between 2005 and 2007, and it also examines how the socio-cultural, socio-political situation affects teachers adopting new practices. Given that we see mathematics learning as taking place in a socio-cultural, socio-political environment, discussions about teachers changing their practices, as though it is an individual choice, are not appropriate. However, this is often how research into professional development portrays teachers (Meaney, Lange, & Valero, 2009). We therefore see this as another rarely discussed issue, which would benefit from being explored in a way that acknowledges the different influences on teachers who are considering how to change their practices, so that their students' learning can be improved.

Pedagogical practices are at the heart of the teaching/learning process. However, they are often reduced to discussing how best to support students to construct appropriate mathematical knowledge. Although this is a valuable component of the teaching/learning process, in a *kura kaupapa Māori* it is not sufficient. At Te Koutu, it has involved investigating how students also come to see themselves as Māori mathematics learners. We, like Māui, do not just want to accept the situation as it is. Instead, we feel that a situation can only change and improve if we – researchers, teachers, community members, and students alike – take an active role in investigating what is currently happening. In this part we describe what has been learnt about how students acquire the mathematics register, what are considered to be Māori pedagogies, and what influences teachers to adopt new teaching practices.

We describe in Chapter 13 the process of meeting challenges using the case studies in the earlier chapters. Like the story of Māui, we do not feel that we have finished all of our investigations. There is still more work to be done with more challenges to be uncovered and resolved. Research is not like a Western fairy tale with a nice happy ending; it is rather an invitation to continue thinking and moving forward as occurs with many Māori legends.

Chapter 10
"They Don't Use the Words Unless You Really Teach Them": Mathematical Register Acquisition Model

In Part II, we looked at issues related to language being a mechanism for thinking about mathematics. Intimately connected to this is the teachers' role in supporting students to acquire this tool. There has been little specific advice available to teachers on how to support students' acquisition of the mathematics register. In regards to mathematical writing, Doerr and Chandler-Olcott (2009) noted that the NCTM (2000) "*Standards* offer little sense of how writing activities might fit together or how students' writing might develop across tasks and over time" (p. 286). With little indication of how this could be done in English-medium schools, the challenge for us was to document the processes that the teachers used at Te Koutu to open up discussions about improving students' learning opportunities. This was an internal challenge, and having recognised its importance, we took it on ourselves. It also built into our thinking about Māori pedagogical practices, which is the subject of Chapter 11.

The strategies that teachers used are described and categorised according to the model for mathematics register acquisition (MRA). The MRA was first described by Meaney (2006a) and was based on a research conducted in an English-medium secondary classroom. The ongoing research at Te Koutu has refined and developed our understanding of the model.

We see the mathematics register as consisting of technical terms, diagrams, and grammatical constructions, such as logical connectives. As students progress through school, as well as meeting new terms, diagrams, and grammatical expressions, they are introduced to further layers of meaning for the terms and expressions they already know. For example, a student in his or her first year of school will have a limited understanding about a triangle compared with a student at the end of high school who is conversant with trigonometry. The shape of the triangle has not changed, but the meaning given to it has gained many layers. Teachers need to help students to understand when to use each of these specific meanings. Leung (2005), in regards to vocabulary learning in mathematics, summarised the outcomes of her paper by stating:

T. Meaney et al., *Collaborating to Meet Language Challenges in Indigenous Mathematics Classrooms*, Mathematics Education Library 52,
DOI 10.1007/978-94-007-1994-1_10,

- Learning vocabulary, whether in a technical domain or in everyday use, means learning both formal and semantic (core and non-core meanings) features of words in a variety of contexts.
- Learning vocabulary involves thinking with and through the concepts associated with the word/s involved; this means exploring limits and boundaries of word meaning, generalising and extending meaning from one instance to another, and these thinking and negotiating processes are mediated through informal everyday language.
- Learning vocabulary, particularly in terms of its associated concepts and linguistic properties, is an incremental activity; the meanings of an item of vocabulary can develop and expand as part of meaning making. (pp. 133–134)

There have been a number of calls for the explicit teaching of the language needed in mathematics. For example, Solomon and O'Neill (1998) suggested, "[c]hildren who are not given the opportunity to develop a confident command of written genres and an awareness of their functions and variation are disadvantaged" (p. 211).

Nevertheless, learning the mathematics register is not considered to be an easy activity. In discussing the science register, Halliday (1988) suggested that "[b]ecause it is the language for an expert, it can often be problematic for a learner" (p. 176). However, this complexity can go unrecognised. Schleppegrell (2007) proposed that "[i]f mathematics concepts are not introduced and explained in oral language that moves from the ordinary language that students already understand to the more technical language that they need to develop for full understanding of the concepts, student learning suffers" (p. 156). Yet, when students are in a second-language situation, their ordinary language may need to be developed concurrently with the technical language. Therefore, bridging the gap between conversational language and official maths language (Herbel-Eisenmann, 2002) is not straightforward and requires the teacher to be much more innovative than simply relying on oral discussion as the vehicle for the transfer.

Early research on learning and using the mathematics register focused on modelling and scaffolding (Bickmore-Brand & Gawned, 1990). Modelling is when a teacher uses mathematical language appropriately. For example, if a student provides a response to a mathematical task in everyday language, a teacher might rephrase it in more appropriate mathematical language (see Chapman, 1997). In Doerr and Chandler-Olcott's (2009) research "students needed to have models of good writing before they could be expected to write such responses independently" (p. 294). Scaffolding is when a teacher provides significant support structures that enable a student to use new mathematical language. Wood, Bruner, and Ross (1976) originally described the scaffolding of an adult as that which "enables a child or novice to solve a problem, carry out a task or achieve a goal which would be beyond his unassisted efforts" (p. 90).

Scaffolding and modelling are two-way interactions. Rogoff (1988) showed that students had an influence on the types of scaffolding and modelling, which were offered to them. Yet, as the major holder of power in the classroom, the teacher has the most control over what language-learning opportunities are provided in the mathematics classroom. In regard to writing, Doerr and Chandler-Olcott (2009) commented:

> the teacher plays an important role in selecting writing tasks for students and in framing them in ways that attend to audience, purpose, and genre. The teacher also plays a role in responding to students' work, especially that of students who are struggling with written expression, in ways that support students in achieving greater clarity and more coherence. (p. 287)

A teacher would be expected to gradually reduce the extent of scaffolding and modelling that they provide. However, as Williams and Baxter (1996) stated, this transfer of responsibility fails to occur in many classrooms: "Edwards and Mercer pointed out that handover, or the process of gradually shifting control of learning from teacher to student, was missing in the classrooms they observed" (p. 25).

In the research projects at Te Koutu, our assumption has been that the teachers were effective in supporting students to acquire the mathematics register. From 2005, we video recorded teachers giving mathematics lessons. The appendix to this chapter provides details from five Year 6 lessons in 2005. Between 2005 and 2006, we concentrated on how oral language was developed, whilst in 2007 we focused on how the teachers supported students' writing in mathematics. Although we had these foci, we found that teachers moved between different language forms constantly during a lesson. After the lessons had been recorded, the teachers explained to a researcher, generally Tamsin Meaney, what they saw in the videos. Uenuku Fairhall was one of the teachers whose lessons were recorded. In this way, the teachers could discuss in a one-on-one situation the reasons for using a particular strategy. Then at staff meetings, teachers could describe to others what they had done to others.

Mathematics Register Acquisition Model (MRA)

The MRA model is based on Gass' (1997) model of second-language acquisition. Registers such as the mathematics register are not the same as languages because registers are derived from natural languages and thus cannot be considered independent of them (Halliday & Hasan, 1985). Nevertheless, there are advantages in viewing the acquisition of a register as being similar to that of a second language. Students are likely to use the knowledge of what they know about a natural language and how they learn it, when learning a register, in a similar manner to how they learn a second language.

Gass' (1997) detailed model has several distinct stages. In the first stage, Gass acknowledged that not all the language, which is available to learners, is utilised. Characteristics of the learner, like their mood, and also features of the language-learning situation, such as how often a new term is used, will influence what is noticed. Gass described the next stage of apperception as the "process of understanding by which newly observed qualities of an object are initially related to past experiences" (1997, p. 4). Once a new term has been noticed, the learners can work with it so that it is comprehended. This may include relating it to what they know about their first language and also to the parts of the second language that they have already learnt. Gass suggested that comprehended input became intake if it was processed further, but this depended upon the level of analysis done by the learner. By testing out their ideas of when and how a new term or expression is used, the learner gains feedback, which tells them if their ideas need modification. The context of the interaction influences the learner's understanding of how and when a new term or expression is used. Once this understanding has been stabilised, the term is integrated. The final stage, output, is when the new language is used fluently by the learner. Whether or not a learner chooses to use the new term or expression will depend on his or her mood, the situation that they are in, and the expectations of other participants.

The MRA model is based on the ideas in Gass' stages and is outlined in Table 10.1. It has four stages: noticing, intake, integration, and output. In the initial stages, the teacher is very much in control of the interaction as new language is introduced, and students are encouraged to use the language in restricted ways. Scaffolding and modelling feature heavily in these stages. In the final two stages, students take over the control of choosing how and when to use the language. The teacher's role becomes one of providing opportunities for students to use the newly learnt features of the mathematics register. The appendix for this chapter provides a description of a set of lessons where it is possible to see who initiated and controlled the interaction during different parts of the lesson.

Table 10.1 Mathematics acquisition model

TAUMATA (Stage)	WHAKAMĀRAMATANGA (Description)	
KITENGA NOTICING	Ka kitekite i ngā kupu me ngā kīanga hōu me ako. Ka kitekite i ngā wā e kōrerotia ai. Taka huirangi ai te kōrero i ngā kupu me ngā kīanga hōu.	Students have to notice that there is a new language to be learnt and when it is used by others. With prompting by others, students will use the new terms and expressions.
AKORANGA INTAKE	Ka kōrero i ngā kupu me ngā kïanga hōu i ngā āhuatanga rere kē kia akoako pai ai i ngā momo wā me kōrero.	Students start using the terms in a variety of situations. Feedback, both positive and negative, helps them to refine their understanding of when and how to use the terms and expressions.
TAUNGA INTEGRATION	Ka rite te kōrero i ngā kupu me ngā kīanga hōu.	Students will use these terms consistently except when the situation is challenging, and they may revert back to simpler terms.
PUTANGA OUTPUT	He wāhanga pūmau ngā kupu me ngā kīanga o te reo tātaitai o te ākonga, ā, ka kōrerohia i ngā wā e tika ana.	Students are using the terms fluently even in the most demanding situations.

The MRA stages have similarities with the three stages of the model for gradual release of responsibility that Doerr and Chandler-Olcott (2009) described for supporting students to become mathematical writers.

> The gradual release model shifts the responsibility for performing a task from resting entirely with the teacher to being taken up by the students in their independent performance of a task. The model includes three stages. The initial stage is characterized by teaching approaches that include teacher modeling, explanation, and demonstration of how to perform a desired task. The second stage is that of guided practice, where the teacher gradually gives students more responsibility for performing the tasks and provides scaffolds that support and guide the students' attempted performance. The final stage is that of independent practice and application to new situations. The gradual release of responsibility model provides a way for teachers to think about explicitly scaffolding instruction for increased student independence over time. (p. 289)

In the next sections, we describe the strategies that the teachers used according to the four stages of the MRA model. Occasionally a strategy seemed to straddle two stages. Generally, we allocated it to the stage, which seemed to be closest to the teachers' aim for that part of the lesson. It was not unusual for teaching strategies from each of the MRA stages to be present in most lessons, with sets of strategies being applied in relationship to different terms and expressions. Students rarely went through all four stages in one lesson for the same set of terms or expressions. Rather students were expected to develop fluency with specific terms or expressions across a unit of work.

In each of the sections on the different stages, we list some of the strategies that teachers employed and include a few examples in more depth. More details of these strategies can be found in Meaney, Fairhall, and Trinick (2007) and Meaney et al. (2009b).

Kitenga/Noticing

The Noticing stage is when the teachers introduce new terms or expressions or add extra meanings to ones that students are already familiar with. The teachers do almost all of the cognitive work by designing the activities in which they use the new terms frequently and then entice students into using them. As one teacher put it, "they don't use the words unless you really teach them" (Year 6 Teacher, Staff Meeting, November 2008). The teachers repeated new terms and expressions many times, often associating them with the physical activities that students were engaged in.

Some of the strategies that were identified for this stage were:

- providing opportunity for the new terms to be used appropriately
- using linguistic markers to highlight what was to come
- using intonation to emphasise a correct term after students used an incorrect one
- repeating new terms and expressions several times in appropriate places

- rephrasing the expressions by using other terms
- writing the new term in an equation which is related to what has just been discussed
- giving definitions in a variety of ways
- emphasising the relationship between ideas using diagrams or physical materials and words
- modelling a new term/skill (idea) as it is being explained
- after teacher explanation, having students say back the new term
- having students repeat the final answer after the teacher has modelled finding the solution
- relating new terms to already known ones
- using a set of leading questions so that students are channelled into using a particular term
- using fill-in-the-blank sentences orally
- acknowledging the difficulty of learning some terms (ideas)
- providing a rationale for the need to learn a new term (idea)
- requesting students' attention before introducing a new term
- describing a new term as being important in a subsequent lesson.

The teachers of young children often started lessons by having students read unfamiliar words. In the following extract, the Year 1 teacher had *āhuahanga* (geometry) written on cardboard. The children read it with her. She then drew different shapes and had the children name them and connect them to the new term. Spacing how often a term is repeated has been noted as important in vocabulary acquisition in second-language learning (McNaughton, MacDonald, Barber, Farry, & Woodard, 2006).

Year1 Teacher:	Ko tō mātou mahi i tēnei rā, kia ako he kaupapa hou – ko te āhuahanga. Koutou katoa…	T3:	Our work this day is something new – it is geometry. All of you..
All students:	Āhuahanga	All students:	Geometry
Y1T:	Āhuahanga	T3:	Geometry
All:	Āhuahanga	All students:	Geometry
Y1T:	Āe. Anei te kupu. Kōrero mai.	T3:	Yes, here is the word, say it.
All:	Āhuahanga	All students:	Geometry
Y1T:	Āhuahanga. Anei ngā āhuahanga. titiro. Ko te kupu āhuahanga e pā ana ki ēnei mea (kei te tuhi i runga i te papatuhituhi)	T3:	Geometry, here is some geometry. Look. The geometry word that relates to these things (draws on board)
All:	Tapatoru – Tapaono	All students:	Triangle (three sided) – hexagon (six sided)
Y1T:	Ko te aha ēnei?	T3:	What are these?
Students:	Porohita – Taimana	Students:	Circles- diamonds

At this stage, students needed to hear or see the new vocabulary or grammatical expressions frequently and to gain meaning from it/them. The meaning often came in the form of a definition. Research into reading showed that "fourth graders can acquire new vocabulary from listening to stories if there is a brief explanation of the new words as students encountered them in the stories" (Brett, Rothlein, & Hurley, 1996, p. 419). If this can be extrapolated to mathematics vocabulary learning, then providing background to new terms in the context in which they naturally arise would be one way of increasing students' awareness of new terms. Teachers also gave rationales about the importance of the mathematical idea, and these provided another kind of meaning to the new aspect of the register that they were highlighting.

At the *kitenga* stage, teachers modelled writing in several ways. These were the writing of words, symbols, or diagrams as a part of a focused discussion; the modelling by the teacher of the mode of writing that students would do as part of participating in an activity; and the modelling of an extended piece of writing that students then would be expected to copy into their books.

Figure 10.1 shows an example of the teacher writing something on the board, which was copied by students into their workbooks. Students then could use these copied notes as a model if they were called upon to do their own writing or drawing at a later stage. When students did this independently, they would be working at the *taunga* (integration) stage. If students did not have to refer to the model at all, but could write or draw it fluently, then they were at the *putanga* (output) stage.

As an introduction to the diagrams or symbols needed for writing, some teachers involved the students kinaesthetically. *Kanikani Pāngarau* is described in Chapter 5 and is one example of these activities. Figure 10.2 shows a teacher with her students engaged in another activity around shapes.

In this lesson, the teacher had students previously manipulate concrete examples of the different shapes. Using their bodies to make the shapes is a move away from this manipulation, but continues to highlight certain features of the shapes.

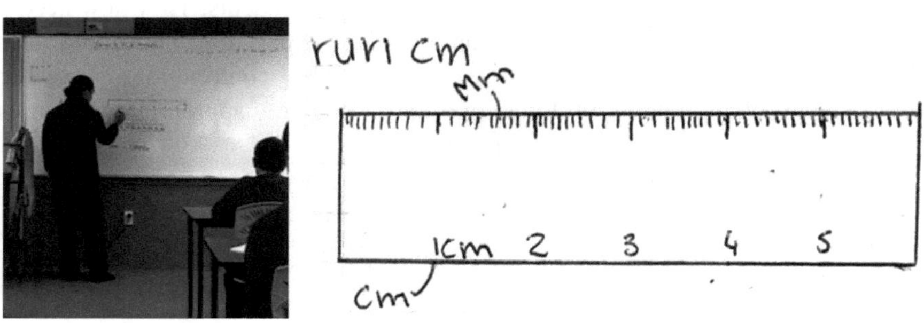

Fig. 10.1 Teacher writing on the board; copied into students' books

Fig. 10.2 Making shapes with the body

Akoranga/Intake

The function of the Intake stage is for students to form understandings of when and how new aspects of the mathematics register are used. This is an important part of this stage because "words don't just represent what we claim to know already, they also allow us to make observations and to formulate novel meanings within a negotiated range of acceptable/accepted possibilities or limits" (Leung, 2005, p. 131). Effective strategies are ones that support students to explore the appropriate contexts for using these new aspects of the mathematics register. This support includes providing students with both positive and negative feedback about their experimentation with the new aspects. This then provides them with an understanding of the limits in which this term can be used appropriately.

By this stage, some cognitive load has shifted to the students. They now need to give definitions and examples, rather than just being expected to notice and interpret those provided by the teacher. However, the students' contributions are usually short, thus providing them with little opportunity to give inappropriate responses. Consequently, the teacher is still very much in control.

The strategies that teachers used at the Intake stage of scaffolding students' acquisition of mathematics terms and expressions were:

- ensuring choral responses from the students
- having students as a group do choral responses
- giving the first syllable of a term so that students are reminded of the term and then completing it
- asking students for names, definitions, or explanations of terms

- having students model the use of terms/skills (ideas)
- asking students for examples of a term
- using the similarities between concepts (e.g., 7 + 3 and 70 + 30) as an entry into having students reflect on the differences
- having students draw their own diagrams or use materials to show a particular term
- repeating or having students repeat appropriate responses
- elaborating on students' responses in words and with diagrams
- asking further questions to help students reflect on what they were describing and to check on what they know or have done
- having students provide a rationale for what they are learning
- ignoring inappropriate answers and just acknowledging appropriate ones
- querying students' inappropriate responses
- suggesting that students' inappropriate responses are close
- having students work backwards from an inappropriate answer to the question which was asked
- using specific amounts to illustrate a general rule (idea)
- focusing students back onto the main idea being discussed to help solve a problem
- using student-devised terms in giving an explanation
- going over an activity which requires the use of the new language as a whole class, before expecting students to do the activity as individuals
- showing students the relationship between what they already know and can do and the new language term or skill
- having students answer a series of closed questions to lead them to using the new term/skill (idea)
- after the teacher models how a new term or skill is used, students repeating the action
- recording in writing what had been discussed or done
- students querying obvious errors by the teacher or another student.

Teachers often involved students by having them contribute words either orally or in writing. This could begin with students interpreting what they were reading, such as a diagram, to others. In the following extract, the teacher had a student explain what was happening when two lines met on the graph. The student went up to the whiteboard and was helped in the explanation by suggestions from other students and from the teacher. If the student had been able to draw the diagram without help, he would have been at the Output stage.

Y11 Teacher:	Inanahi, i tuhi au i ngā rārangi e rua me te pātai ki a koutou kōrerohia mai te tutakitanga o ngā rārangi e rua. Nō reira, Student1 haere ki te tuhi i ngā rārangi e rua	Y11T:	Yesterday I drew two lines, and asked you for the intersection of the two lines. Therefore, Student1 go and draw the two lines
Student1:	E ai ki tōku mea	Student1:	According to my thing...
Y11T:	Oh, koinā tāu e kī ai he rerekē	Y11T:	Oh, that's why you said it's different

Student1:	Whā ripeka, oh, māku e tuhi [Student1 stands up and goes towards whiteboard]	Student1:	Four crosses, I will draw them [Student1 stands up and goes toward whiteboard]
Y11T:	Student2, kei te pai kē mehemea i tino pango te rārangi o waenganui, he uaua te kite	Y11T:	Student2, it's better if the line between is darker, it's difficult to see
Student2:	Oh	S2:	Oh
Student1:	Oh he aha tēnā?	Student1:	Oh what is that?
Student2:	Oh	Student2:	Oh
Student:	Whā kei runga, rua ki te taha	Student:	Four above, two at the side
Student:	I whakaaro au i tuhi au e rima	Student:	I thought I drew five.
Y11T:	Koinā te tūtakitanga, nē?	Y11T:	That is the intersection, okay?

The Year 3 teacher, in 2007, also used students' contributions in writing. She told in a staff meeting how she transcribed some students' contributions, because they were too slow to write their ideas down, which impeded what else had to be done during the lesson. At the beginning of the next lesson, the Year 3 teacher asked them about those ideas and whether they still agreed with them and if they wanted to add anything to them. She found that doing the writing for these students resulted in their ideas being valued by other students. If they had to write sentences, they rarely got anything else done in the lesson. Thus, a tension existed between ensuring that students' ideas were valued and ensuring that they gained writing skills. As a strategy that was used occasionally, it had advantages but there was a risk that the students would continue to rely on the teacher to do their writing for them.

On another occasion in 2007, the Year 4 teacher described how she used various strategies to build up the students' own writing. In the first example in Fig. 10.3,

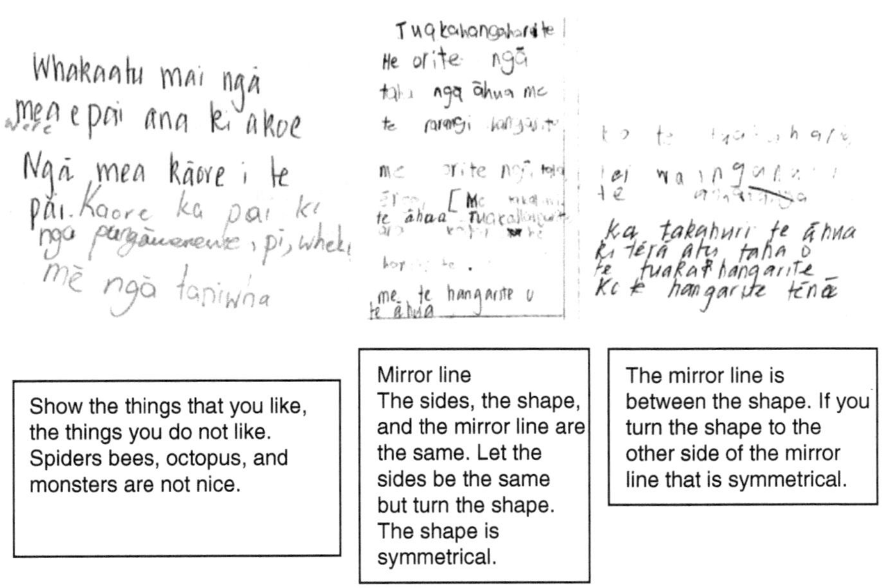

Fig. 10.3 Examples of the teacher working with Year 4 students' writing

she began the writing and then had the student complete it with a sentence. In the second example, the teacher had corrected the student's narrative and in the third example she had the student interpret what he had wanted to write and then wrote it for him. The teacher actively monitored the students' work whilst they were doing this writing, so that she could activate the most appropriate strategy.

The teacher's work with students to improve the quality of their writing involved students reflecting on the process of writing, not just the end product. This stage concentrated the students on finding out when and how they should use the new vocabulary or expressions.

Taunga/Integration

By the Integration stage, students have a good understanding of the new aspects of the mathematics register. However, they may need to be reminded that they have good skills and knowledge, so that they can use them in the activities. The students have the major responsibility for making use of the new language. If the students are unable to operate at this level, the teacher is able to supply support quickly. The teachers reminded the students of what they knew through:

- using commands and linguistic markers to highlight for listeners that they need to pay extra attention to what they are hearing and doing
- encouraging students to make contributions to the teacher and to each other
- reminding students to think about what they already know
- asking a student to repeat a good response
- if a slight correction is needed, the teacher repeats the response correctly
- summarising what a student has said
- if a slight correction is needed, the teacher can model doing the action so that the students self-correct their own response
- prompting in a general way for more details
- having students write a summary of, or record as a diagram what they have learnt
- facilitating an environment where students will correct each other
- asking students to say whether an answer/term is correct
- repeating the question if the students appear to have responded to a different one
- having students complete appropriate actions as they respond to questions.

In the following example, the teacher used a general class discussion at the beginning of the lesson to remind students about the features that they needed to include in a map about the places in the movie *Shrek*. She did this by asking the children questions and then recording their answers as a list (see Fig. 10.4). The children drew their maps based on this list. An example of the map can be seen in Fig. 10.5.

Fig. 10.4 Teacher recording students' answers on the board

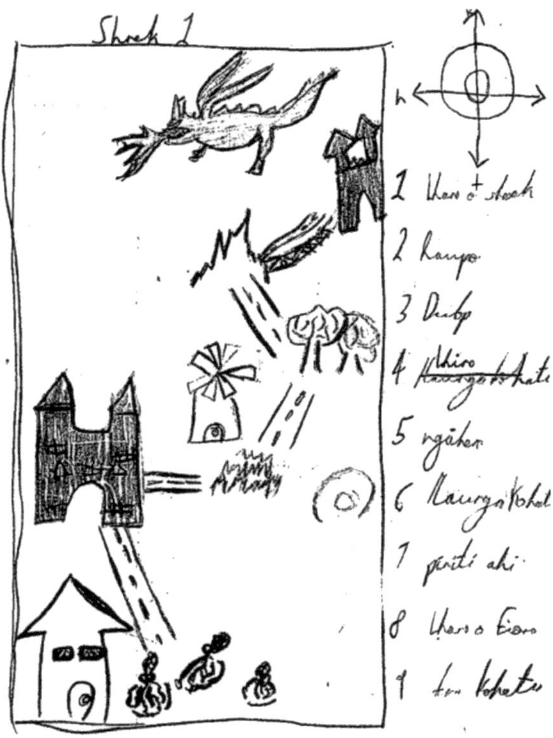

Fig. 10.5 One child's map of the places in Shrek

This student has included a compass and placed different places in relationship to each other. Although it does not fulfil the requirements of a conventional map, for a Year 4 child it shows considerable understanding about the role of a map. At a later time, the concept of a map will be given more layers of meaning such as needing to be from a bird's-eye perspective.

Another very common strategy at this level was for teachers to collect in students' work and check it for accuracy. This is a common strategy in all schools. Yet, by being delayed, it lacks the immediacy of feedback that was seen in many other strategies from this stage. Sometimes, the students wrote an initial draft, made changes, and then produced a final version. In Doerr and Chandler-Olcott's (2009) research, the middle school teachers had found that "the editing of student work began to yield improvements in the quality of students' written responses" (p. 294). The immediacy in this feedback would contribute to the students being able to gain the most from the "teaching moment".

In Fig. 10.6, an earlier version of the sentence can be seen faintly underneath the final sentence. This is most obvious in the writing of Kahuri. When writing is to be displayed, the teacher closely supervised the work of students. Teachers returned to using the *akoranga* strategies if students started to flounder.

One way to provide immediate feedback was through the use of computer programmes. In one recorded lesson, students used MS Word drawing functions to produce tessellating patterns. Winch, Johnston, March, Ljungdahl, and Holliday (2004) suggested that students find revision of narrative pieces of writing much easier if they can use word-processing programme. The use of computer technology to alleviate some of the physical demands of writing has been available in mathematics classrooms for some time. Brown, Jones, Taylor, and Hirst (2004) found that students were more able to engage with a problem about the diagonal properties of quadrilaterals using Geometers Sketch Pad, whereas this had not been the case when using pencil and paper. It may be that students can use computers to replace the tediousness of some parts of mathematics, such as drawing tessellating patterns and graphs and so engage more willingly with these topics.

Figure 10.7 shows the development of a pattern using a translated shape. Others in the class rotated their shapes to form their patterns. The software allowed a very quick development of a complicated pattern that would have taken many hours to draw by hand. The first picture shows the student choosing a shape. He then drew the original shape, copied it, and pasted several examples onto the document. The

Fig. 10.6 A student's written explanation of rotation

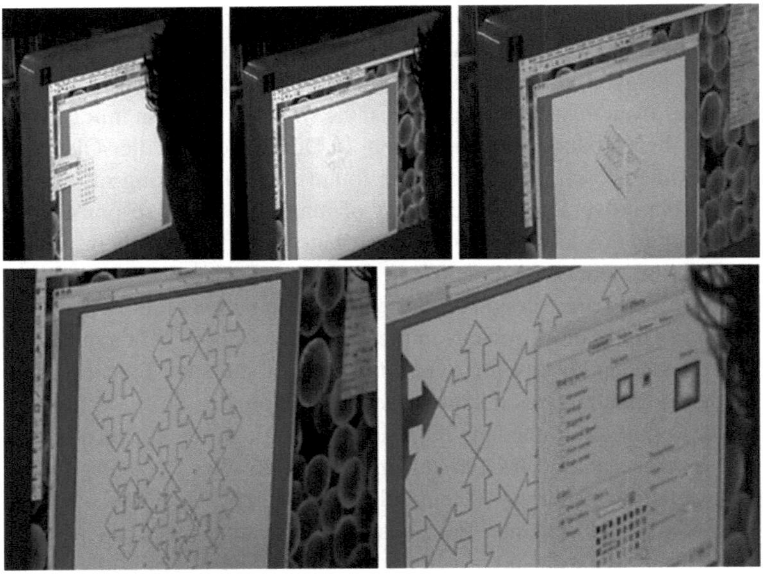

Fig. 10.7 Stages in developing a tessellating pattern using translation

student slid (translated) the copies of the shape around the page to form a pattern. The final picture shows the student choosing colours to shade the shapes in the pattern.

When students did mathematical writing on playgrounds or on whiteboards, they also displayed their fluency, but not in the same way as the static posters put up around the classrooms, of which Fig. 10.6 was an example. Public writing was done quickly and was only available for immediate scrutiny and discussion, as the work would be removed at the end of the lesson, if not earlier. If students had produced something that was incorrect, then there were opportunities for clarification to be made. Well-presented pieces of work were also discussed. These activities were part of the *taunga* stage because they allowed for feedback about the appropriateness of the response.

In the senior classes, students were regularly expected to present their ideas on the white board. An example of this can be seen in Fig. 10.8.

In asking students to display their knowledge, it is assumed that they have the skills to do so and that the classroom environment was supportive of them if this was not the case. This supportive environment was used to remind the students of what they already knew, and if they were unable to resolve the clarification issues, then the teacher could intervene by using strategies from the *Akoranga* stage.

Fig. 10.8 Presenting an explanation of the length of the hypotenuse of a triangle

Putanga/Output

In the final stage of the MRA model, students show what they know and can do without support from the teacher. The teachers provide activities where the new aspects of the mathematics register would arise naturally. The following two strategies were the most common ways that teachers did this:

- providing opportunities for students to use their acquired aspects of the mathematics register between themselves and with the teacher, and
- providing an environment in which the students can query the language use of the teacher.

Assessment tasks tested students' fluency as well as their mathematical understandings. A teacher in Doerr and Chandler-Olcott's (2009) research had given students the same writing prompt at the beginning, middle, and end of a unit. The students' work had provided her with insights into how their understanding had grown whilst completing the unit. In the second year of that project, the teacher had encouraged the students themselves to look at the work and consider how to improve it.

In 2007, two teachers at Te Koutu, the Year 8 teacher and the Year 4 teacher, asked students to write about a topic both at the beginning and at the end of a unit of work. This enabled not only the teacher but also the students to see improvements. In the September staff meeting, the Year 8 teacher described why she had students produce the two pieces of writing about transformations. The following comes from the minutes of the meeting.

The Year 8 teacher mentioned that in her group she has some students who struggle with writing generally. She saw in the examples of writing about transformations (reflection, rotation and translation) that some students appeared to have played safe. For example, to show reflection a student chose the letter "T". Although this was reflected, it does not actually change, which would have been the case if she had chosen something like the letter "K". A good piece of writing on this topic gave the explanation generally through diagrams. However, some students would have done better by providing a longer written text. Students need strong *te reo Māori* if they are to produce good narrative texts. Even with good mathematical vocabulary they also need good general writing skills.

Having students present their understanding of a topic means that they have to make some independent choices about what they are going to do. The teacher felt that it shows her where they are at.

This teacher had students complete a second piece of writing on this topic by having them choose a *kōwhaiwhai* [repeating] pattern and then describe the transformations within it. In this case she felt that she provided more explanation about what she was wanting than she had with the earlier piece. The first piece was in some ways a diagnostic test to see what students knew about the topic.

Figure 10.9 shows the two pieces of writing from a Year 8 student. The prompts for the different pieces of writing were not the same. After the students had produced the first example, the teacher felt that they had used simple shapes that were easy

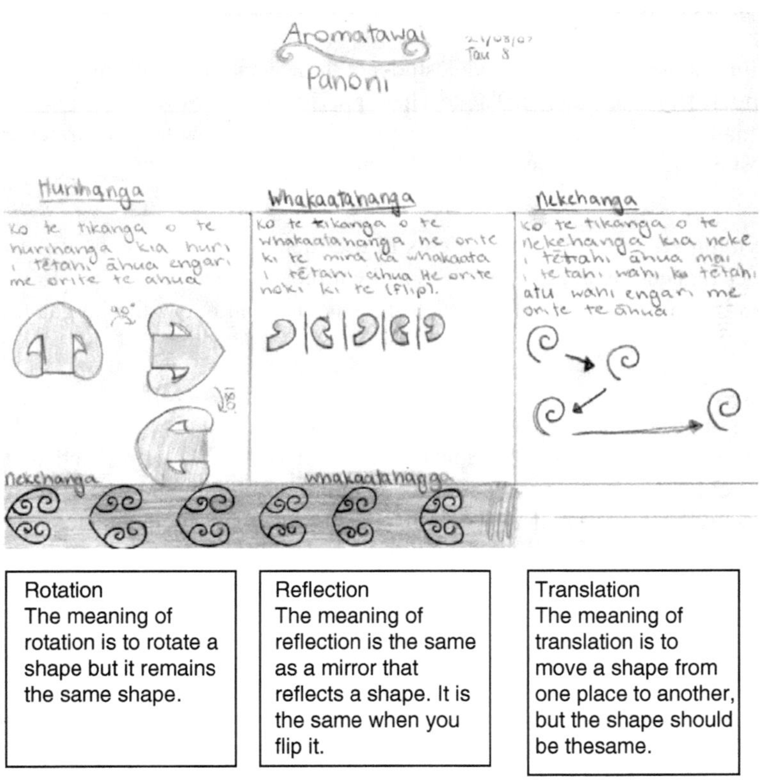

Fig. 10.9 Transformation assignments by a Year 8 student

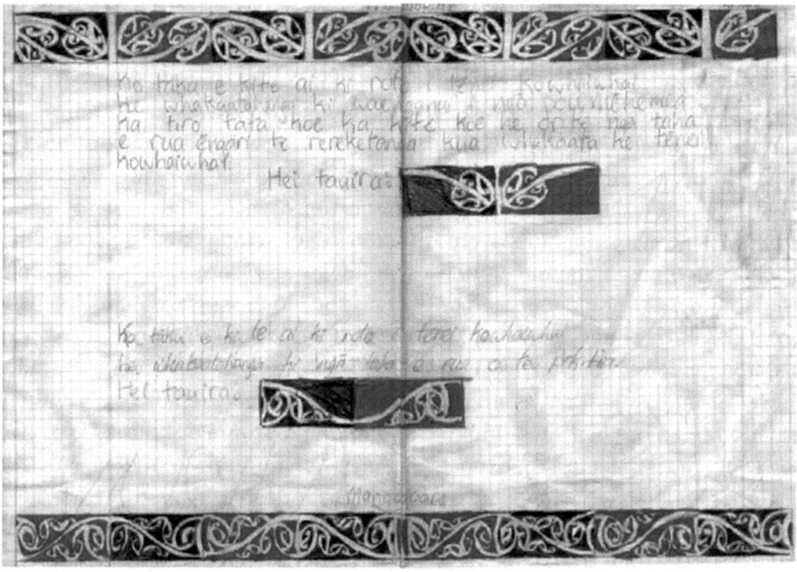

What I see in this *kōwhaiwhai* is a reflection between the columns. If you look closely you will see that both sides are the same. The difference is that this *kōwhaiwhai* is reflected.

For example

What I can see in this *kōwhaiwhai* is a reflection on both sides of the diagram For example Hammerhead shark (the *kōwhaiwhai* pattern)

Fig. 10.9 (continued)

to transform. For the second piece of writing, she asked them specifically to choose traditional Māori shapes and describe how these had been formed. This was a more difficult task, but still provided students with a large amount of freedom to choose shapes that they felt comfortable describing. Although the students' actual writing appears not to be more developed, what has been attempted is more difficult because it is not just a definition, but an attempt to explain how the patterns were formed using different transformations.

Combining Strategies for Effectiveness

During staff meetings, our discussions were often about the effectiveness of various strategies. It was clear that what constituted an effective strategy had to be considered in regard to its role in acquiring the mathematics register. A strategy that was effective at the Noticing stage may not be an effective strategy at the Integration stage, because it could not fulfil the functions of both stages well.

Previous research in mathematics classrooms, with English as the language of instruction, had focused on dialogical structures and their contribution to students' mathematical understanding (see Nathan & Knuth, 2003; Bill, Leer, Reams, & Resnick, 1992; Moskal & Magone, 2000; White, 2003; Tanner & Jones, 2000). Consequently, some structures were criticised or recommended according to how successful they were in prompting students' mathematical thinking. For example, Wood (1998) criticised the use of leading questions where students provided one-word answers to questions that were difficult to get wrong. She stated, "although the teacher may intend that the child uses strategies and learn about the relationship between numbers, the students need only to respond to the surface linguistic patterns to derive the correct answers" (p. 172). She suggested that an alternative pattern, which she labelled "focusing", would be more effective in promoting learning. "A high level of interaction between the teacher and students creates opportunities for children to reflect on their own thinking and on the reasoning of others" (p. 172).

However, the strategies that the teachers used to support students to acquire aspects of the mathematics register are likely to be different from those used to promote thinking. Having students repeat an answer, after the teacher has gone through an explanation, would be labelled by Wood (1998) as an example of the "funnel pattern". She saw such strategies as being of limited value to students learning mathematics because the teacher is the one who does the cognitive work. Yet this strategy may have some value if it is used in introductory tasks, which help students to use, both through interpretation and production, new aspects of the mathematics register. If this is one strategy of many, which raise students' awareness of these new aspects, then it has some value. In each of the lessons, the teachers always used more than one strategy. Therefore, it is difficult to dismiss a particular strategy in isolation as being ineffective. Combining a range of strategies, therefore, seems to be part of what makes effective support for students who are operating at different stages in respect to the same set of new terms and expressions.

Language Acquisition Strategies and Year Level

Wells (1999) stated that "in the hands of different teachers, the same basic discourse format can lead to very different levels of student participation and engagement" (p. 169). Consequently, there is a need to better understand the contexts in which the strategies were used, both individually and combined, to support the acquisition of the mathematics register.

Although there seemed to be a progression in how aspects of the mathematics register were acquired, the relationship between the strategies that teachers used and the year level that they taught was much less clear. Table 10.2 shows the strategies used by at least two teachers in the lessons, which were recorded in 2006. We did not capture all of the strategies that each teacher employed, especially as the teachers all had different amounts of lessons filmed. Nevertheless, the strategies that were recorded seemed to be independent of age.

At the Noticing stage, *writing the new term in an equation* and *using fill-in-the blank sentences* were the only strategies used by teachers from just one area of the

Table 10.2 Strategies used by teachers of different student year levels in 2006

	Junior primary			Senior primary		Intermediate	High school
	Y2	Y3	Y4	Y5	Y6	Y7/8	Y10/11
Noticing							
Providing opportunities	√		√	√	√	√	√
Using intonation	√						√
Repeating new terms and expressions			√		√	√	
Rephrasing the expressions					√	√	
Writing the new term in an equation				√	√		
Giving definitions				√		√	√
Emphasising the relationship between ideas	√			√	√		√
Modelling a new term/skill	√			√			
Using a set of leading questions	√				√	√	
Using fill-in-the-blank sentences						√	√
Intake							
Students do choral responses	√	√		√			
Asking students for definitions					√	√	√
Having students model use of terms	√			√		√	
Asking students for examples of a term	√					√	√
Having students draw or use materials	√			√	√		√
Repeating appropriate responses	√	√		√	√	√	
Elaborating on students' responses	√			√	√	√	
Asking further questions	√			√	√	√	
Having students provide a rationale					√	√	√
Querying inappropriate responses					√	√	√
Illustrating a general rule with amounts						√	√
Focusing back onto the main idea	√						√
Relationship to new language	√	√			√	√	
Answering a series of closed questions	√				√	√	
Having students repeat the action	√	√		√			
Recording in writing	√				√		
Integration							
Using commands and linguistic markers						√	√
Encouraging students to contribute	√		√	√	√	√	√
Reminding students to think				√	√	√	√
Summarising what a student has said				√			√
Prompting for more details				√	√	√	√
Having students write a summary				√		√	√
Facilitating an environment where children will correct each other	√	√					√
Asking whether an answer is correct	√		√	√			
Repeating the question	√	√				√	
Having students complete actions	√	√			√		
Output							
Providing opportunities	√	√	√		√	√	√
Students can query the language use of the teacher			√			√	√

kura. Teachers working with students in the later years of primary school used the first of these strategies. However, given that students needed to learn how to write mathematics symbolically at all year levels, this was unlikely to be an age-dependent strategy. The second strategy, where teachers began a sentence that students then finished, was used only by the teachers of the older students. Once again, it is difficult to accept that this was an age-dependent strategy. These strategies are related to the teacher-student exchange that has been documented in many classrooms and is known as the IRF (initiation – response – feedback) exchange (Mehan, 1979). The teacher asked a question by leaving the sentence unfilled. The students provided the response and then the teacher gave either explicit feedback, through affirmation or negation of the response, or indirect feedback by asking a new question. Given its prevalence in English-medium classrooms, it seems unlikely that this was a strategy that only teachers of older students used.

At the Intake stage, the teachers of the older students were the only ones to use specific amounts to illustrate a general rule. They could have been connected to the introduction and use of algebraic equations to this age of students. The relationship between general rules and specific amounts may not be so important in the earlier year levels.

The two teachers teaching the intermediate and high school mathematics classes were the only ones to *use commands and linguistic markers* as a scaffolding strategy to encourage students to make use of the language skills and knowledge that they had. It is difficult to know whether this is an age-related strategy. For students to take advantage of linguistic markers, they must have good skills in everyday *te reo Māori*. It may be that this knowledge comes later in the acquisition of *te reo Māori* and therefore is not available to young students. It may also be related to the fluency level of the teachers.

Commands such as "to listen carefully" are useful to older students because they have the skills to interpret them. Younger students may need more explicit directions about what they are expected to do, especially when the activity that they are working on is made up of several components. The Year 1 teacher in 2008 seemed to confirm this by stating in an interview:

> I mean with my kids, because they are only little, I really have to break it down, because it is a bit hard for them and I always start off with – do they understand, with groups, like what is a group and could they group the class, and group the boys and then group the girls, some with jerseys some without jerseys, things like that. Just so they could understand what they will be learning over the next couple of weeks. I always start off the whole lesson with the whole class doing that so they've all got the same information. (Interview, June, 2008)

The Output stage contained no strategies that were used solely by teachers of one age group of students. This may possibly be because there was only a limited number of strategies at this stage.

In summary, the overwhelming impression from Table 10.2 is that there were almost no strategies that seemed to be related to the year level that students were in. The possible exceptions for this would be the strategies of *illustrating a general rule with specific amounts* and *using commands and linguistic markers*. However, given the small sample of videoed lessons that we collected, it may be that even these two strategies were used sometimes by teachers of younger students.

The Effect of the Newness of the Topic on Strategy Use

Although the age of students did not influence the strategies that teachers used, the newness of the topic did. When the topic was first introduced, the strategies tended to be from the Noticing stage. As the unit of work continued, the strategies tended to come from the later stages.

The Year 6 teacher's strategies, from a series of lessons in 2006, are categorised by their MRA stage in the appendix to this chapter. The relationship between the strategy stage and the point in the unit of work is illustrated through different fonts. Noticing strategies were only used in Lessons 1, 2, and 4. At the Intake stage, strategies were used in Lessons 1, 2, 3, and 5. Strategies from the Integration stage were used in Lessons 1, 2, 3, and 4. The Output stage's strategies came from Lessons 1, 2, and 5. This suggests that there was no clear pattern in the distribution of strategies across the lessons. However, scrutiny of the information in the appendix shows that the situation is complex.

The topic for the unit of work was on the properties of arrangements of cubes as part of a larger unit on three-dimensional shapes. At the beginning of the first lesson, the strategies were categorised as being from the Output stage and related to the class discussion on the previous lesson. A student queried the teacher about what had been written up on the board at the end of the previous lesson. The strategies from the Integration stage in this early part of Lesson 1 concentrated on having other students gain an understanding of what the original student was explaining. Although this student was operating at the Output stage, the teacher seemed to believe that some of the class were still at an earlier stage. She cued them into listening carefully through the use of linguistic markers but also used *rephrasing* and *fill-in-the-blank sentences* to introduce this vocabulary. Concurrently, she went over students' knowledge of three-dimensional shapes by having them provide examples.

In the rest of the lesson, the teacher introduced ideas about the faces that are on cubes. She had put dots on each face and then placed several cubes together. The aim for this part of the lesson was for students to use an equation to work out the number of dots that could be seen. The majority of the time was spent in the teacher explaining how to work out the number of dots. To do this, she used a series of questions that led the students to understand what was required. Then, she had students work on different arrangements of blocks and provide explanations of what they had done. The teacher moved on to introducing the *whāritenga* (equation). With the aim of the lesson being to develop and use the equation, it is understandable that the majority of the lesson revolved around strategies from the first two stages. The strategies from the Integration stage encouraged students to use existing skills and understanding to support them to recognise when an equation was appropriate and how to use it.

Lesson 2 also continued with identifying the number of dots, but the aim was for the students to be able to give clear explanations. The strategies from the earlier stages were used to ensure that the students were familiar with terms such as *huapae* (horizontal) and *poutū* (vertical) that were needed in the explanations. In this lesson, students were also expected to draw sketches of their block arrangements and thus provide a written description of their experiences.

Over time, the emphasis in the lessons shifted from "learning" to "using" aspects of the mathematics register in relationship to the arrangements of blocks. Lessons 3 and 4 continued this shift, and Lesson 5 involved the students in providing descriptions of the block arrangements to their peers. It was the feedback from their peers that illustrated to the speakers whether or not they had been clear.

When the new topic began, there was a concentration on introducing new aspects of the mathematics register. At the same time, there was a need to ensure that there were opportunities within each lesson for students to use aspects of the mathematics register with which they had fluency or near fluency in. In this way, students connected new information to what they already knew, whilst at the same time giving them understandings about the contexts in which the new aspects would be relevant. Acquiring the mathematics register involved students gaining fluency in speaking, listening, reading, and writing. Moving between strategies, from different stages, also showed how a teacher used the students' ability in one language skill to support them gaining fluency in another.

Meeting the Challenge of Documenting How Teachers Supported Students to Acquire the Mathematics Register

Once a decision had been made to teach mathematics in *te reo Māori* and a mathematics register developed, the challenge for teachers was and continues to be one of finding the best methods for having students learn this register. Recognising this as a challenge involved the teachers working with researchers to document what was already being done so that this information could be used to think about what to improve.

The mathematics register is not something that is used frequently away from the context of the classroom by most non-mathematicians. This is because its specific function is to discuss formal mathematical ideas. For second-language learners, when the home language is different from the language of instruction at school, there are even fewer opportunities to use the mathematics register learnt at school. Therefore, it is essential to understand more about how students learn the mathematics register at school. In some ways the development and refinement of the MRA model, as a result of this documentation, provided a good basis for a discussion about the strategies that the different teachers were using. Teachers could describe what they were already doing as well as consider whether strategies used by other teachers also might be worth trialling. In 2006, many teachers adopted the practice of providing rationales to students about the language that they were learning. This was a strategy that they heard about from the Year 6 teacher and Uenuku Fairhall.

An outcome of the research was a much more thorough understanding of how teachers supported students' acquisition of the mathematics register. This is contributing to wider understanding in other education systems about how to teach the mathematics language as part of mathematics lessons. At Te Koutu, the teachers who were part of the original group that contributed to the development of the MRA model are able to discuss it. However, the considerable change of staff in the last few years means that it is no longer an established part of staff discussions. The ability to sustain the development of understandings that inform teaching practices is a continuing challenge for the school.

Appendix: Year 6 Teacher's Scaffolding Strategies

T1	Teacher initiated	Student initiated	Comments
Noticing	• Use of "arā" to mark that a definition will follow • Use of kē to mark that information following may be unexpected		Language devices in Māori to highlight/alert the need to listen
	• After going over a new rule, the teacher begins a sentence for students to complete with a one-word answer		This fill-in-the-blank sentence has only one possible response
	• "he momo koeko" is rephrased as "te whānau koeko"		
	• Teacher begins by asking how many dots can be seen. She then clarifies through a series of leading questions what is meant by "seen" in this context		
	Teacher asks a series of leading questions which have clear, one-idea answers that build towards the equation (te whāritenga) which is what the teacher originally expected the students to provide	• Students make (wild) guesses about the number of dots that can be seen	
	• Teacher reminds students of the relationship between poutū and te pou poutū (vertical and a pole)		
	• Teacher uses the term and describes it more fully	• Student offers "ele" (L-shaped) as the name of a configuration of blocks	Teacher accepts the term offered but then describes it more fully so that there is a shared definition of the term's meaning
	• Teacher asks whether the number of dots is different for several configurations of the blocks	Students keep saying that the number of dots is the same	This exercise reinforces what "ōrite" (same) means in this situation. This then leads to the modelling of the sentence about the most dots that you can have
	Teacher then says "22 ke te mea rahi rawa i tēnei wa" (the most dots you can have is 22)		
	• *Teacher rephrases "he rerekē te nuka" as "tirohangā"*		
	• *Teacher rephrases "te whakautu" as "te otinga kimi" and repeats "te whakautu"*		
	• *Teacher rephrases "nuka tere" as "huarahi tere"*		
	• *Teacher uses "i te mea" in several explanations*		By repeating the expression, the teacher would be modelling its use in explanations

T1	Teacher initiated	Student initiated	Comments
Intake	• Teacher asks for different three-dimensional shapes (which can be related to Euler's rule)	Student offers pororua (cylinder)	This exchange starts with a more general request than those seen in the Noticing stage. However, when a student suggests a three-dimensional shape to which Euler's rule cannot be applied, the teacher limits the students' choices
	Teacher then restricts students' choices to the family of pyramids	Students offer different suggestions of types of pyramids (porotapatoru)	
	• Teacher asks how many faces and vertices there are on a "koeko tapatoru"	Student responds with 10	
	Teacher then asks for the number of sides	Student responds with 8	Students are once again given more option to show their understanding of the terms. The teacher's counting reinforces that they were correct
	Teacher counts the sides and asks again for the number of sides before asking them to add on an extra 2, following Euler's rule		
	• Teacher begins an explanation of how to work out the number of dots, which students need to complete with one-word answers		Could be noticing, except "tango" has not been used by the teacher previously. Students are channelled into using this term that they already know in this new context
	• Teacher asks for the names of different shape configurations (huapae – horizontal, poutū, ele – l-shaped)		
	Teacher goes over the need for the blocks to be face-to-face (mata ki te mata)	• One group of students has a configuration where the blocks are separate	
	• Teacher asks a student to explain fully how she got the number of dots in her block configuration	Student gives a short answer	
	Teacher adds extra words that the student repeats	Student then provides other details	This part of the exchange is probably more like the Integration stage but with the reversion to Intake when it is clear that the student's answer does not fulfil the teacher's requirement of a building which shows fewer than 22 dots
	Teacher has the student repeat what she said so that other students who were talking could hear the explanation. Teacher then has the student count the missing dots to show that the amount is not less than 22 which is what the problem was		
	Teacher prompts to get an explanation of how the number was achieved (even offering a calculator for the student to use to work it out)	• Student responds to question about the number of dots	
		Students respond with numbers (sometimes inappropriate numbers)	
	Teacher starts to repeat a student response when the needed answer was given	Students complete the repeated sentence	
	Teacher rephrases the student's response as a generalisation (without all of the specific amounts)	• Student gives an explanation using specific numbers	In this exchange, the teacher is not suggesting that the student's response is wrong, just that there is another (more appropriate) way of expressing the explanation
		• Student uses term "huapae" ↓ **Another student then uses a mispronounced version, "ruapae"**	

T1	Teacher initiated	Student initiated	Comments
Integration	• "āta whakaaro kao" (command to understand)		Students are credited with being able to understand but the teacher is aware that some might miss the opportunity to do so
	Teacher asks for repetition by repeating the initial past tense participle ("e")	• Student provides explanation Student completes repetition	Students are again expected to understand others' contributions but the teacher's intervention highlights the need for students to understand
	• Teacher makes a formal request for a student to repeat what they said	• Student requests clarification of task requirements "ngā mea o raro?"	
	• Teacher asks the students to explain their strategies for working out how many dots there are	Students provide strategies using/giving specific amounts in their explanations and the use of "i ngā mea"	
	• Teacher commands the students to provide full descriptions of how to work out how many dots (not just saying horizontal or vertical) are in the configurations of blocks. She then provides an example	Student then gives an explanation which is not terribly clear	This exchange has parts where the teacher is encouraging students to use the language they already have (to recognise that they can give an equation rather than just use a counting strategy) but is restrictive at times, such as would be more typically seen at the Intake stage
	Teacher moves the students from counting to using a more general equation/strategy using subtraction to work out how many dots can be seen. Students are prompted to use what they already know		
	Teacher focuses students back on to the original question (ko te pātai tonu – the question was) This discussion of the most dots is then turned around to ask students to think about a block configuration with the least number of dots showing	• Students go off track in responding to the teacher's questions	
	Teacher asks about the number of dots which can't be seen	• Student describes the arrangement of blocks	
	Teacher queries this suggestion Teacher then commands the student to think before speaking again	Student suggests that they should be added Student gives a fuller explanation of adding 4	The student is fairly competent but needs prompting similar to that in the Intake stage to provide a full explanation
	• Teacher asks students to make isometric drawings of their block configurations and to explain their strategies for determining the number of blocks (writing equations is given as a suggestion for doing this)		
	Teacher prompts for more details	• Student gives an explanation of how he got 24 blocks for his drawing Student provides details when prompted so that he gives a fuller explanation	
	• *Teacher uses "nē" and "neha" as requests for interaction*		
	• *Teacher reminds students of what was covered in the previous lesson (e hia ngā ira and location words)*		

T1	Teacher initiated	Student initiated	Comments
Output		• Student explained a problem with the numbers given in a table for Euler’s rule	The student raised this problem with the teacher at the end of a previous lesson. The teacher introduced this in a new lesson and asked the child to state what the problem was
		• Student queries whether the number of dots that have to be seen is 22	
	• Teacher commands students to draw the different configurations of blocks		
		• **Students use location words in giving an explanation describing the amounts in blocks and in describing the arrangements of different coloured blocks**	

Times New Roman – Lesson 1, Comic Sans - Lesson 2, Arial Lesson 3, *Italics – Lesson 4*, **Bold – Lesson 5**

Chapter 11
"Māori were Traditional Explorers": Māori Pedagogical Practices

In the literature on *kura kaupapa Māori*, there is much mention of Māori pedagogy, often considered to be the practices used for teaching traditional skills and knowledge (Hemara, 2000). The discussions include little if any reference to how relevant these pedagogical practices were to the teaching of Western knowledge domains, such as mathematics. In this chapter, we explore how teaching and learning at Te Koutu are related to a concept of Māori pedagogy. Given that Te Koutu's primary aim is to support Māori language and cultural revitalisation, there is a need to better understand how this objective is supported in the mathematics classroom.

At Te Koutu, the importance of better understanding teachers' pedagogies and its connection to Māori culture and values have been recognised as a challenge. As was noted in Chapter 3, parents had concerns that teachers may be so tainted by their Western teacher education that they may not be able to move out of this mindset to adopt a more culturally appropriate way of teaching. As Smith and Cram (1997 cited in Pihama, Smith, Taki, & Lee, 2004) stated, "there is a growing body of literature regarding *Kaupapa Māori* theories and practices that assert a need for Māori to develop initiatives for change that are located within distinctly Māori frameworks" (p. 10). This is linked to "puristic ideologies" that have underpinned the modernising of the language (see Chapter 2) for teaching and learning and is more to do with the status of *te reo Māori* and people's attitudes to its status (Smith, 1997). The challenge is to document and extend pedagogical practices currently being used in the school in order to develop a clearer understanding across the school of what Māori pedagogy could be.

Debate about and critique of content and epistemology, most often bundled as "Māori knowledge", has been reflected in the dialogic practice at Te Koutu since its early days. One of the debates centred on what traditional Māori mathematics knowledge and practice should and could be included in the Te Koutu curriculum. In discussions from this time, concern was raised about whether the worldview of Māori ancestors had been lost or changed, and the effect this had on traditional practices being seen as mathematics. As well, Uenuku Fairhall was unsure if teachers, as adult second-language learners, would know how best to facilitate the mathematical learning of their students who were more fluent speakers of *te reo Māori*.

T. Meaney et al., *Collaborating to Meet Language Challenges in Indigenous Mathematics Classrooms*, Mathematics Education Library 52,
DOI 10.1007/978-94-007-1994-1_11,

> Things have gone together in their heads. We may be teaching maths in such a way that it's not going to make any sense to them, it's not the way for them to get there with the tools that they've got in there so that's why I'm saying, we've got to go beyond it and think what could be, what is the world view that's expressed in the language and the other way round in the past. Now we can talk quite easily about flax weaving in English, that's how Peter Buck [Māori anthropologist from the early twentieth century] could write a book in English about flax weaving. But what's going on inside the head of the native speaker, when they are doing the flax might not be in that form at all. If they are seeing the flax as being the manipulated, as more important than the person manipulating the flax, that creates a whole balance between you and the flax, you know, the relationship, and that's the world, the physical world. Māori *tūpuna* [ancestors] saw themselves in a quite a different relationship with the world than we did. (Uenuku Fairhall, Meeting, 1999)

On the other hand, there was also a belief that the way that children learn might be universal. In an earlier meeting, a parent stated:

> Because if sequence of student learning which is maths, has probably been developed, I don't know, against a backdrop of all that human growth knowledge, all that Piaget and all that sort of stuff over years and years and years and lots and lots of lived experiences of teachers say this works and that doesn't work and all the rest of it. To suddenly change it and say well we will do it so it is more culturally appropriate, for sequence? How do our children learn differently to all that other stuff? (Parent4, Meeting, 1999)

The tension between believing that their students were similar to their non-Indigenous counterparts and also wanting them to be distinctly Māori remained throughout the meetings in this first project from 1998 to 1999. Smith (2003a) suggested that one of the strengths of Māori education is:

> That teaching and learning settings and practices are able to closely and effectively 'connect' with the cultural backgrounds and life circumstances (socio-economic) of Māori communities. These teaching and learning choices are 'selected' as being 'culturally preferred'. (p. 12)

However, over the course of the research, we have been querying whether there is a definitive set of teaching practices that could be constituted as uniquely Māori pedagogical practices in contemporary Aotearoa/New Zealand. Rather, it may be better to think of a combination of practices which when used together in a lesson, or classroom or school, could be denoted as Māori pedagogy.

Alternatively, there are other influences that mitigate against the likelihood of even the earlier-mentioned interpretation of Māori pedagogy being a possibility. For example, government-funded schools are required to implement and thus to conform by default to mandated curricula and so "[p]edagogical practices are therefore expected to be aligned to curriculum requirements documented in the curriculum statements" (Rau, 2001, p. 4). In discussions on literacy development, Rau (2001) suggested that this stranglehold of government priorities on Māori-medium education has limited the ability of *kura kaupapa Māori* to determine their own epistemological and pedagogical priorities. In July 1999, Uenuku Fairhall made a similar comment:

> But I must admit, you know, that for the last few meetings, I've been asking myself how effectively different are we? We're different in the way the school is run and we are different in the language that we use and we're different in what we expect to be normal behaviour

> in some cases, but curricula-wise it's a Western corpus of academia and curricula that has been quite whole-heartedly accepted.

In the next sections, we discuss what pedagogical practices might be before turning to Māori pedagogical practices. We then explore the pedagogical practices that have been used at Te Koutu and what value there may be in labelling them as Māori. We structure this discussion around the main points from *Te Aho Matua* (Kura Kaupapa Māori Working Group & Katarina Mataira, 1989), the guiding philosophy document for *kura kaupapa Māori*.

What are Pedagogical Practices?

Pedagogical practices like many educational terms are ill-defined. Carr et al. (2005) equated pedagogy with the attributes of a good teacher. More frequently, pedagogical practices are linked to theoretical understandings of how children learn. For example, Draper (2002) stated the following:

> Constructivism offers educators a way to think about how people think and come to know. Constructivist pedagogy requires that teachers take into consideration what students know, what they want to know, and how to move students toward desired knowledge. (p. 528)

Teachers' pedagogical practices do have a direct impact on students' meaning construction (Gutstein, 2003). Walshaw and Anthony (2008) suggested that "[t]he harsh reality learned from the OECD [Organisation for Economic Co-operation and Development] study is that mathematics pedagogy affects learners in disproportionate ways" (p. 134). For many Māori students, mainstream mathematics pedagogies have resulted in lower educational outcomes than their non-Māori peers. For example, results from the National Education Monitoring Project showed that Māori students in mainstream schools did not perform as well as their European background peers (Flockton, Crooks, Smith, & Smith, 2006). As well, Forbes (2005) found that "Māori students were markedly less likely than non-Māori students to continue in mathematics in all years" of senior high school and that "the accumulated mathematics attainment of Māori students, as a group, was less than two-thirds that of non-Māori students" (p. 3). In particular, teacher expectations have come to be seen as a cause of Māori student underachievement:

> deficit theorising by teachers is the major impediment to *Māori* students' educational achievement for it results in teachers having low expectations of *Māori* students. This in turn creates a downward spiralling, self-fulfilling prophecy of *Māori* student achievement and failure. (Bishop, Berryman, Tiakiwai, & Richardson, 2003, pp. 4–5)

Thus, it could be expected that pedagogies that align more with the ways that Māori students learn are likely to produce improved academic performances. This would include having teachers with high expectations of students.

Walshaw and Anthony (2008) described effective pedagogies in mathematics as those where the teaching facilitates the learning of diverse students and stated, "pedagogy is linked not only to achievement outcomes, but also to outcomes relating to affect, behaviour, communication, and participation" (p. 135). They outlined the following principles for effective pedagogy that focus on both teaching and learning practices:

Mathematics teaching for diverse learners:

- demands an ethic of care
- creates a space for the individual and the collective
- demands explicit instruction
- involves respectful exchange of ideas
- demands teacher content knowledge, knowledge of mathematics pedagogy, and reflecting-in-action (Anthony & Walshaw, 2007).

However, there appears to be no one set of practices that are effective with diverse student populations, but rather it is how individual teachers implement these principles that make them effective.

> Precisely because pedagogy encompasses elements characterised not only by regularities but also by the uncertainties of practice, it has to take into account the physical, social, cultural, and historical space in which the teaching is embedded. (Walshaw & Anthony, 2008, p. 146)

The pedagogical practices used by the teachers at Te Koutu are aimed at supporting Māori children to learn mathematics. Nevertheless, these practices would vary from teacher to teacher because of the influence of their own education and understanding of mathematics, as well as their beliefs about children's previous mathematical experiences and future needs. For these practices to count as Māori pedagogy depends on whether there is a consistent belief about how Māori children learn, which is then manifested in a set of teaching practices designed to facilitate this type of learning.

What are Māori Pedagogical Practices?

All pedagogy is culturally based (Lipka et al., 2005), and to designate certain practices as Māori may obscure the fact that mainstream practices are also culturally based. In deciding what are Māori pedagogical practices, there is a need to also determine why it is valuable to label practices in this way. Therefore, Lipka et al.'s (2005) question about pedagogical practices – "whose cultural knowledge and practices are they based on?" (p. 369) – could be rewritten as "from whose cultural knowledge are decisions about cultural practices being made?" These issues are also connected to whether it is sufficient for a *kura kaupapa Māori* to have children achieve well in mathematics, or whether this achievement needs to be as a result of the implementation of Māori pedagogies.

As was the case with the discussion about pedagogical practices mentioned earlier, there is a range of views about what Māori pedagogical practices might be. It has been suggested that teaching in *kura kaupapa Māori* is different because "teaching and learning occur[s] within a Māori framework, spiritual dimensions of the learners are given important consideration, and Māori is the medium of instruction" (McMurchy-Pilkington, 2008, p. 619). Definitions of Māori pedagogies often are connected to the traditional Māori worldview (Rau, 2001).

> Writer after writer indicates that Māori pedagogy is not new, but is derived within a long and ancient history of tikanga Māori [Māori culture] and is informed by mātauranga Māori [Māori knowledge] that is sourced in thousands of years of articulation and practice. The ability and commitment to look to the past for answers to present (and future) Māori educational developments is perhaps the most critical factor to Māori educational achievement. (Pihama et al., 2004, p. 53)

Smith (1997) noted that the term *kaupapa Māori* as in *kura kaupapa Māori* is underpinned by the practice and philosophy of living a "Māori" culturally informed life. Pohatu (1996) argued that cultural underpinnings of *whenua* (land) and *whakapapa* (genealogy) are imperative to ensure cultural transmission and acquisition. For Te Koutu, this was most obvious in that it takes its name from the geographical location of the school, and the children and teachers are predominately from the same *hapū* (sub-tribe) (see Chapter 3). This reinforces Lipka et al.'s (2005) opinion that it is important that "historically silenced knowledge of indigenous peoples such as the Yup'ik is privileged alongside traditional academic discourses" (p. 369). However, the need to emphasise the legitimacy of Māori pedagogy can lead to a romanticising of the situation so that the reality of what actually happens in classrooms is not recognised for what it is. There has been much theorising of Māori pedagogical practices but little evidence from research in classrooms themselves.

For some, the fact that mathematics is being taught in *te reo Māori* to Māori children by Māori teachers means that it is being done using Māori pedagogical practices. For example, Christensen (2003) felt that Poutama Tau was different to the mainstream Numeracy Project because it was set within Māori-medium education:

> Te Poutama Tau, like the mainstream Numeracy Project, is responsive to the Ministry of Education's strategy for improving levels of literacy and numeracy in New Zealand schools (see Ministry of Education, 1999). Unlike the Numeracy Project however, Te Poutama Tau is also firmly located within the overall context of Māori development, which includes the maintenance and revitalisation of the Māori language. (p. 9)

However, the Poutama Tau project adopted many of the Numeracy Project pedagogies. The attraction of the Poutama Tau project to many Māori-medium schools and teachers was the focus on developing students' "oracy" as a means to developing mental strategies to solve mathematical problems. Te Koutu's commitment to the revitalisation of *te reo Māori* meant that because Poutama Tau provided activities and resources in *te reo Māori*, it was adopted by teachers. Programmes that supported students to become good speakers and writers of *te reo Māori* were considered valuable in fulfilling this primary aim of *kura kaupapa Māori*. Yet there remains a question about whether encouraging language use is a sufficient requirement for a practice to be included as Māori pedagogy.

It may be that the long-term dominance of teaching and learning in formal school systems has limited Māori educators' perceptions of what is possible.

> Power shapes people's perceptions, preferences, and thinking such that they accept their role in the existing order of things. In this dimension conflict is unlikely to arise because people think there is no alternative, or that their state is natural and unchangeable, or divinely ordained and beneficial (Freire, 1972). (McMurchy-Pilkington, 2008, p. 625)

Even when there is an acceptance that Māori-medium education has been influenced by Western ideas about learning, the emphasis remains with the contribution made by Māori knowledge and culture.

> Te Kōhanga Reo [Māori-medium early childhood centres] and Kura Kaupapa Māori [Māori-medium primary education] provide opportunity for Māori parents, working within national curriculum guidelines, to change the rules that have previously excluded Māori language and culture from recognition as cultural and linguistic capital in schools, and beyond. Māori knowledge and language competencies thus come to frame, but do not exclude, those of the dominant Pākehā [European] group, and they are themselves the subject of negotiation and change. The stated outcomes of Kura Kaupapa Māori clearly highlight this process of mutual accommodation, with their emphasis on bilingualism and biculturalism. (May, 2004, p. 34)

Rau (2001) stated that "[c]ontemporary definitions of Māori pedagogy are being shaped through efforts to successfully blend traditional Māori views of learning and teaching with modern principles and practices evolved directly from those valued by the colonising, hegemonic culture in this country" (p. 2). Combining practices to form modern Māori pedagogical practices is positioned as being led by Māori understandings of the world. This, thus, continues to ensure that what occurs in Māori-medium education is perceived to be different to what is provided in English-medium education.

Hemara (2000) completed a literature review on Māori pedagogy by examining records about traditional teaching, learning, and child-rearing practices as well as education practices since the institution of formal, Western education. He identified a number of themes from this review. Although some of these themes, such as hooking new learning to the familiar, suggest a general theory of children's learning, other themes such as streaming and gender-specific learning indicate that not all children were expected to learn the same things in the same way. Traditionally, all children would not have had access to the same education. In the twenty-first century, these ideas are unlikely to receive credence.

At Te Koutu, there has been ongoing concern that a rush to use particular pedagogical approaches because they were more "Māori" might be detrimental to the students whose backgrounds they were supposed to be connecting to. The following extract comes from a meeting in 1999 when this issue was raised:

> Uenuku: You know, there's been a lot of talk that Māori might be more, who are the ones who are analytical and linear, is that left-brain? And so the talk is that the right brain, Māori are generally more right brained but our kids, I just can't see that, I see quite, you know, I see analytical linear thinkers. I see cluttered thinkers, I see lateral thinkers.

Senior Primary Teacher: There really is a real mixture.
Uenuku: They are all very different from each other.

At Te Koutu, Māori pedagogy is considered to be those practices used in Māori-medium education that fulfil the joint aims of *kura kaupapa Māori* of revitalising the language and culture and supporting students to fulfil their potential, including achieving academically. Therefore, the staff can draw upon traditional understandings of Māori teaching and learning as well as understandings from Western education about how to teach mathematics education. Like the effective pedagogies described by Walshaw and Anthony (2008), the practices that are seen in any particular classroom or school will be different as teachers respond to the individual needs of their students in the ways that resonate with their own understandings of pedagogy.

Te Aho Matua

Teachers at Te Koutu follow the guidelines set down in *Te Aho Matua* (TAM) (Kura Kaupapa Māori Working Group & Katarina Mataira, 1989) when planning and implementing their lessons. TAM was written only a few years after the first *kura kaupapa Māori* was established. Smith (2003b) suggested, "its power is in its ability to articulate and connect with Māori aspirations, politically, socially, economically and culturally" (p. 10). As a philosophical base for curriculum planning, TAM sets out the principles to ensure students receive a well-rounded education. These points contribute to centring the student's education around Māori language and culture, in a way that supports students achieving at the highest level. The focus is on developing the whole child rather than just his or her career options.

TAM has six parts, with each one focusing on different aspects that are deemed as essential for educating children (Kura Kaupapa Māori Working Group & Katarina Mataira, 1989). These parts are:

- *Te Ira Tangata* – the nature of humankind
- *Te Reo* – language for communication
- *Ngā Iwi* – the communities who socialise the child
- *Te Ao* – the world
- *Āhuatanga Ako* – principles of teaching practice
- *Te Tino Uaratanga* – the outcomes from the education process.

Written in *te reo Māori*, this set of principles has been interpreted and applied in a range of ways. The following summary is one interpretation, and many more details are available in the document itself.

Te Ira Tangata focuses on the physical and spiritual properties of children and the need to nurture these. To do this educators and parents work together to develop a holistic education that will lead to children respecting themselves and other people.

Te Reo is concerned with the mastery of *te reo Māori* and also English. The *kura* should make its own decision about when to introduce English, with *te reo Māori* being the language of instruction for all subjects. The two languages should be separated, and code switching between them should be avoided.

The third part, *Ngā Iwi*, is about the wider community's role in educating the children. This includes the students' families but also anyone else who is associated with the *kura*. There is also a call to respect and nurture the various tribal affiliations of the students and their families. As well, the provision of teacher education to meet the needs of the *kura* is discussed.

Te Ao considers knowledge of the wider world that surrounds the child and how it informs the child's learning. Children's learning should be based on a Māori understanding of this world.

Āhuatanga Ako describes the types of learning practices that will contribute to the child-centred, holistic education that revolves around the Māori world. It also acknowledges the role of the national curriculum in the provision of an appropriate education.

The final part, *Te Tino Uaratanga*, identifies the outcomes that the children need to gain so that they become fully functioning human beings. These outcomes recognise the need to foster the children's individual talents and interests to ensure that they achieve at the highest level.

Pedagogical Practices at Te Koutu

In the following sections, we provide information about the philosophies informing the beliefs of parents or teachers about why a particular approach is important and then elaborate with an example of a matching teaching practice. The information comes from interviews, meetings, and lesson transcripts. We do not present them as a definitive description of all the pedagogical practices at the school. Instead, we have looked for information in our data sets about practices that seem different to those discussed in mainstream education. As can be seen from examples of teachers' practices in earlier chapters, many pedagogical practices were similar to those used to teach mathematics in mainstream schools. However, in this chapter, we have looked specifically for practices that have the potential for being described as Māori. We have examples for each of the specific aspects of TAM, but many of the examples illustrate a range of aspects. The separation under the following headings is more for the convenience in discussing the aspects than because there is a real distinction.

Te Ira Tangata

Nurturing is not often a term used to describe educational practices in schools. Yet for some teachers at Te Koutu, it was a strong focus in forming the relationship that they had with their students. In a set of interviews with a teacher who had started halfway through 2006 with Year 0 class and then kept them as a Year 1 class in 2007 and as a Year 2 class in 2008, there was much discussion about how Māori values, such as nurturing children, impacted on her teaching.

Tamsin: How in your own teaching, how do you work with the little ones? What's in your mind when you are preparing lessons and when you are implementing lessons?

Y2 Teacher: I think about, this is where I'm at with them. Where do I need to be taking these children? Because Māori children are brought up to be nurtured and loved and that's a conditioning in *Te Aho Matua*. It's with our Māori people as well and those *tikanga*, those customs, have come down from ages ago and when I am planning and when I am having to do work for my kids, I've got that *whakaaro* [notion] in me as well. And I am thinking, well I've got this child in my care where do I need to be taking this child as a mother, as a grandmother, as a Māori woman, and as a Māori teacher? What will this child look like for me when she reaches their time, because they will be going on? And what do I need to be doing for this child, what can I put into this child in order for this child to be a huge success in life? So I think about my culture. I think about our language. I think about my people. I think about my friends and colleagues and those are all part and parcel of the teaching and nurturing of these children. Because what I'm thinking about is all those good role models, Māori role models that are out there and what I want to transmit through to these *tamariki* [children] for when they move on and grow. Yeah, so it's trying to bring all those things into them. (Interview, September, 2008)

She described how she put this philosophy into practice in relationship to two boys. Although the pedagogical practices that she employed were reactive to the boys' behaviour, her approach was part of a general philosophy of teaching. As has been described in earlier chapters, most children were second-language learners of *te reo Māori*, and so the impact of their fluency on their learning was an issue that the teachers had to deal with:

Y2 Teacher: Yeah, they are quite boisterous, the two young boys, and it's been difficult getting them on task and getting them to learn, really difficult. They are typical boys and that's something I've got to think about in our classroom structure. What do I do, how do I deal with them as a Māori woman and teacher, how do I deal with *tamariki* [children] like that?

Tamsin: Mmm

Y2 Teacher: Because you've got one, the boy that's the most boisterous of all, doesn't have Māori in his home, doesn't have the *reo* in his home.

...

and consequently that particular child, he sets the whole tone in that classroom because he doesn't have *reo* at home and while his *reo* has come along really, really well, he is not as articulate as the others.

Tamsin: Yeah

Y2 Teacher: So he creates all this...

Tamsin: Because he can't get his point across?

Y2 Teacher: Yeah...

Tamsin: So this is a way of getting attention ...

Y2 Teacher: Which is what he is doing all the time. Yeah, so I've got to find a way of fixing that. (Interview, September, 2008)

In the extract, this teacher acknowledged that the child's lack of *te reo Māori* had resulted in him becoming boisterous and disruptive in mathematics and other lessons. Although she had not determined a way forward, she accepted that a broader approach was needed than just improving his language skills. It involved her deeply considering her relationship to the child, so that she would work with him rather than trying to change him to fit her expectations.

Nurturing of students in *kura kaupapa Māori* is a culturally appropriate practice. According to Salmond (1997), early ethnographic records note that Māori were traditionally over-indulgent of their children. This view is affirmed by Pere (1997) in her definition of the word *tamariki* (children) when she stated, "children are the greatest legacy the world community has" (p. 4).

Two principles which underpin the nurturing of children are *mana* and *tapu*. All children and teachers have *mana*. Pere (1982) defines *mana* as "psychic influence, control, prestige, power and associated beliefs" (p. 32). *Mana* is "multi-form" and can be ascribed and acquired (Pihama et al., 2004). In terms of teaching practise and pedagogy this means that children's *mana* (and the teachers') must be respected and not trampled on. For example, if someone (teacher or child) is standing to speak, only that person speaks. This encourages other children's attention and the speaker to experiment with ideas without fear of ridicule or indifference. Figure 11.1, from 2005, shows a student providing an explanation at the board, while Uenuku, the teacher has taken a seat in the class.

Associated with *mana* is the concept of *tapu*. *Tapu* functions at many levels in the school and acts as a corrective and protective principle. Figure 11.2 shows students in a Year 1 class using their bodies to form a square.

In Māori traditions, heads are *tapu* and cannot be placed near feet. Although only five or six years old, these students arranged themselves so that in forming a square their heads were together and their feet were together, thus showing an understanding of *tapu*.

Fig. 11.1 Year 11/12 student providing an explanation at the board

Fig. 11.2 Year 1 students using their bodies to make a square

Te Reo

Te Koutu was founded on the principle of revitalising *te reo Māori* but celebrated being multilingual, not just bilingual. English, the dominant language in mainstream New Zealand, is not taught formally until students begin Year 7. However, Spanish is introduced as an additional language when children start at the *kura* at the age of five. Spanish was chosen as an additional language because it does not threaten *te*

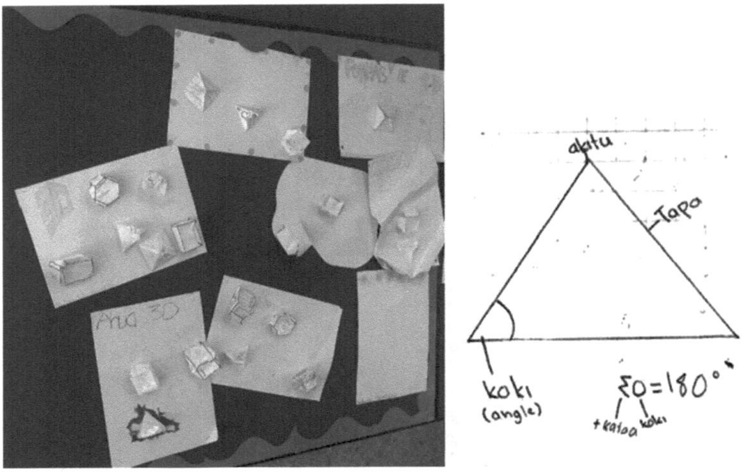

Fig. 11.3 Student work from Year 7, 2007

reo Māori, in the way that English does, and there are many sound systems in common in the two languages. Although TAM recommends that code switching should not occur, in some mathematics lessons, on rare occasions both English and Spanish have been used to clarify ideas presented in *te reo Māori*. Figure 11.3 shows work by Year 7 students in 2007. Some of the posters about three dimensional shapes give the title in *te reo Māori*, *Ahua 3-D*, and some in Spanish, *Formas de 3D*. It also shows a copy made by a student of blackboard notes about the features of a triangle. *Koki* has the English translation, "angle", underneath. The more general situation was for every lesson to be presented only in *te reo Māori*. It was believed that students who were taught in a full immersion situation would have stronger language understandings than if they were given initial insights from their other languages, especially English which for most students was their primary language.

Much of what has been written in this book has been about students' command of the mathematics register in *te reo Māori*. The views of the teachers about the necessity of being able to fluently use the mathematics register is summarised in the following comment:

> I do think it's of huge importance that kids are able to communicate efficiently in whatever context they are dealing with but in maths, yes, it's hugely important. Out in that big wide world there's a different form of language for every situation you find yourself in. (Year2 Teacher, September 2008)

As has been discussed in previous chapters, the teachers consciously tried to improve students' fluency in the mathematics register. Students also recognised that their fluency levels, in that they had very few people other than their classmates to discuss mathematics with, did have an impact on their learning. Figure 11.4 shows a graph from a school-wide student survey completed in 2007 in *te reo Māori*.

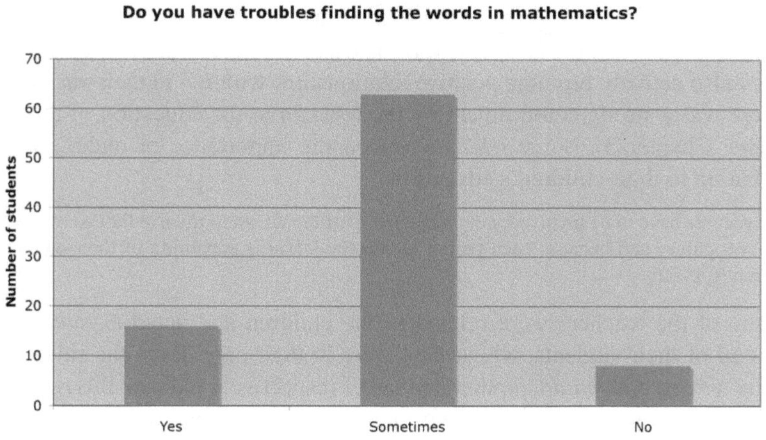

Fig. 11.4 Students' responses to survey question about knowledge of mathematics vocabulary

It shows that many students sometimes struggled with being able to describe their mathematical ideas in *te reo Māori*.

Having fluency in the mathematics register was a concern for the whole school community. Parents had chosen to send their children to Te Koutu, and they expected that their children would learn mathematics through *te reo Māori*. The students were aware of the effect of their fluency on their mathematics learning. This focus affected teaching practices and students' learning of mathematics. The same amount of emphasis and awareness about fluency levels is unlikely to be seen in most mainstream classrooms, especially high school classes, where language often is not seen as something that needs to be focused on.

Concern about fluency in a language is a concern for the cultural traditions of that language. However, acting upon this concern by implementing pedagogical practices does not necessarily make those practices culturally appropriate. On one hand, the use of other languages, such as Spanish, within mathematics lessons is not supported by *Te Aho Matua*. On the other hand, the support by the whole school community for making the students multilingual could contribute to making the students comfortable in their participation in school, including in their mathematics lessons.

Ngā Iwi

Over 50 percent of children at Te Koutu have strong *whakapapa* (genealogy) connections to Ngati Whakaue, one of the *hapu* (sub-tribes) of Te Arawa. There has been migration into the area from other tribal areas (particularly rural) for work and educational opportunities. However, the *kura* recognises and promotes all children's connection to their *iwi* (tribe). The community, including parents, children, and caregivers, are included in the *whānau o* Te Koutu (extended family of Te Koutu).

A number of staff are actively involved in tribal affairs and work hard to build relationships between the *kura* and various tribal entities. A number of parents and *whānau* also actively promote positive relationships with *iwi* in their various roles.

There was a strong commitment by the *whānau* to the education of all the children (see Chapter 3). Some *whānau* voiced the importance of making an active contribution to their children's education:

> The parents have to be involved, not so much to participate but to involve themselves within the curriculum and because each parent knows the learning capability of their own child. (Parent9, 1999)

Many of the teachers were related to the children and acted as surrogate parents to all of their students, whilst they were in their care. Even the titles given to teachers, *whaea* and *matua*, mother and father respectively indicate this relationship. If a student was unhappy at Te Koutu for whatever reasons, then the teachers discussed with the student and the caregivers what alternative school was most likely to support the student to do well.

In the discussions in 1998 and 1999, the parents felt that Māori students would only do the academic work if they had a good relationship with their teachers.

> You know, if there is a thing around that the kids don't like it's the teacher and therefore I won't learn because I've got this teacher and you are trying to say you know, it's science you don't have a problem with, science is actually quite important. (Parent3, 1998)

When students and teachers viewed each other as part of the same family, the issue of liking or disliking between teachers and students is not a distinction that can be made. Māori families put effort in ensuring that relationships are long term and beneficial to all parties.

Research in mainstream schools has shown that parental support for their children's education results in improvements in academic results (Anthony & Walshaw, 2007). However, this usually implies that caregivers must support what is done in the school and the school does not need to learn about what occurs in the home. Yet as Brenner (1998) stated:

> there is substantial evidence that the participant structure in a traditional classroom, that is, the roles and responsibilities assigned to the different persons can act as an inhibiting factor to children who come from a culture that stresses different participant structures from those found at school. (p. 215)

Adults at both home and school worked with students to provide an appropriate education and it may well be that this is a different scenario than what Māori students experience in mainstream education. The involvement of *whānau* is an integral part of *kura kaupapa Māori*, as it is in *kōhanga reo* (Pihama et al., 2004).

Te Ao

TAM advocates that children should learn about the world from a Māori basis. Caregivers and teachers also recognised this as something that they wanted for the children.

Tamsin: What would you see as a role for a curriculum in a school? What sort of thing do you think you'll be developing at this point?

Parent2: Realistic applications of things our kids need to learn as opposed to things that are probably irrelevant to our children's lifestyles ... One good example I have is of where my daughter went to a country school in the middle of nowhere and where they have normal electives like metal work, wood work and things like that, her electives were making an *hīnaki* [eel trap] which is something that you catch eels with and also cutting up a cow, killing a cow and cutting it up. That's what was useful for them in their area. Because that was the lifestyle that they lived in at the time and plus they had no metal work facilities anyway. So things that are useful for our children. (Interview, August, 1998)

Y2 Teacher: Right from when our kids are born, there's always a maths world going on around them. We go to the *marae* [traditional meeting area], there's their mothers and fathers in the back there, doing all the menial tasks of getting *kai* [food] ready. There's measuring, cutting up – are all maths type things in their everyday life. I suppose so they are exposed to it right from birth and throughout until they actually get to school, where it becomes a focus and they realise "oh, that's maths" and so I think they have this idea of maths happening in their world in any case but I think nowadays there's this push to ensure that maths is more than just doing sums and it's a way of life. (Interview, September, 2008)

This belief in making connection wherever possible to cultural traditions and children's own lives resulted in geometry activities in which children described how shapes were transformed from car symbols to traditional Māori designs. In Fig. 11.5, a Year 4 child has completed a worksheet about how rotation, reflection, and translation are used in various traditional designs as well as in drawing their own design using two types of transformation. As the child has used only one transformation (reflection), the teacher has provided feedback on how translation also could be used.

One activity that Uenuku Fairhall has used successfully with various groups of students and adults is that of land division (see Meaney, Fairhall, & Trinick, 2008). Traditionally, land belonged to extended family groups. With the advent of the Māori land court, a person's share in the land depended upon the number of generations since the original title was recognised. Generally family members are unclear of how they came to have their proportion of land. In 2009, Uenuku worked with his Year 9 class to have each person represent his or her family tree and then connect it to the proportion of land that he or she would gain. The students produced a blog in *te reo Māori* on their process which at publication was still accessible at: http://www.tekoutu.com/2009/10/matua-tau-9-year-9-parents.html. This blog provided information to parents and other school community members but was also

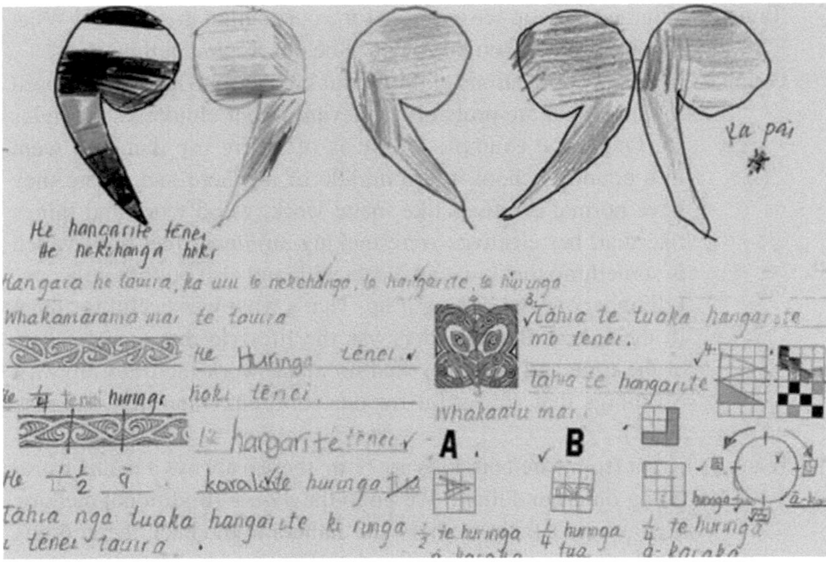

Fig. 11.5 Year 4 child's work on transformation

accessed by other people across the world, which was documented by a map attached to the site. The use of technology, such as blogs, has been noted as one way that successful home–school relationships can be maintained (Bull, Brooking, & Campbell, 2008).

Figure 11.6 shows two posters that illustrate how students were required to draw their family trees and then use a hypothetical 1 ha as the original land size to work out what their own proportion would be. The complexity of their calculations depended upon the individual family. The first figure shows a reasonably straightforward family but still requires the student to be able to work to four decimal places. The second poster shows a much larger family and consequently much smaller proportions.

Students were asked to decorate their posters with photos that represented their artistic understanding of land division. As can be seen in the two examples in Fig. 11.6, students did this in different ways but were quite clear in their reasoning for what they had done. For example, in the bottom left corner of the paper, strips of different photos have been integrated together like a piece of weaving to represent that although one piece of land was divided into many family shares as the student was a descendant of many different ancestors, they were connected to a number of different pieces of land.

Use of traditional Māori understandings could provide a basis for developing students' mathematical understanding. However, there was awareness that mathematics could not colonise the cultural practice by making the mathematical understanding seem more important. Instead, there was a need for a two-way adding of value so that mathematics could give extra meaning to the traditional practice, and the traditional practice would also give extra meaning to mathematics (Meaney & Fairhall, 2003).

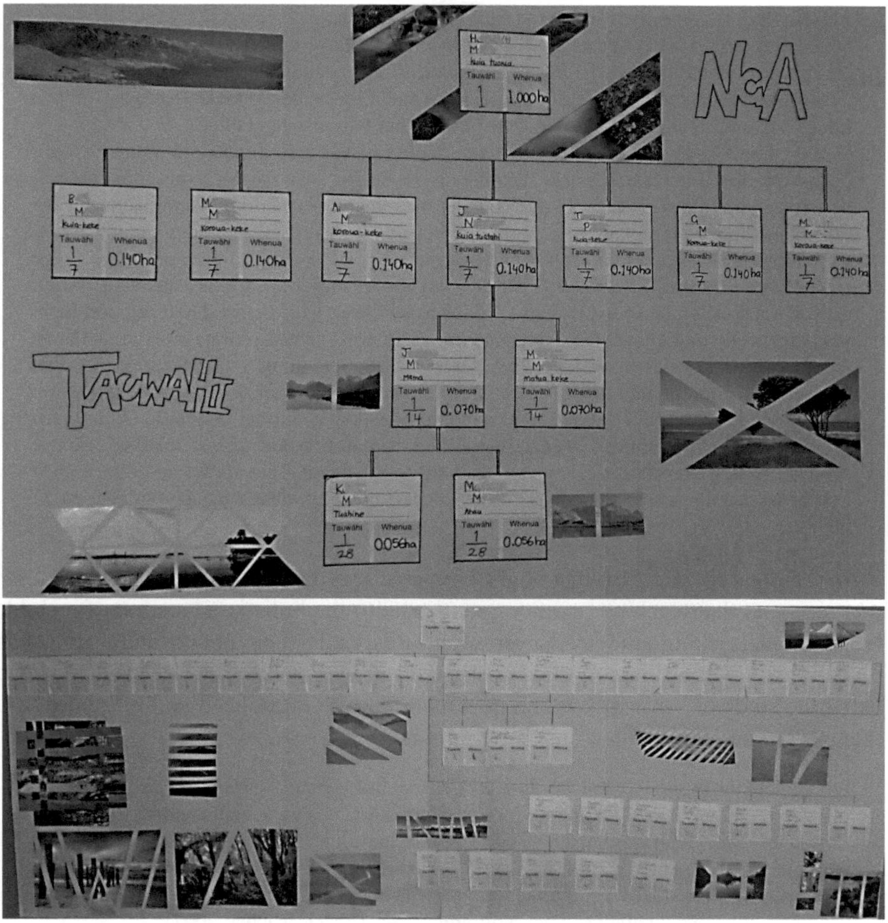

Fig. 11.6 Two Year 9 students' family trees showing their proportion of land entitlement

Consequently, mathematics learning that is based on traditional practices is clearly a culturally appropriate approach for teachers to adopt. It supports students' comfortable participation in the classroom by starting from culturally known content but also assists them to move forward in their understanding of the wider world of mathematics.

Āhuatanga Ako

Āhuatanga Ako concerns learning while at a *kura kaupapa Māori* and beyond. Te Koutu's teachers and *whānau* (extended family) are concerned about the future lives and career pathways of students. This can be seen in the following extract from an interview with a parent in 2008.

> My daughter is currently wanting to be a forensic scientist . . .
>
> For her, that's what she is aiming for and, as we have already explained to her, the requirement would be that she has to have a clear understanding of maths, that she has to have a clear understanding of different sciences and also she has to look at things in quite a linear way and so we spend a lot of time pushing her to understand things . . .
>
> For the likes of my daughter, she's ten, it was just simple things and although she's been explained it in class, it's also showing her all the lines are the ones, tens, hundred and how to add up things, so that the one is under the other . . . or multiply them one under the other and although it took us a lot of arguing in that eventually she figured out that what we were doing was also correct. It took an hour and a half of arguing and her statement to her teacher "Dad thinks he knows everything", but her teacher looked at her and said, "Well, do you know what, he possibly does". So the aspect from us is being able to support them mathematically and although I don't state to be any great professor in regards to maths, in fact I am not that great, at the stage they are at, if they are willing to take the help then that gives them a good ground to ask for help. So for us, the school allows that as well, because we will send her back if we don't understand the question or she can't translate [from *te reo Māori* to English] it in a way we can understand the question and what is required, we just send her back to the class and say you just have to ask again. That allows her to be able to ask the question again so she gets a clearer understanding of what the maths subject might involve.

In previous chapters, a range of teaching practices has been described. *Āhuatanga Ako* suggests that teaching practices contribute to the holistic education of a child. For example, asking students to represent artistically their understanding of land division would contribute to this holistic education. For teachers, it was also about using the students' strengths when new material was introduced and to ensure that they had ownership of what they were doing.

> I think with the group I've got, because they're real independent readers and writers, I think I can start really zooming in on the story problems. And we are actually learning to add on from five, so maybe if I use that concept. Because they don't like to write either. Little kids don't, not a whole heap of stuff, anyway. They do if they've got their own ownership of it. If it has come from them, you know. If you are trying to put the ideas into their heads and they think it's yours then it becomes a different story. (Year1 Teacher, 2007)

As was noted earlier, there was concern that some traditional teaching practices may not be appropriate for children growing up in the twenty-first century, as the societal demands on children are different than when traditional teaching/learning practices had been employed. For example, it was believed that in traditional Māori society, certain knowledge was rarely questioned; however, in modern society children were expected to query knowledge.

> Regardless of the cultural practices in the past, I see huge advantages in the questioning child . . . in a closed society, what you were doing was replicating and there was huge value in replicating things and so there was also a hidden value on innovation and so Māui was the cultural hero. He broke the rules and they admired him for that but they also would not approve of that at a personal level, in someone around them, the "Māuis" were always told off. So that seems to me to reflect that unquestioning, what you ought to do is to observe and learn and take in. It's like an old transmission model, everything is going to get transmitted but the modern world is not like that, the society is no longer closed and the whole way that the global society is, you push the boundaries in all sorts of ways and even questioning some of the replicated things which may even lead to increased pollution and use of resources,

> so we have to make a decision on whether we're going to override that cultural norm of non-questioning. (Uenuku Fairhall, Meeting, 1999)

Parents saw the need for their children to be able to ask questions about what they were learning because it contributed to them being confident in themselves.

> But the way the teachers teach here gives them [the students] enough scope to ask questions without feeling stupid and that adds to the strength they have inside because being able to go out there as global people is about an inner strength and an inner confidence (Parent3, September, 2008)

There was some discussion about how praise, as a teaching practice, was traditionally given to children. This was compared with Western educational beliefs about praising effort and ability. There seemed to be an acceptance in this school community to use Western education beliefs about praise.

> Uenuku: Ah, that's a question that has to do with cultural things because I've heard that in some cultures, the absence of praise means that there is praise. Whereas it seems that sometimes to me that a lack of praise, not saying anything is almost equal to disapproval, because especially in modern educational theory, you're trying to be so positive that it's hard to be negative so you don't do anything so the children soon learns that
>
> Parent6: Oh, okay it's like a disapproval (Meeting, August 1999).

Assessment was used not only to inform teachers about students' understanding of material but also to share student progress with parents.

The example, from 2006, in Fig. 11.7, comes from an assessment booklet. These booklets were sent home to caregivers of primary children at the end of every term two and four of the school year. Although there are ticks alongside each answer to show the child what questions they had correct, there is also information for the parents about what the work shows. As noted at the top of the sheet, parents can see across the year how their child had improved and how this related to what was expected of students at their level.

The use of these assessment booklets, although extremely time-consuming for teachers to produce, provide rich information to parents as well as to the children themselves about what they can do. They also showed to all participants in the learning how "[n]ew knowledge, skills and activities were related to preceding and following lessons" (Hemara, 2000, p. 10). Praise that is given to students then becomes something directly related to academic progress rather than just given for general performance.

Students were also actively encouraged to make their own assessments of the quality of their work. The following extract is from an interview with a Year 6 teacher about the strategies he used to improve students' written explanations.

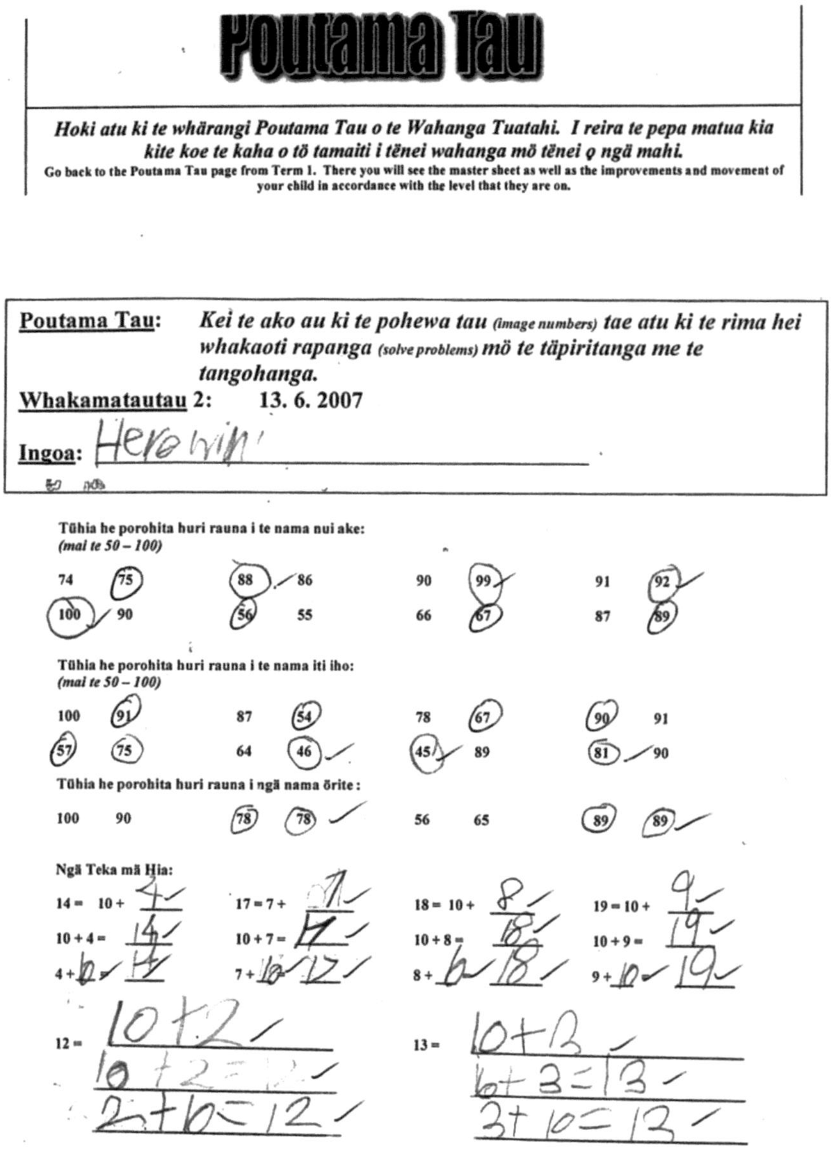
Poutama Tau

Hoki atu ki te whärangi Poutama Tau o te Wahanga Tuatahi. I reira te pepa matua kia kite koe te kaha o tö tamaiti i tënei wahanga mö tënei ǫ ngä mahi.

Go back to the Poutama Tau page from Term 1. There you will see the master sheet as well as the improvements and movement of your child in accordance with the level that they are on.

Poutama Tau: *Kei te ako au ki te pohewa tau (image numbers) tae atu ki te rima hei whakaoti rapanga (solve problems) mö te täpiritanga me te tangohanga.*

Whakamatautau 2: 13. 6. 2007

Ingoa: Herewini

Tūhia he porohita huri rauna i te nama nui ake:
(mai te 50 – 100)

74 75 88 86 90 99 91 92
100 90 56 55 66 67 87 89

Tūhia he porohita huri rauna i te nama iti iho:
(mai te 50 – 100)

100 91 87 54 78 67 90 91
57 75 64 46 45 89 81 90

Tūhia he porohita huri rauna i ngä nama örite:

100 90 78 78 56 65 89 89

Ngä Teka mä Hia:

14 = 10 + 4 17 = 7 + 10 18 = 10 + 8 19 = 10 + 9
10 + 4 = 14 10 + 7 = 17 10 + 8 = 18 10 + 9 = 19
4 + 10 = 14 7 + 10 = 17 8 + 10 = 18 9 + 10 = 19

12 = 10 + 2 13 = 10 + 3
10 + 2 = 12 10 + 3 = 13
2 + 10 = 12 3 + 10 = 13

Fig. 11.7 Year 3 student's work

What I've been producing and role modelling and then getting them to design their own. Then they read all their stuff out and one boy read his out and his answer was quite good which was surprising for him. He said he took it home and his Mum helped him out … in terms of explaining what he was trying to say. Straight away in the question, it was a good example for everyone to follow. Read his out and I made everyone look back at their own to see similarities between theirs and his. The whole of them said no there were no similarities. … Even they could see the value of it to work at their own level. (Y6 Teacher Interview, November, 2007)

Finally, students in Years 7 and above are given the chance to appraise their teachers, in relationship to the delivery of the curriculum to class management and attitudes to students. Each year, the students fill out an assessment known as the Tauwāwā that consists of 25 questions relating to teaching practice. The results inform individual teachers of their perceived strengths and weaknesses, which in turn inform their personal development plans.

Teaching practices illustrate teachers' beliefs about how children learn. The practices of the teachers at Te Koutu focused on providing students with a holistic education. This holistic education was not simply about connecting students' learning across subjects, but also about informing all participants in the education process about the learning that was being undertaken. These practices are also culturally appropriate. Students are expected to be active participants in the learning, and this is reinforced through informing and involving the rest of the family. By actively involving the children in the learning process, there is a need to connect new knowledge to what children bring with them to the learning situation. If this is not done, it is difficult for students to be active learners as the alternative is to passively receive information provided by a teacher that may or may not have anything to do with what they already know.

Te Tino Uaratanga

Te Tino Uaratanga, a distillation of all the other sections, highlights the knowledge and skills that students need in order to become whole human beings. Parents were especially clear about what they wanted for their children in order for them to become successful adults. The following extract comes from a parent who had had three children at Te Koutu for more than ten years. When she was interviewed in 2008, two of her children had gone out into the workforce. She, therefore, was able to reflect on what they took with them from their schooling.

> I think maths is a part of a whole philosophy, because of Uenuku and the way he teaches. My oldest son, he was a maths whizz so the way Uenuku taught just extended him. Every day, he was excited to come to school. Uenuku's classes, no matter what they were, he was the first one there because he wanted to be pushed to the limit, whether it is maths or Spanish and whatever he was teaching because that's the way he does it. Whereas the second one who's not quite so bright enjoyed [another teacher]'s method, which was more laid back and a bit slower. But maths here has been taught well for me as a parent. It has been taught at different levels to suit the kids. I don't know how it adds to their global view, I know when [another teacher] was teaching [second son], he went from hating maths, hating school, hating coming, to loving it, to loving school, to loving coming. [Another teacher] has a way of making them feel valued and I like the two polar views of teaching maths here. It's not one size fits all and I honestly don't know how it fits into my global view but it fits into a system of making the kids stronger as people, but every subject does that. I believe that every subject, they are taught, teaches them intellectually what they want to know

Individual needs are catered for at Te Koutu. In this case it was through the provision of different teachers. However, Te Koutu is a small school with a limited number of teachers of mathematics so there is always a need to look for different ways to individualise the instruction, which does not require a different teacher.

Interviewer: Koina tāu mahi tātai i whirihia?	Interviewer: Is that the maths you chose?
Student: Ae. I whiri e au i ngā koki tapatoru. He rawe te mahi nā te mea he momo whakaaro – me whakaaro koe kia tae atu ki tētahi whakautu. Arā he rawe aua mahi ki ahau. Ko te tino mea – me kii – ngaro nei, ko nga whakaritenga me tae atu ki tēhea pātai – tēhea whakautu – he momo huarahi maha nei ki te tae atu ki tētahi whakautu. Ara, he rawe.	Student: Yes. I chose the angles of a triangle. The work is fantastic as it's kind of like thinking – you need to really think to get the solution. You know, I really like that sort of work. The main thing, really, the hidden bits are the steps needed to achieve the solution – whichever answer – it's like there are many paths to arrive at the answer. It's great.
I: He aha ōu maharatanga i te wā i mahi i taua mahi – he aha tāu e mahara ana mō aua mahi?	I: What are your thoughts when you're doing the work – and what do you think about the work?
Student: Ki ōku nei whakaaro me mau i ngā tino kiko o te mahi kia tae atu ki te whakautu i nga mea māmā– me whakaaro whānui hoki engari – he pai ki te mau i ētahi mahi, nā te mea, he maha nga huarahi arā, ka taea koe te whiri i tēhea mea	S: I believe you need to keep a hold of the main points in order to solve the easier parts – you need to think broadly – [but] it's good to concentrate on some of the work, because there are several paths to follow, and the trick is to choose the best one.
I: I waenganui i te tau – ka hoki atu koe ki nga mahi kua mahi kētia? Ngā mahi i mahi koe i te Wāhanga Tuatahi, Wāhanga tuarua rānei?	I: During the year – do you go back to work you've already done? The work you did in the first or second term?
Student: Ae	S: Yes.
I: He aha ai?	I: Why?
Student: Nō te mea, ko te nuinga o nga mahi he mea ka rata i te tangata, a, ka rata au, he rawe ki te maumahara pai.	S: Because, most of the work you'll find interesting, and I like it. It's wonderful to be able to retain it.
I: Kei te mōhio koe he aha tēnei mea te tuhinga kōrero i roto i te tātai?	I: Do you know how to express your mathematics in writing?
Student: Ae, engari ki te mau pai i te tikanga, me mau pai ki tau e mahi ai, ko tōku nei whakaaro ka mau i ngā tikanga arā, ka mahara.	S: Yes, but you need to understand the principle. You need to grasp it in order to do the work. I think that if you grasp it, then you will remember it.
I: He aha nga momo tuhinga i roto i te tātai?	I: What sort of things do you write in maths?
Student: He roa nga tuhinga, kia tae atu ki tētahi whakautu me haere huri rauna kia whiwhi i tētahi "mea" kia tino kitea i tau e whakarite ana nā tērā ka mōhio te tangata maaka, nō hea taua pātai.	S: There's a lot of writing. In order to arrive at the answer, you need to go all over the place to find what you used to solve [the problem] whereby the marker can see how the question was worked out.
I: He aha nga momo tuhinga e pai ana ki a koe?	I: What types of writing do you like?
Student: Ngā mea roa	S: The lengthy ones.
I: He aha ai?	I: Why?
Student: Nō te mea, ka tino whānui ōku whakaaro	S: Because I have a lot to think about.
I: He aha nga momo tuhinga kāre e pai ana ki a koe?	I: What sort of writing don't you like?
Student: Ngā mea paku	S: The short ones.
I: He aha ai?	I: Why [not]?
Student: Nō te mea, he māmā ki te mahi (jumps) ngā momo wero	S: Because it's too easy. [I like] a challenge.

Students also saw that mathematics education was not just about getting correct answers but about improving their thinking and providing them with skills for when they became an adult. The following interview is with a Year 10 student from 2007, and was primarily about writing in mathematics.

I: Ki ōu whakaaro, he aha te take me tuhi i roto i te marau tātai?	I: Why do you think you should write in the maths class?
Student: Nā te mea ko te tātai tētahi o nga mea matua i te wā e pakeke ai koe ko tērā tētahi o nga...	S: Because maths is one of the main things you need to know for your adult life, it's one of the ...
I: Engari he aha ai – me tuhi....	I: But why do you need to write...
Student: Kia tautoko i a koe e whai ake nei.	S: To help you later on.
I: Ka tautoko i te aha?	I: To help with what?
Student: Ki tāu e hiahia ai	S: Whatever you want to do.

For this student, learning was about thinking hard about what you had done. As a student, she could choose which solution path to take, but this meant that she had to extend herself by learning the different ways. She also saw that writing longer pieces in mathematics was useful in helping her to improve her writing. Her overall understanding about learning seemed to be that the teacher provided her with opportunities, but it was up to her to make use of these opportunities. Skills such as writing in mathematics would help her in her life as an adult, and so it was up to her to learn them.

Setting goals for students was a shared endeavour by the whole *whānau* or school community. For each child to reach his or her potential, it was not just the teachers who needed to set goals, but parents and caregivers had to be in agreement with these goals. Students also needed to accept these goals and to work with the adults to achieve them. This whole school effort made these practices culturally appropriate. To be a fully functioning adult in the modern world was not just about knowing about traditional Māori practices, but also learning how to be a Māori, global citizen who could face any challenge with confidence knowing that he or she has the skills to work through any issue. As the parent stated in an earlier chapter, "Māori were traditionally explorers". The pedagogies used at Te Koutu were allowing them to become explorers of the modern world.

Meeting the Challenge of Working Within Māori Pedagogy

Consideration of what Māori pedagogical practices are and their impact on the teaching of mathematics at Te Koutu is not simple to conceptualise in regards to meeting a challenge. The insistence by Māori theorists of the existence of Māori pedagogies and their usefulness in *kura kaupapa Māori* perhaps can be understood as a resistance to meeting the challenge of making the concept functional. Without proper documentation about what these pedagogies look like, or the underpinnings about how children learn, they become constraints rather than opportunities for educational development. The need to make Māori education appear different and to some degree superior to Western education restricts the possibilities for discussions about what these pedagogical practices could be and could become.

In this chapter we have attempted to open up the activity space around Māori pedagogical practices by discussing what they might be. To do this we have used the principles in *Te Aho Matua* as a de-facto theory of how Māori children learn. In some ways this is a circular argument because by providing a description of the principles for children's learning, TAM is describing pedagogical practices rather than a learning theory. However, it does give a starting point for discussing differences that are solely connected to the Māori background of the students. From looking at how each principle from TAM was realised in teaching strategies at Te Koutu, it is possible to consider what makes Te Koutu different to mainstream schools. Although its students also learn a Western curriculum of mathematical knowledge, the way that this is done involves the connecting of all participants in the educational enterprise through the agreement of a similar set of goals. Therefore, it can be said that at Te Koutu, the fundamental difference in pedagogical practice is the relative stability of the aims for the mathematics education of its students within a wider holistic education that regards the students as needing a strong Māori background, in order for them to go out and take their place in the wider world.

Chapter 12
"And That's What You Want to Happen. You Want the Shift in Classroom Practice"

In order for a school not to remain static, teachers themselves need to continue to be learners and make changes to their teaching practices. Certainly in the early days of Te Koutu, the parents were quite vocal about their expectations in this regard (see Chapter 3). In this chapter, we outline some of the factors that contributed to teachers at Te Koutu developing as teachers. We consider changes to teacher practices to be the outcomes of learning. Teacher learning, like student learning, is not a purely cognitive activity that is located solely within a person's head. Rather it is something that happens within the constraints and opportunities arising out of interactions with the wider society. Thus, changes to teachers' practices are not likely to occur simply because of an increase in knowledge, but because they are related to how teachers perceive themselves, as well as the opportunities they have to implement different practices. By recognising the complexity of the situation in which teacher learning occurred, we are able to better illustrate how teachers dealt with the challenges in trying to improve their practices. We do not specifically consider the relationship between the changes that teachers made to their practices and student outcomes. However, given that teachers' main reason for changing their practices was to improve students' achievement in mathematics, it is the background for the discussions in this chapter.

Teacher learning or teacher development has been the focus of much research for some time (Garet, Porter, Desimone, Birman, & Yoon, 2001). On the whole, this work has concentrated on identifying the features of professional development that were likely to contribute to improved student achievement. For example, Timperley, Wilson, Barrar, and Fung's (2007) review of professional development research collated lists of features in professional development programme that were associated with increased student achievement. Their model for ensuring effective professional development, shown in Fig. 12.1, suggests that there is a direct link between professional development and changes in teacher practices. Therefore, teachers need to consider only themselves and their students when identifying how to improve their practices. According to this model (Timperley et al., 2007), context is irrelevant both in relationship to what the students need to learn and in how the teachers design and implement learning activities. Yet Duncombe and Kathleen (2004) suggested that such a narrow model of professional development does not respect the complexity

T. Meaney et al., *Collaborating to Meet Language Challenges in Indigenous Mathematics Classrooms*, Mathematics Education Library 52,
DOI 10.1007/978-94-007-1994-1_12,

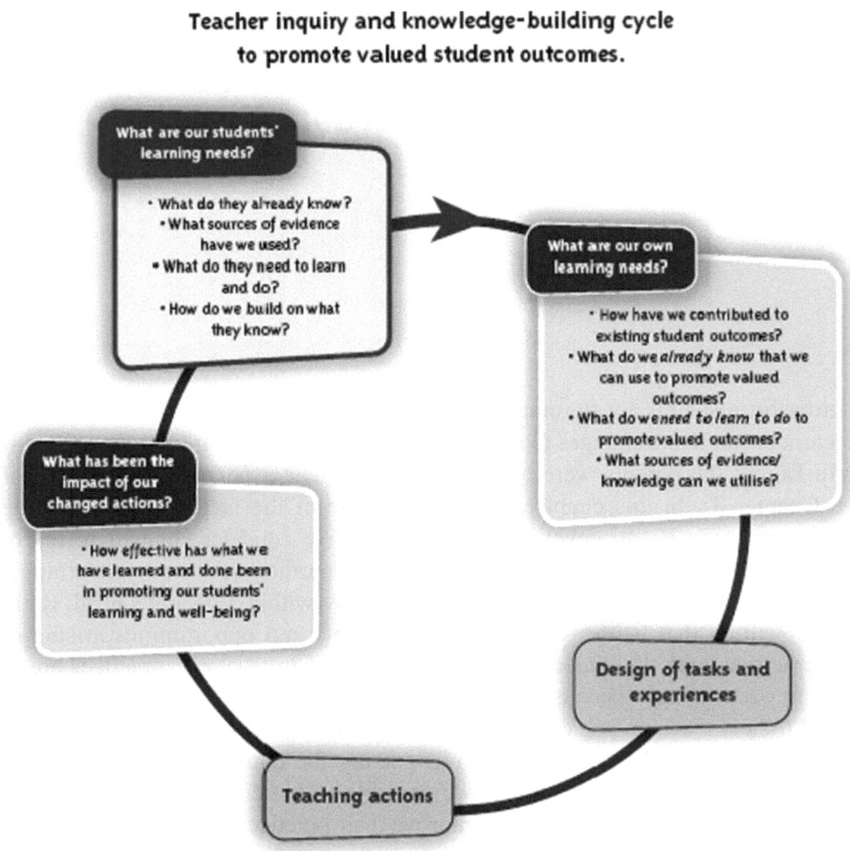

Fig. 12.1 Teacher inquiry and knowledge-building cycle from Timperley et al. (2007, p. inside front cover)

of why and how some teachers make changes to their practices. Historically, professional learning models in New Zealand/Aotearoa did not value culture (Macfarlane, 2004), particularly the complexities of Māori-medium and Māori ways of seeing the world, but rather tended to take a Eurocentric focus (McMurchy-Pilkington, Trinick, Dale, & Tuwhangai, 2009).

As outlined in earlier chapters, the situation at Te Koutu was recognised as being complex. Therefore, in order to investigate why and how teachers changed their practices we needed to find a theoretical framework, which respected this complexity.

Generally, professional development has focused on changing teachers' practices through increasing teachers' knowledge about either pedagogy or mathematical content (White, Mitchelmore, Branca, & Maxon, 2004). For example, Jaberg, Lubinski, and Yazujian (2002) felt that there was a need for teachers to construct "their own perspectives on change and [to be provided] with information from research about how students learn" (p. 3). Focusing on overcoming predetermined gaps in teacher knowledge has contributed to a lack of valuing of the context beyond the classroom. Yet, it is known that changing teachers' practices is difficult (Jaberg et al., 2002).

Trying to understand the influence of context is difficult, especially as there has been limited research about the specific concerns of teachers in Māori-immersion schools. Even a professional development programme such as *Poutama Tau* that focused on teacher practice was initially built on research concerning English-medium teacher learning and non-Māori students. This is not to suggest that Māori-medium teacher professional development cannot be informed by English-medium professional learning theory. As documented in this book, some of the challenges facing Te Koutu teachers may be the same as those facing teachers in mainstream school. However, there is a range of issues unique to *kura kaupapa Māori* which must be considered. The impact of these differences needs to be better understood if the challenges to teacher learning are to be resolved in ways that are beneficial not just to students' mathematical learning, but to teachers' beliefs about themselves.

Thus, we considered teacher change to be an outcome of teacher learning rather than an outcome of professional development. For Meaney, Lange, and Valero (2009), teachers changed their classroom practices as a result of teacher learning, rather than professional development. Their model, seen in Fig. 12.2, recognises that teachers have lives away from school, and these experiences contribute to their understandings about what occurs in their classrooms. It incorporates the ideas of Radford (2008); Skovsmose (2005a, 2005b); Alrø and Skovsmose (2002); Alrø, Skovsmose, and Valero (2009); and Kemmis and Grootenboer (2008) to illustrate how the teacher development process is affected by the teachers' interpretations and awareness of the educational context, their individual teaching situations, and the other people with whom they interact.

Drawing on ideas from each of the preceding sets of researchers, the relationship between an individual and the socio-political nature of teacher learning is highlighted at each level of the model. From Radford, at the centre of the model teachers are seen to reconceptualise who they are as a consequence of the new meaning that they acquire. The middle part of the diagram is influenced by the ideas of Skovsmose and colleagues who provided details of how the circumstances are not add-ons, but essential components of the learning process. On the outside of the diagram are Kemmis' ideas of practice architectures that suggest how teachers could be constrained by circumstances that surround them. By combining all of these ideas, it is possible to more fully understand the complexity of teacher learning.

The three tiers of the model represent the nested set of intersecting components. Our focus on the complexity of the situation is driven by "relational ontology that seeks to explore connections between all elements of a system, in contrast to an atomistic ontology in which 'objects' are defined discreetly, and in non-relational ways" (Wheelahan, 2007, p. 189). At the subjectification/objectification level, the model considers how teachers connect their learning to their perceptions of themselves or their identities. At the meso level, the focus is on how an individual teacher might perceive the context in which they learn, including their own background

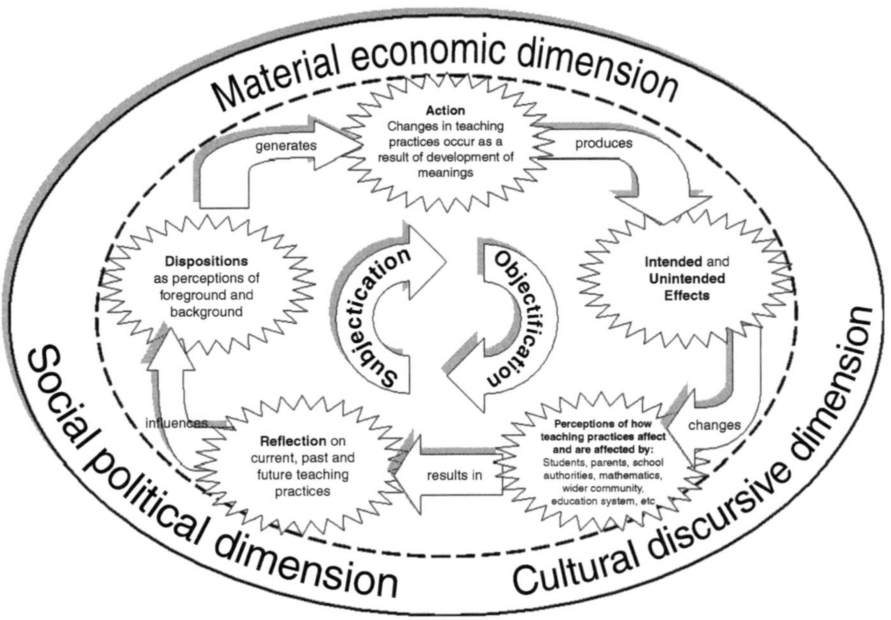

Fig. 12.2 The mediation of teacher learning

and opportunities for their futures. The macro level consists of the wider societal influences on what teachers perceive as opportunities or constraints with regard to their learning. Each layer deals with how individual teachers are connected to society whilst engaging in learning, but emphasises this relationship in different ways. The layers and the connections between them are contested spaces, which enable different practices to become possibilities that teachers can adopt. In this chapter, we explore teachers' descriptions of their participation in the projects at Te Koutu to show how the three-tiered model contributed to their learning and subsequent changes in their teaching practices.

The Teachers' Experiences of Learning

For this chapter we predominantly used teachers' comments from interviews, meetings, and surveys carried out between 2005 and 2007. During the different projects, teachers watched their own videoed lessons and then discussed with a researcher what they saw in these lessons. Regular meetings provided opportunities for teachers to share what they had done and their reflections about them. Teachers were not "taught" new activities or practices but rather became aware of possibilities by watching their own teaching and by talking with others about what they wanted to achieve. On rare occasions, specific resources were introduced into the discussions by the researchers.

Table 12.1 New teaching practices in 2007

Teacher and year level	What were the new practices that were tried in 2007?	Why were these practices tried?	How do you know if they were effective?
Year 1 Teacher	Explained and clarified what was expected of them in more simple and easier-to-understand language.	Children weren't really getting the gist of what was required from them.	Children gave and showed clear understanding or better understanding using pictures, words, numbers, and symbols.
Year 2 Teacher	I try to explain things more clearly, and I also write it out for them. We also write achievement objectives for the lesson for all of us to see.	I thought it would make things easier for the students as well as be good modelling.	By this term (term 4) the students are used to writing out their explanations with less help.
Year 3 Teacher	Presenting questions so that they would write their understanding. State what we are writing.	Was one way I thought may assist.	Not sure that it did other than they did start writing their ideas down.
Year 4 Teacher	Words around the classroom, new vocabulary in books, ideas of how to write things.	To see if by making sure that there was a buildup of vocabulary, explanations would be easier.	As the year has gone on, children can write more. Those that can only do a sentence, there is more depth in it.
Year 5 Teacher	Making them justify answers orally and then through writing. Making more displays and adding explanations.	First was done following suggestions. Second was adopted as a simple way to try to encourage writing.	* Feedback * *Reo* was being used in everyday situations following a unit. Their recall was stronger.
Year 6 Teacher	Creating a set writing plan to help students.	The need for explanations to be more specific and descriptive and also to see inner strategies of students.	Seeing students being able to write at length and then being able to present.
Year 7 Teacher	Writing with words, nets.	Initially because of the project and later so they HAD to clarify their ideas.	Explanations they gave, and understanding gained.
Year 8 Teacher	Not much more.		
Year 11 Teacher	Using models for instructions including numerals, arrows, etc., to supplement words.	To ensure that the writing task didn't drown the understanding required, i.e., in case the task got too big.	Students inform me that they return to their "instructions" or at least I direct them there.

As teachers came and went through the projects with different levels of experience, information about the relationship between teacher learning and changes to teacher practices varied quite considerably. Not all teachers felt that they made significant changes from being part of the different projects. This variation can be seen in Table 12.1, which collates some of the answers to the survey done at the end of the 2007 project on writing in mathematics. Some of the questions from the survey are given at the top of table.

In Table 12.1, it can be seen that some teachers felt they had made significant changes to their practices, whilst the Year 8 teacher did not feel that she had made any changes. Of the teachers who had made changes not all were clear about why they had made changes, and others were unclear if the changes had had any impact on their students. In the following sections, we consider how teacher learning was connected to changes in teachers' practices by describing some of the influences on their learning.

The Wider Societal Structures as Influences on the Teachers' Learning

Schooling as an historical institution operates within wider social-cultural-political expectations about what content should be covered and how it should be done (Faure et al., 1972). In New Zealand, being funded by the government as an educational organisation requires a school to conform to certain regulations, including using the mandated curriculum (see Chapter 3). However, the origins of *kura kaupapa Māori* outside of the mainstream school system meant that they also needed to implement a set of principles, *Te Aho Matua*, for the holistic development of Māori children, which was in line with traditional values (see Chapter 11). For many Māori teachers, societal expectations about the academic ability of Māori students also formed a backdrop in considering what was an appropriate teaching style. This knowledge was available from teachers' lived experiences, although it had also been documented in many studies. The cultural-discursive, material-economic, and social-political dimensions interact together to set up the parameters, which bind the practices that teachers engage in. When aspects of one dimension are contested, then this has ramifications in the other dimensions.

Culturally discursive preconditions mediate teacher learning by influencing what practices teachers believe should be changed and how they could be changed. Kemmis and Grootenboer (2008) stated that "[i]n the case of the education of the educator, the disposition of *epistēmē* – aimed at attaining truth – is formed and developed through engaging with, and coming to one's own conclusions about, the different knowledge and traditions that have shaped and formed education in the past, and the perspectives of different educational theorists that inform different approaches to education today" (p. 41). The cultural-discursive dimension, in which Te Koutu teachers worked, meant that they were fully aware that Māori children traditionally under-performed at school (Chapple, Jeffries, & Walker, 1997).

This knowledge was not accepted by the teachers, but instead was contested so it could be changed.

> Because out there, there's this perception that Māori are not high achievers. Soon as I walked into the school, I could see that there was this high expectation and you knew, I knew straight away that, by goodness, these kids here, their conversations, their ability to communicate in Māori was just really, really good. They were articulate speakers. The teachers were the same but the whole culture in the school made me see that Māori can achieve highly. (Year 2 Teacher interview, September, 2008)

There was a concurrent perception that mathematics was something that many Māori were uncomfortable with and that this perception also needed to be overcome. In the curriculum development project that we carried out in 1998 and 1999, we had wanted parent participation. For some parents, there was a fear that they had too little understanding of mathematics to be able contribute. This was in contrast to what the teachers felt the parents could contribute to the process.

Primary Teacher:	The parents, I think they need to know what's happening in the curriculum, need to know about it... Whereas I think the Family Maths nights that we've already planned are ...
Infants Teacher:	I had one or two mums saying I don't know too much about maths so it's not worth me coming. (Meeting 7, 15/8/99, p. 72)

As well, Māori teachers were thought to struggle with mathematics and how to teach it. This was given as a reason for them opting out of professional development about mathematics.

> I think they are still scared of maths. Specially, Māori are still scared of the whole maths concept. The talk is too above. With Māori teachers they just want the nitty gritty. How to do it sort of thing and that's realistic. That is it. They want the stuff. When it comes to high-falooting maths talking or anything they start shutting off. That's still sad. (Year 8 Teacher interview, November, 2007)

Māori-immersion mathematics teachers' fear of mathematics had also been noted by others. For example, Trinick (2006) described the change in teachers' attitudes from being involved in Poutama Tau, a professional development project on numeracy.

> Prior to Te Poutama Tau, a number of teachers and students had negative attitudes towards *pāngarau* [mathematics]. Despite their professional training, many teachers still lack confidence, based on memories of their own mathematical learning experiences. In the case study schools, teachers and principals felt there had been significant change over the duration of the project in teacher and pupil attitude to *pāngarau*. Much of the change on the teachers' part was that they could see the positive outcomes and thus felt more inclined to change their practice. (p. 90)

Within the cultural-discursive dimension in which the teachers at Te Koutu operated, Māori children were labelled as not being able to achieve highly and Māori teachers as being afraid of mathematics. This societal discourse was accepted as being a valid interpretation of the situation, even by some Māori. Bishop (2007), using the ideas of Michel Foucault, described how teachers could become trapped by an acceptance of these norms:

> That is, we are not *of* the explanations but rather, by drawing on particular discourses to explain and make sense of our experiences, we position ourselves *within* these discourses and act accordingly in our classrooms. The discourses already exist; they have developed throughout history and are often in conflict with each other through power differentials. Most importantly for our desire to be agentic, some discourses hold solutions to problems, and others don't. (p. xviii)

Māori are of course not the only ones who do not expect to do well in mathematics:

> Pupils who encounter difficulties and poor results are led to believe that they lack ability, and this belief leads them to attribute their difficulties to a defect in themselves about which they cannot do a great deal. So they 'retire hurt', avoid investing effort in learning which could only lead to disappointment, and try to build up their self-esteem in other ways. Whilst the high-achievers can do well in such a culture, the overall result is to enhance the frequency and the extent of under-achievement. (Black & Wiliam, 2001, p. 6)

The situation is exacerbated for Māori because experiences of failure in mathematics are connected to ethnicity, and this affects the teachers' considerations of how they can change the situation. The need to overcome stereotypes about Māori educational achievement was reflected in discussions about the types of practices that the teachers at Te Koutu wanted to implement. In the following two interview extracts from November 2007, Māori underachievement is not mentioned explicitly. However, there is discussion of the importance of having high expectations of the children's learning.

> Now I am just for kids, Tamsin, and every child making progress, you know, whether they are Māori or *Pākehā* [European New Zealanders], I don't care, but for them to do well in life and to be knowledgeable and to be able to impart their knowledge and to be able to share it, all those sort of things. If I can do any little bit to make a child move forward in their lives I am happy, yeah. But I know it is important how we do it. (Year 1 Teacher Interview, November, 2007).

Teacher in charge of the primary section of the school (TIC): At the *hui* [meeting], a lot of people were saying my kids can't do that but then they go to Year 1 Teacher's class for writing, not maths, some draw a picture then orally they are giving her a paragraph so why can't they do that for maths? We aren't making the connection between things that can happen as well in the maths.

Tamsin: The probability stuff from last term where she did that, they drew a picture and then they used the probability words in a sentence.

TIC: There is no reason why they can't, what are our expectations? Can our new entrants only draw a picture, then don't worry about them orally giving you anything? You make assumptions from that. Is a sentence not enough? In a sentence, they are just warming up. I think our expectations are too low. (TIC interview, November, 2007)

Material-economic orders and arrangements are to do with the provision of resources to teachers and expectations of the sorts of work they would do as teachers (Kemmis & Grootenboer, 2008). Although it had its origins outside the state education system (see Chapter 3), Te Koutu was expected to operate like a school once it had accepted state funding. For example, teachers taught classes of students in specially designed classrooms. Teachers had to also follow state-mandated curricula. During the time that we were collecting data, *Pāngarau i roto i te Marautanga o Aotearoa* (*Mathematics in the New Zealand Curriculum*) (Ministry of Education, 1996) was the legal document from which teachers had to programme their teaching. When this document was developed, the Ministry of Education had required that the achievement objectives, which guided teachers' planning, were translations from the parallel English-medium document (Ministry of Education, 1992). As well, the NCEA assessments at the end of high school (see Chapter 4) provided a structure within which the school had to operate. However, the fact that the school taught five-year-olds through to 18-year-olds meant that Te Koutu had flexibility that traditional schools, which only taught a narrower age range, did not have.

My wife she says it is very difficult at [her] school because they have this ultimatum, we think this is what they should have when they leave [primary school] and she was talking about the *kura* and she said it is a great opportunity that we have because we've got *wharekura*, year 0 to 13. Now what do we want our stakeholders, what do we want them to look at by the time they reach Year 11 Teacher? Ultimately if we went backwards from what Year 11 Teacher wants down the, each of the years what each person should implement and the steps incrementally as they go up so that ultimately he's not having to try and do basically what should have been done in year 7 or year 5. So what does it look like at that end and go backwards from there then you know what year 0 looks like. (Staff Meeting, November, 2007)

In most primary schools, teachers are not clear about the links between what they teach and the end of high school assessments, so they confine their responsibility to planning from the curriculum for their year level. At Te Koutu, there were possibilities for discussions about what should be taught at what year level so that students had the best possibilities for success. For example, the Teacher in Charge of the primary section of the school stated her belief that the sharing of expectations about what should be taught had to go both ways.

We are all in a unique situation where we can all link together but it can't be *wharekura* [high school] dictated like they hoped because we have our structure in place. That gives the kids coverage. You have to teach the kids all those things. Year 11 Teacher, Year 8 Teacher and Year 7 Teacher, this year have been more than happy to jump on board .... We all did that triangle unit [in 2006] and it worked really well but that's not always going to happen. It doesn't have to happen every term because they are doing unit standards and we're not. We are giving the coverage. We can't have dictation from the *wharekura* and we can't dictate to the *wharekura*. There might be just one unit that as a collective we can all do. (TIC interview, November, 2007)

Consequently, there were differences in perceptions of the possibilities that arose from the material-economic orders and arrangements. As well, teachers at *kura kaupapa Māori* had to follow the guidelines set down in *Te Aho Matua* (see Chapter 11). This also contributed to the contestation of the supremacy of the curriculum and end-of-high-school assessments.

The social-political dimension is concerned with how teachers are positioned in relationship to others. Relationships, such as ones with principals, are historically constructed sets of power relations. If teachers, like students, have no say in what they learn and how they learn it, because power relations have enabled others to determine this for them, then they are less likely to engage in the learning activities provided to them (Bonner, 2006). In 2007, all Te Koutu mathematics teachers attended a conference in the first week of the school holidays. The school paid for the teachers' registration and accommodation, but the principal expected teachers to attend it. Not all teachers were enthusiastic about attendance.

> I wasn't happy about losing a week of the holidays. I thought, "Well, we are here. How can we get the most out of it?" At the end of the day you are going to find something if you are positive. (Year 6 Teacher, November, 2007)

When she was asked in September 2008 about further attendance at the next conference, this teacher stated that she was "uninterested". This was not professional development that she felt was worth making the sacrifices of spending time away from her family. Although she had gained something from her attendance, it was not sufficient to entice her to attend again.

On the other hand, other teachers felt that they gained from attending the conference. For example, the Year 8 teacher felt that she learnt about developing for older students internal assessments, known as unit standards. Her inexperience in teaching these assessments had been her motivation to attend the sessions. Yet her input about her experiences of teaching in Māori-medium educations was also valued by the presenters of the sessions. The sharing of information can be seen as a distribution of power (Gordon, 1980).

> Yes, that was interesting because they are obviously looking for a lot of Māori input into the Unit Standards. I chose that one to see what was it all about. Getting ideas about setting up unit standards for the kids in... This is my first year. There are those who had been doing unit standards for a while so they were a great help. Do this, do that. I gained a lot out of there. In the end, I was approached to go down to their next Unit standards *hui* [meeting] because of the Māori input. That was one. (Interview, November 2007)

The official status of the curriculum meant that teachers were legally required to plan their lessons using the learning objectives and for this planning to be available for evaluation by appropriate authorities, such as the Educational Review Office. In the school, teachers had to present their programme to a member of the school executive in the first couple of weeks of every term – "I expect teachers to be planned. They are given more than enough time to plan. For me there is no reason as to why they are not planned" (Teacher in Charge of primary section of the school, interview, November, 2007). The teachers had accepted that in order to work in a state-funded school, they must conform to these legal requirements, and the school's role became one of ensuring that this was achieved. By conforming to the material-economic dimension of their work as teachers, the teachers also had to conform to the existing power relationships within the social-political orders and arrangements.

The wider society influenced the teachers' sayings, doings, and relatings about the practices that they considered to be valuable. The cultural-discursive, material-economic, and the social-political dimensions set up expectations about what teacher practices should be. Although constrained by these practices, teachers were not restricted by them. For example, they worked to overcome expectations about Māori students' lack of achievement in mathematics. Thus teachers' learning could result in contesting of societal norms. These dimensions influenced teachers' learning and their decisions to change their practices.

Perceptions of Themselves Within the Immediate Context as Influences on Teachers' Learning

At the meso level of the model, teachers negotiate the local influences on their teaching such as the awareness of their students' needs and expectations of parents. This negotiation was connected to their reflections on the intended and unintended outcomes of their teaching, or to their responses to professional development activities. If teachers are not invited to engage in meaningful learning acts in these professional development activities, the field is left open to all sorts of other meaning productions. In relationship to students' learning, Alrø and Skovsmose (2002) discussed *underground intentions*, which "refer to the students' zooming-out of the official classroom activity ... partly setting an alternative scene for what is going on in the classroom" (p. 158). If adapted to teacher development, these intentions could result in teachers learning amongst other things how to manoeuvre themselves out of the development opportunities, sit next to the right person, or complete other work surreptitiously.

On the whole, being involved in the research projects provided opportunities for learning that many teachers seemed happy to engage in. The discussions and reflections that they did on their teaching practices provided intentionality to the activities that they then chose to adopt. Alrø and Skovsmose (2002) took "intentionality" as a defining element of action, thereby separating action from mere activity. If the learning situation allowed the active involvement of teachers, the resulting learning process could be one of action. When teachers identify with the intentions of the development activity, joint ownership and shared perspectives between teachers, facilitators, and others could develop. Thus, intentional learning acts constitute forms of learning that can be described as action. The meaning ascriptions resulting from learning-as-action would be different to those where teachers do not engage willingly.

Videoing lessons and then having the teachers talk through what they saw happening often made them reflect about issues that they may not have otherwise been aware of. In 2006, the Year 2 Teacher stated:

> I have never really thought of maths as a language. I knew it on the surface but it hasn't really sunk in until we started the project. I see it as a whole new language to learn. Very cool. (Survey, November, 2006)

In the 2005 lessons, she had seen herself giving four different explanations for the same mathematical idea and felt that this would have confused the children. She made changes to how she explained concepts as a consequence. Reflection on her teaching in conjunction with the discussion in the project meetings gave intentionality to her subsequent actions in which she became more thoughtful about how she explained mathematical ideas to the students.

In the 2005–2006 project, the Year 4 Teacher also began focusing on aspects of the mathematics register in her lessons.

> Previously the lesson focus had been on the maths concept with an expectation that a five-minute discussion of language was sufficient for children to gain it. As well the language given in an explanation might have been wordier and therefore more difficult for students to understand. Being aware of the issue of language meant that [I] was now simplifying how [I] explained ideas. (Survey, November, 2006)

She felt these changes resulted in the students gaining the language more quickly. By watching what she did in the 2006 videos, she was able to see how doing something differently had come across more clearly to the students. She was able to identify what had worked well and think about why this had worked.

For the principal, Uenuku, the main gain from having the teachers engage in the projects had been that he now had staff members who were happy to discuss mathematics teaching. In his survey at the end of the 2007 project, in answer to the question "what had been the most interesting thing for you about being involved in the project", he wrote, "enjoying the development of staff, leading to more conversations about maths". Timperley et al. (2007) described the necessity of having institutional support if a project is to be sustained beyond the initial intervention period. For teachers, their reflections on their teaching are more likely to be of a deeper nature if done in conjunction with others. They are also more likely to result in changes to teaching practices if others are also making similar changes. It may be that teachers, who initially reluctantly join professional development programmes, can become active participants through these joint discussions.

The projects also contributed to teachers seeing themselves as only one of many contributors to the students' mathematical learning journeys.

> We have a little bit of flow about where we want to get our kids up to year 6 so that we are covering specific things like. So, like in year 2 we cover this, so that they will know that by year 3. But we, I don't know where my kids need to go to get up to Year 7 Teacher's stages. I've no idea what they do up there. And they're like "oh, you were in high school", yeah like ten years ago. So if we could set those things up. (Year 2 Teacher Interview, November, 2007)
>
> I think with writing, it's what you were expecting of the kids. You were doing a lesson, this is what you have to do and that is how they do it. Something like the probability strand. It's where it's common sense and you can't see where their thinking is. It is very hard for many. It's probably a process that needs to be worked up from the bottom, all of a sudden you have probability. The last time you had probability was in year 5. You are in

year 10. It is quite a long way, you know, the way of thinking. (Year 8 Teacher Interview, November, 2007)

Discussions in what to teach across the school turned into discussions about effective teaching practices. The Year 3 Teacher stated:

If we don't hook on to some strategies that we believe will work and change classroom practice then we won't do anything. You won't see the change that we want. And that's what you want to happen. You want the shift in classroom practice. (Interview, November, 2007)

Effective teaching practices were considered to be those, which supported students to gain the intended outcomes from the lessons. Although outcomes usually were connected to the curriculum or other policy documents, teachers were more focused on their students' needs.

Year 2 Teacher: I discovered something today that I did take for granted. I thought that they knew the word "add".

Tamsin: *Tāpiri*

Y2T: *Tāpiri*, I really did think that they knew. That they just seemed to regurgitate the answer to you. But until I went deeper and started using the five-finger method with them. And then they couldn't get that *tāpiri*, you had to actually use the five. They were coming up with all sorts of answers. And I thought "oh, my god they don't actually understand this *tāpiri*". So I had to go right back and explain what it was. Show them what, use five and if I *tāpiri* this amount, how much will it be? You know, how much will the lot be? (Interview, April, 2008)

In 2006, the Year 6 Teacher started to expect his students to give more structured responses. In the *Poutama Tau* programme, students were expected to provide explanations of their thinking. However, the teacher was aware that he had not stringently adhered to the requirement to ask "how do you know" and even when he had done so, he had accepted responses such as "just because I know" and "because it just is" from his students. Being part of the project meant that he understood better the importance of his students giving explanations. Consequently he changed his teaching practices to insist that students explained what they had done.

One approach the teachers had learnt about at the Mathematics Teachers conference was RAVE. RAVE was a mnemonic, which stood for "rewrite" the question, "answer" the question, use mathematics "vocabulary", and use "examples". One of the keynote speakers at the conference, Helen Doerr, described how teachers in a project she had run on writing in mathematics had supported students to use this to structure their responses. Although the teachers at Te Koutu could see that RAVE had value, they understood that it was beneficial only if it was adapted to meet their students' needs.

Year 3 Teacher: Helen had some great things to say. There was only a core bit of her *kōrero* (talk) that related to us or that could be used. RAVE is what we have taken from it, getting used to justifying themselves.

Tamsin: That isn't part of RAVE; that's the movement here.

Year 3 Teacher: We've not really taken the RAVE, we are making up our own. And we have to be aware of that. RAVE is a guideline. We must be careful not to get stuck into that. (Interview, November, 2007)

In evaluating its usefulness, teachers considered the needs of their students. Contestation was part of the discussions. The Year 2 Teacher felt that RAVE would need to be modified to suit the needs of young students.

> It sounds like it would be good for the seniors, like even the year 6s sound like they were doing really well with it and the year 5s do well, I think. Yeah but I don't think my babies would be able to, unless we simplify it. (Year 2 Teacher, Interview, November, 2007)

At the meso level, teachers combined what they knew about their students with the mathematics learning they wanted their students to gain.

At this level of the model, the focus for learning shifts to collective units of teachers and students, in this case linked by being part of the same school. Being part of a collective unit constrains what possibilities are considered in the same way that that the cultural-discursive, material-economic, social-political dimensions do at the macro level of the model. Although the constraints are different, they remain open for contestation. By collectively reflecting on what they knew about teaching mathematics and identifying what they wanted to find out, the teachers were able to focus their learning on their own needs as well as those of their students. Perceptions about what were appropriate new learning and approaches to this learning developed from discussion with others. Consequently, connections between the individual teacher and societal structures are visible.

Teachers' Sense of Self as an Influence on Their Learning

Radford (2008), in developing his cultural theory of learning, considered that within a culturally mediated experience a learner or subject comes to understand an object in "the dynamic and ever changing cultural-normative sphere of knowledge" (p. 225). In this way not only the object comes to the notice of the learner, but the learner himself/herself becomes part of what is to be understood in the learning process. Subsequently, learning is seen not only as a process of knowing but also as a practice of becoming. Learning about teaching mathematics means learning about being a teacher of mathematics. It is tied up with teachers seeing themselves as being successes or failures and so is connected to emotions. Therefore, the tacit knowledge that teachers possess becomes part of the learning.

Many researchers have commented on the important role that reflection has in professionals' work. Castle and Aichele (1994) felt that in regard to the work of teachers "[r]eflective insights provide a deeper and richer understanding of what it means to teach, thus contributing to professional knowledge used to make autonomous decisions" (p. 5). Schon (1983) described this as reflection-in-action where research is happening in context and thinking is not separated from doing:

- There are actions, recognitions, and judgements which we know how to carry out spontaneously; we do not have to think about them prior to or during their performance.
- We are often unaware of having learned to do these things, we simply find ourselves doing them.
- In some cases, we were once aware of the understandings which were subsequently internalized in our feeling for the stuff of action. In other cases, we may never have been aware of them. In both cases, however, we are usually unable to describe the knowing which our action reveals. (Schon, 1983, p. 54)

The discussions that the teachers engaged in contributed to their understanding of the collective responsibility at Te Koutu for students' mathematics learning, as well as to an understanding of themselves. Teachers can have knowledge about themselves and their teaching practices that is so ingrained in what they do that it remains unrecognised and unvalued. Yet this knowledge could contribute to learning if brought into the awareness of the individual teacher or of other teachers (Duncombe & Kathleen, 2004). The discussion based on what they had seen of their own teaching in the videos contributed to the teachers becoming aware of their reflection-in-action. Regarding this, Kemmis (2009) stated the following:

> The process of gaining experience is a process of self-formation, especially when a person becomes 'experienced' in a deep and reflective sense. The identity of the practitioner who lives in and through familiar passages of practice is similarly shaped and formed by practice – the 'skin' of the practice is not external to the practitioner's identity but part of it. The practitioner is an *agent* and *subject* of the practice; her or his *subjectivity* is reflexively formed and transformed by living through both familiar passages and new and surprising ones that call for new ways of working or living within the practice. (Kemmis, 2009, p. 11)

When the first project began in 2005, almost all of the teachers found being filmed daunting. It made them see the "reality" of themselves as teachers rather than the perceptions that they had built up from the stories that they had told themselves about their teaching (Sfard & Prusak, 2005). The apprehension that the teachers had about seeing themselves on video had been one of the reasons behind the decision that sharing of the videos with other teachers would not be part of the staff meetings. This apprehension had shaped our project, whilst at the same time the project had shaped teachers' learning. As the Year 5 Teacher described it, once teachers realised that the filming was not "big brother", they were able to use the videos to improve their teaching, but still did not want to show them to others.

At the end of the 2005–2006 project, two teachers admitted to thinking more about what they were teaching when planning the lessons that would be videoed. For one of these teachers, this was because he was aware he would watch them with a university researcher and was conscious of not wanting to present himself poorly. As someone new to teaching as well as to the school, it had been quite a shock to find himself being filmed in his first term as a teacher. The other teacher found that having her lessons filmed forced her to reflect on the progression of the lessons that she was presenting.

> If you were teaching a similar one to the previous lesson, [the videoing] made you think about why it needed to be similar. If it was different it made you think about why and how you had moved the focus on. This meant that you were thinking about how you were moving the children on mathematically. Sometimes when you were busy with your teaching, you did not think too much about ensuring that the children were moving mathematically on. (Year 3 Teacher, Interview, November 2006)

This teacher's comments show how the process of being videoed had in itself made her reflect upon those actions that she would normally carry out spontaneously. Much of the apprehension about being video recorded was linked to whether the teachers would see themselves as failures as teachers, and this was tied into their subjectification of their learning. The Year 7/8 teacher had the last of her 2006 videoed lessons be of two of her students giving and receiving a set of instructions for a geometric construction. She had wanted this, because she felt it would provide her with information about whether she had succeeded or failed in her teaching. For the students to be "able to be videoed on their own, giving instructions and following instructions means that the students need to have the vocabulary and the drawing skills" (30 August, 2006, feedback on teaching). The following extract comes from her written summary and comment on the entire unit of work.

> I am absolutely pleased that this class (despite the noisy background at times) had all completed or met this objective. The students of *Tau* [year] 7/8 have a very wide ability range, and although there was fine work done by the two students, which is seen in the last movie, the samples show that all the students know how to construct correctly a network using the mathematical materials. Very fine samples. I could have chosen another couple of students in place of the two who were videoed, but clarity of language was needed. A great activity where *te reo Māori* (of the students) came to the forefront with minor mistakes. I enjoyed immensely this activity and believe the students did too.

When students failed to learn, teachers often saw this as being their fault. In an interview with the Year 4 teacher at the end of 2007, there was a sense of confusing her children because she herself was not sure of the correct terms. She had introduced some terms for clockwise and anticlockwise, but was unsure if they were appropriate. After discussion with Uenuku as he walked passed the classroom, she then introduced alternative terms:

> Clockwise was easy… *me he karaka* which means like a clock. Then he told me, *a karaka*, so that was easy to change we just decided anticlockwise was *ka huri tua* just opposite …he gave me a whole different phrase that didn't relate at all so it was a bit tricky. I think I confused them in the end. (Interview, November, 2007)

On the other hand, introducing new teaching practices which seemed to support students' learning contributed to teachers feeling successful. This often can be seen in the emotive language that teachers employed to describe different activities or practices. The 2007 writing project made the Year 2 teacher think about mathematics and the role of writing in it.

> Thing is, it's made me think about how much writing we do in maths. Because before, I'd never actually thought of it as writing. Golly gosh, it's maths sort of thing, I'd never really thought about it as being a written language or anything. So for myself as a *kaiako* [teacher], it has made me think about it. (Interview, November 07)

At the meeting in November 2007, the Year 3 teacher described why she felt that the whole school should introduce RAVE, as a way to support students' writing:

> I think we would be silly not to introduce the RAVE thing. It gives you a bit more direction and some of us have introduced it this year and it does make a hell of a lot of difference as to giving you a bit more direction, not so much to your teaching, but your end result of how much your kids produce. I have seen a great change in my kids and they are only year 3s. I am amazed with some of them the amount of words they learned just in a three-week period and the amount of writing they did. It might only be a sentence, but it is rich. (Meeting, November, 2007)

Like many of the other teachers, the Year 5 teacher also discussed her reasons for wanting to implement RAVE. She described her initial involvement in the project when she was trying to adopt better practices as something like being in a lolly shop.

> actually I was plucking any out of the air things that I thought might work, while I feel I am someone who has been to the lolly shop and I know exactly what it tastes like and I want to try it. (Meeting, November, 2007)

However, she was unsure that she was having an impact on students' explanations.

> I thought I was being a part of something, I didn't know if I was contributing, but if we went with that [RAVE] I could see myself contributing. I could make a difference. I am not sure I was really making a difference. I was teaching the best I could but I didn't have any new strategies that could help that writing component. (Year 5 Teacher, Interview, November 2007)

Although this teacher recognised that she wanted to improve her practice, this did not lead her immediately to adopting any different teaching practices, even when they were suggested by others. She needed to discover a teaching approach that resonated with her current teaching practices. RAVE became this approach.

The contribution to a sense of purpose in their teaching, as being connected to a sense of their own identities as teachers, can be seen in the following extract:

Tamsin:	It's full-on here. It's absolutely full on.
Year 0 Teacher:	I see it as a good thing.
Year 1 Teacher:	No, it is a good thing.
Year 2 Teacher:	It is a very good thing. It also lifts your expectations of yourself. Because everyone is working really, really hard to ensure that the kids achieve and achieve highly, and achieve well. You can't help but follow that same sort of thinking. It really helps also, to know that your leader, is a leader. He's not coming from here. He's taking you by the hand and he leads you, he shows you

Y0T: He's passionate, *āe* [yes]. It's a good thing. And even Teacher-in-Charge-Junior-Section (TIC) is a leader of this section. She's really great. Oh I've just been so amazed with her – that support. But TIC's just full on, makes sure that the staff are all happy. That's really good to see. Never been involved in that sort of thing.

Y2T: And it is all about supporting each other. (Meeting, September, 2009)

Learning about new mathematics teaching practices did have an impact on how teachers felt about themselves as teachers. It was not simply a cognitive activity because by internalising the new knowledge and practices, teachers also reconceptualised themselves as teachers. How they saw themselves as teachers was also mediated by what they knew about their students, the school, and the wider school community. In particular, the success or confusion of their students in learning mathematics contributed to their sense of being successful or failing as teachers. The practice architectures that set up the norms by which teachers make judgements about their own success or failure constrain the likelihood of learning taking place.

Meeting the Challenge of Changing Teachers' Practices

In this book, we have documented many of the challenges that were faced by different groups connected to the school – parents, to teachers, to children, and to the wider community – in relation to mathematics and language. In this chapter, we have dealt with a challenge to us as researchers, as we grappled with describing some of the influences on teachers when changing their practices. The main aims of Te Koutu are to revitalise *te reo Māori* and to ensure that students gain academic results. In order for these aims to be met, it was necessary for some teachers to change some of their practices. However, given that the school was doing well in meeting both of these aims, there was no point in having teachers change their practices unless it was likely that improvements would result. In fact, when we began our language in mathematics research in 2005, the project aim was to document the practices already in place. Given the paucity of previous research in Māori-immersion mathematics education, we, as researchers, have had to think carefully about the way that we framed what we were documenting. The refinement of the mathematics register acquisition model was one of the frames that we used (see Chapter 10). We were unhappy with previous models about teachers changing their practices as a result of professional development, because of the lack of acknowledgment of the role of context. Our work at Te Koutu had emphasised how context had a large influence on what teachers did. We, therefore, drew on the model developed by Meaney, Lange, and Valero (2009).

We have considered changes to teachers' practices to be the outcomes of teacher learning. Previously, teacher learning has been considered to be a cognitive activity, and so professional development only had to provide appropriate knowledge for teachers to change their practices. In this chapter, we adopt a more complex view of teacher learning. Using a three-tiered model to show how learning as a practice embedded an individual within their society, we were able to emphasise different aspects of the relationship between the individual and the society to illustrate how teachers at Te Koutu came to adopt different practices. This gives a more rounded understanding about why teachers change their practices. From this it is possible to make suggestions about appropriate support for teachers' learning.

Teachers' reflections enable them to contest perceptions of their role as teachers and question established patterns of behaviour. Yet the complexity of the context in which teachers operate means that often aspects from the different tiers and within each tier can have conflicting influences on teacher learning. For example, a teacher may adopt a new practice, such as having the students explain their thinking because he/she felt that it would improve their understanding of mathematics. However if it confuses the students on the first attempt, then the teacher could stop doing it because its implementation may make him/her feel like a failure. Teachers, in order to persevere, would need to ignore negative emotive responses through a strong belief that the new approach will bring benefits to the students.

Another challenge that Te Koutu, like most schools, has had to face is the constant changing of staff. Each year, new staff join the school, sometimes at the beginning of the year and sometimes later in the year. Whenever this occurs, there is a need for new staff to be initiated into the reflection process. Joint reflection meetings are only successful if there is a critical mass of teachers who accept that this is an important part of being teachers. This remains an ongoing challenge that the school faces.

Teacher workload in Māori-medium contexts is identified as an area of concern in Māori-medium contexts (McMurchy-Pilkington et al., 2009). Teachers at Te Koutu have added responsibilities over and above those of teaching. Many are involved in local *kapa haka* groups and tribal management. Additionally, as the majority of *kura kaupapa Māori* are small, there is a smaller community of practitioners to carry out the responsibilities of the *kura*. This has implications for teacher involvement in professional learning outside of school hours (McMurchy-Pilkington et al., 2009).

Supporting and documenting teacher change have affected a range of groups connected to the school in different ways. As researchers, we have had to find a model that showed how teacher change occurred within the complex contexts in which they operated. In documenting the changes that were made to teaching practices, teachers were challenged to be video recorded and then to decide for themselves what needed to be changed. The discussions that they engaged in allowed them to reconcile competing understandings about what they saw in their classrooms and what they wanted to see. At the wider school level, these discussions have challenged the whole school to consider more carefully how they can utilise an across-the-school approach to the benefit of the students.

Chapter 13
Meeting Challenges

Māui

Māui returned to his pigeon form and fluttered towards the hole. Folding in his wings, he plummeted into the darkness ringing the distant circle of light. The older Māui sat down to wait the pigeon's return, grudgingly beginning to admire his prodigal brother.

Meanwhile, Māui, even in his pigeon guise, had to use all his skills to avoid the rocky outcrops and crumbling sides of the tortuous tunnel. Finally he emerged from the other end. What a wondrous sight awaited him. This new world, sky and all, seemed complete, if more idyllic than that above.

Māui's sharp eyes and ears discerned a group of people sitting under a large *miro* tree not far away. He was not concerned about the noisy flapping of his wings; no one knew who he was, and the party was earnestly engaged in conversation.

Māui perched on a branch directly above where his mother was seated. She was explaining to the rest, especially a middle-aged man sitting opposite her, the reason for her lateness and the news from the other world. It did not take long for Māui to realise that the man opposite his mother was none other than his father, who was showing great interest in his newfound son's doings.

Māui, stretching out, plucked off a nearby *miro* berry in his beak and threw it down. It hit the man on the head, but the berry being so light, he absent-mindedly brushed it away. Māui threw another, and another, until the man finally stood, peering up amongst the branches.

Another man called out, "Look up there, it's a pigeon!"

Several people, including Māui's father, picked up stones and began firing them at the bird. Māui scrambled about, avoiding every stone, but finally, by design, he was hit right between his eyes by one of his father's stones. Whilst blood vessels burst, turning the eyes bright red, the injured, inert bird tumbled to the ground to a resounding cheer from the party below.

The cheering turned to amazement, for as soon as the pigeon reached the ground between his parents' feet, he turned back into his human form. Taranga

T. Meaney et al., *Collaborating to Meet Language Challenges in Indigenous Mathematics Classrooms*, Mathematics Education Library 52,
DOI 10.1007/978-94-007-1994-1_13,

let out an anguished scream and hugged her youngest son, pulling back his hands from his bloodshot eyes.

Seeing he could do so, Taranga raised Māui to his feet and turned to her husband, "Mākea-tūtara, here is your youngest son, Māui-pōtiki, whom I have named Māui-tikitiki-a-Taranga".

Mākea gingerly pressed noses with his son, conscious of the injury he had caused, but already proud of the wonderful skills he apparently had. Holding Māui by the shoulders, the father stood back to assay his son. "I believe you'll never lose the redness in your eyes".

And so it was to be, Māui, never handsome, now looked even less so. And the *kererū*, though a handsome bird, still has startlingly red eyes to this day!

Mākea, wanting to make amends, said to his son, "Let me now do what I should do. As your father I wish to perform a *tohi* rite whereby you will be even stronger and protected from danger and injury".

Māui gladly followed his father to a sacred pool which they entered naked. His father dipped a chaplet of leaves in the water and began chanting, concentrating on getting each and every word correct and in order. However, his emotions got the better of him, and he missed out a couple of words. He quickly recovered, anxiously carrying on and finishing without further error.

Unfortunately the damage had been done. Yes, Māui had found his parents, he was now even stronger, both physically and spiritually. He would go on to do wonderful, amazing things, but his initial rejection and those few unuttered words were to follow him to the end.

For us, this is a book of hope. Despite all the challenges facing Te Koutu, the teaching of mathematics in the medium of Māori is now a "normal" component of the school curriculum. In the wider world, international languages have come to be considered the languages of power, so that the teaching of mathematics in English has become commonplace, even where it is not the primary language of the learners (Setati, 2002; Lim & Ellerton, 2009). A rationale given for using English as the language of instruction is that governments and parents want children to have access to this language of power, and using it in the mathematic classroom is an acceptable way to do this. The fact that *te reo Māori* is used in the teaching of mathematics is a contradiction to this trend and shows that parental aspirations for their children can be realised in a range of different ways, including through the medium of an Indigenous language. In Aotearoa/New Zealand, *te reo Māori* has become the valorised language amongst Māori who want schooling to do more that pass on Western knowledge and values, especially when academic results proved that even this was being done poorly. In this book, we illustrate how it is possible for a community to utilise their human resource strengths and the strengths within a non-traditional language, for teaching mathematics to create something unique to service their needs. Like Māui's investigations of his mother's disappearances, a determination not to be thwarted by challenges has enabled the aspirations of Māori to be realised. However, to do this has required innovative thinking.

In the last eleven chapters, we have presented case studies of different aspects of using *te reo Māori* in mathematics classrooms. These illustrate not just the range of challenges that have been overcome, but also those challenges, which continue to be worked through. Challenges can bring people together, when they are perceived as requiring an appropriate level of engagement. Challenges that can be solved easily do not need in-depth communication amongst different participants, because the challenges are not intriguing enough. On the other hand, challenges perceived as being too complex could be shelved as too difficult to solve. Rather than bringing different participants together, such challenges can drive people apart as the frustration, from not being able to agree on any alternative actions, can result in a need to blame others for the lack of a resolution.

Initially, the primary aim of parents in setting up Māori-immersion schooling was to maintain and revitalise *te reo Māori*. More recently the focus had been on student achievement. Without Māori-immersion schooling, the language would not be currently in such a strong position (Spolsky, 2004) but, like Māui's father's invocation, the revitalisation has not been perfect. Māori-immersion education may be the *karakia* (prayer) that strengthens and enriches the language through teaching subjects such as mathematics. Yet, there remain concerns, especially amongst native Māori speakers, about this elaboration of the language and its relationship to English. However, the complexity in developing and using a new mathematics register was never accepted as being so problematic that this challenge could not be overcome.

Similar concerns have been expressed about the revitalisation of other languages. The use of Gaeilge in predominantly English-speaking areas of Ireland was championed mostly by second-language learners. Their version of Gaeilge has been described as "a jargon of the middle-class, incomprehensible to native Irish speakers" (Hindley, 1990, p. 142). Commenting on the different movements to revitalise Breton, McDonald (1989) highlighted the immense differences between the Breton transmitted through families and the Breton spoken by urban second-language learners, which was considered to be strongly influenced by French.

Yet, the very success of Māori-immersion schooling in producing cohorts of *te reo Māori* speakers may have lulled the community into believing that the language is saved. This is regardless of the fact that much has not been achieved, for example, a broad base of intergenerational language transmission and the common usage of *te reo Māori* in a variety of workplace domains. These unachieved aspirations for Māori revitalisation can be seen as the missing words within the *karakia*, which ensured that the language was strengthened, but in such a way that the language must continue to contend with new challenges. This book describes one part of a journey in the use of the mathematics register in *te reo Māori*, but like Māui's adventures it is not the complete story.

It will be interesting to follow what the graduates of *kura kaupapa Māori* will do. Currently they still have parents and grandparents to speak to at the *marae* (traditional meeting area), but as time goes on these graduates will probably become the natural speakers for wider *whānau* (families) and *hapū* (subtribe) because of their fluency in *te reo Māori*. If they do, it will be interesting to see how they promote the status of Māori, and whether any aspects of the mathematics register appear in their *marae* speeches. Their fluency with the language will provide them with the resources to be able to be Māori and to participate in the world as a global citizen.

Of future interest also is the effect on *kura kaupapa Māori* of the next generation of students, whose attitudes to the language may be different to those who preceded them. When Te Koutu started, the students came from parents who were ideologically committed to the revitalisation of the language. As well, there was an economic trough, which gave many parents the time to be actively involved in their children's education. Many were inspired to learn *te reo Māori* as a second language, and their commitment to their children's learning resulted in some becoming teachers. Parents' commitment to revitalising the language is still strong, but other commitments, like ensuring that their children achieve academically, have become more prominent in their aspirations for their children. This change in emphasis and changes in parents' employment status have meant that parents are no longer learning *te reo Māori* in the same numbers. The intergenerational transmission of the language so hoped for in the 1980s has not become a reality, except in a small number of cases. A similar situation occurs in some parts of Wales where most parents sending their children to Welsh-medium schools, *Ysgolion Cymraeg*, are native English speakers (Jones, 1998). Thus, schooling has taken on the role of language transmitter. However, the situation may change yet again if graduates of Māori-immersion schooling choose to send their own children to their old schools.

Another reason for this being a book of hope is that the case studies provide information that is useful to mathematics educators in a range of situations, outside of immersion schooling. One of the perceptions that we hoped to overcome is that mathematics education in *te reo Māori* is a translated version of what occurs in English-medium education. Generally, there has been a belief by the New Zealand Ministry of Education that Māori-medium schooling, including mathematics education, benefits only from translated versions of what is designed for English-medium education. The professional development programme, Poutama Tau, is an example of this. We are not suggesting that Poutama Tau is not supportive of Māori-medium teachers. Indeed, there is substantial evidence to show that Māori-medium teachers and their students gain much from being involved in the programme (Ministry of Education, 2009b, 2010). Nonetheless, we feel that much could also be learnt from *kura kaupapa Māori*, and this could feed back to English-medium education. For example, the extensive work done in elaborating and refining the mathematics register acquisition (MRA) model could be valuable for teachers in a variety of situations, including first-language classrooms, who see language development as an integral part of mathematics learning. Another example would be that of understanding how teacher learning in mathematics education is affected by contexts at the societal, local, and individual levels. All too often teacher learning is seen as being in a linear relationship to the amount of professional development that they receive, and there is no acknowledgement of the role of contexts.

In Māori-medium education, the contested spaces within the cultural-discursive, social-political, and material economic orders and arrangements (Kemmis & Grootenboer, 2008) may be more obvious, but they may be just as influential in English-medium education. By looking at the challenges for *kura kaupapa Māori*, it is in some ways easier to identify those features that support or hinder students' mathematics learning, teachers' professional learning, and parents' active involvement in their children's education. This information can then be used to better understand what occurs in English-medium education. Unfortunately, at the present time there seems to be little awareness by those who work in English-medium education that they could learn from working with those in Māori-medium education. There appears to be resistance amongst many mathematics educators, whose native language is English, to investigate and learn from classrooms where English is not the language of instruction.

The Complexity of Factors That Interact When Meeting Challenges

Although we believe that our study has important implications for non-immersion schooling systems, we also acknowledge the importance of individual factors at play on what happens within a school, such as Te Koutu. It is not always possible to directly transfer the features, which support successful outcomes in one school, to another context.

There are many factors that support or hinder particular practices being adopted. In each case study, we showed that every challenge is complicated by other related challenges. Over time the challenges also changed as they become situated in new historical moments. For example, the issues that had faced the development of the mathematics register in the 1980s were different issues than those currently being faced. In the early days, it was a challenge to collate and standardise the different word lists from individual teachers, scattered across the country. At this time, there were a number of strong native speakers, some of whom had had an active role in developing the word lists. In the 1990s, the intervention of the Māori Language Commission resulted in the replacement of many of the transliterated terms that had been included in the word lists. Their mission was to return *te reo Māori* to as pure a state as possible. Currently, the challenges include revisiting some of these earlier choices of terms to determine if they provide an appropriate connection to the mathematical concept. However, there are now fewer of the native speakers to consult who had contributed to the development of the initial word lists. Throughout the last 30 years, the challenge of trying to standardise the terminology was resolved. It may be that in the future, the publication and subsequent use of teaching resources in *te reo Māori* will achieve the standardisation more rapidly then any of the previous initiatives. This will reduce the likelihood that if students move from school to school or even from class to class, they will be faced with having to learn a whole, new set of words. However, a reliance on readily available resources has the risk of teachers not using their creativity to meet the needs of the individual students in their classrooms.

The variability in how different factors interact together in specific situations means that individual teachers or schools will make different choices about how to proceed within their specific contexts. Findings from Te Koutu cannot be transferred readily to other schools or systems. Although other immersion programmes may have faced similar issues to those in Aotearoa/New Zealand, the solutions are quite likely to be different. It is the circumstances in which the challenges are presented, which affect what is perceived as a challenge, and how people respond to it.

Ní Rîordaín (2010) described how Gaeilge-immersion schools in Ireland have become “trendy” in a similar way to how Māori-immersion schools became popular with Māori parents in the 1990s. She described how legislation had first banned the use of Gaeilge in schools and then a change enforced the use of Gaeilge in schools. In some ways, this is similar to the effect of legislation reversal in Aotearoa. Yet, in Ireland it was not until after legislation relating to the use of Gaeilge was lifted that parents began to send their children in large numbers to Gaeilge schools in predominantly English-speaking areas of Ireland. Ní Rîordaín (2010) felt that parents had resisted having the state mandate that their children should be taught in Gaeilge but were not prepared to have the language lose prestige when the legislative requirement was removed. On the other hand, it is only very recently that the impact of teaching mathematics in Gaeilge has begun to be investigated. The legislative responses were arguably directly connected to the historical oppression of the Irish within their own country as were some of the legislative responses in New Zealand regarding Māori. Although the sorts of responses were similar, the actual legislation was different as were people’s responses to it. Other factors, such as Irish people being the majority of the population compared to Māori in the minority, would have contributed to the differing reactions to legislation.

One of the main findings from our long-running project at Te Koutu is how much the context, at both the societal and local levels, influences micro-level interactions in a classroom between individual teachers and students. For example, the belief that Māori students’ academic results are affected by teachers’ low expectations has galvanised many of Te Koutu’s teachers into ensuring that they set high expectations for students’ mathematics as well as language-learning outcomes. How this was achieved in individual classrooms depended very much on how teachers perceived their role. These perceptions were shaped both by their previous experiences and also by aspirations for their future as teachers (Skovsmose, 2005a).

The Stages in Meeting Challenges

We have outlined a number of different challenges ranging from the setting up of Māori-medium education and the development of the mathematics register to examining how teachers and students learn about improving their understandings of mathematics education. Different participants were at different stages in meeting the challenges, and very few of these challenges could be said to have been successfully negotiated by all. Some challenges have been met, such as Te Koutu being established with mathematics classes being taught in *te reo Māori*, but others persist. Often, individual students, teachers, or community members were at different stages in resolving the same challenge. We suggest that there are different stages in facing and meeting challenges.

The first of these stages is to recognise that there is a challenge. We noted that in the teaching of probability there did not seem to be a recognition of the role that cultural activities and first language may have in developing intuitive probability understandings. If the challenge is not recognised, then no investigation or solution will be sought. The reasons for this non-recognition can be varied. In some cases, a situation is accepted as normal practice, and so no alternative practices are contemplated. It was not until our first language project in 2005–2006 had been undertaken that we became aware there was little writing occurring in mathematics lessons. It was accepted as the norm that bookwork consisted of students answering sets of exercise questions with short answers. However, once the issue was raised, it was possible both to document what had been occurring and to consider how writing in mathematics could support students to improve their language proficiency. This led to many teachers at Te Koutu making changes to their teaching so that their students provided more explanations and justifications. Not only was this expected to improve the quality and quantity of students' mathematical writing, but it was also assumed that it would also support students' mathematical thinking and their fluency in *te reo Māori*.

Another stage in the process of resolving challenges is that of resistance. As stated in Chapter 1, we do not see resistance as being negative. Rather, it is part of the contestation needed in order for new ideas to come forth. We acknowledge that resistance can be destructive to the collaboration process as documented by Hynds (2008). Therefore, it needs to be carefully discussed to enable participants to move forward. Silence about the causes for the resistance results in missed opportunities to deal with the underlying causes of the resistance.

For example, the use of *te reo Māori* in everyday settings has been noted as something that was being resisted by non-Māori. Yet if *te reo Māori* is to fulfil a range of functions, more has to be done in the wider New Zealand community so that using the language becomes "trendy", and non-Māori are encouraged to accept it. For conversations in *te reo Māori* to be commonplace, enough speakers with similar levels of fluency have to be in the same place at the same time. If the levels of proficiency are too diverse, it becomes much easier for people to switch to English in which they have similar levels of proficiency. Therefore, there is a need not just to encourage everyday talk in *te reo Māori* but also to support more Māori adults to improve their language skills so that they can participate fluently in conversations.

Another situation of resistance occurred during our research when teachers realised that their mathematics lessons would be video recorded. Although the teachers were and continue to be unenthusiastic about being recorded, many also valued the opportunity to watch what happened in their classrooms and to discuss what went well and what needed improvement. This enabled teachers to determine other courses of action, which they may not have contemplated without seeing themselves teach.

Challenges begin to be resolved once alternative activity spaces are imagined, and then explored. In some ways the setting up of Te Koutu can be seen as an exploration of a previously unimagined activity space. It was a grassroots process, whereby different resources within the community were galvanised to work together to provide an education for children that would suit their needs. For those who had been agitating for change, setting up Māori-immersion schooling became something practical to do. Again, it can be seen that the circumstances affected the possibilities within the action spaces that opened up.

Te Koutu parents knew about *kura kaupapa Māori*, but they had not been able to enrol their children in the local one in Rotorua because of the prevailing perception that *kura* should remain small. This meant that the parents were left with no alternative but to set up their own *kura*, or enrol their children in the local state schools. At that time, the state schools showed no inclination whatsoever to provide anything that would build on the strengths that *kōhanga reo* preschool graduates brought with them to primary school.

Setting up a new kind of school meant that the parents and the teachers had to make many decisions about the sort of *kura* that they wanted for their children. Some of these decisions resulted in some parents leaving the school because they felt that there was too strong an emphasis on academic achievement, and this was detrimental to the revitalisation of the language and the maintenance of culture. Consequently, another *kura* was set up in Rotorua which catered better to the needs of these parents. Such developments were supported by the political impetus to give parents more choice about the schooling for their children. However, if *kura kaupapa Māori* had not been able to be developed because the system had not countenanced it, perhaps activists might have worked more with state schools to provide an education that suited the needs of *kōhanga reo* graduates. It has only been in very recent years that state schools have begun to adapt to the needs advocated by Māori parents. Yet as Hynds (2008) documented, non-Māori parents and teachers have strongly resisted these changes. Perceiving opportunities as contestable spaces is important if challenges are to be resolved in ways that best meet the needs of those involved.

The final stage of meeting challenges is when a new order comes to be seen as the norm. By the early 1970s, it had been over a hundred years since *te reo Māori* was banned from being the language of instruction in schools. The norm was very much that schooling could and should only be done in English in New Zealand. However, the advent of Māori-immersion schooling has put paid to this idea. It is now well accepted that *te reo Māori* can and should be used in schooling and that it is possible for Western domains of knowledge such as mathematics to be taught in this language.

The Features of Collaboration That Support Meeting Challenges

In order for people to meet the challenges before them, it is important that collaboration occurs at each stage of the process. Different components of collaboration come to the fore over the course of meeting challenges. Like the components of the context, these features interact to support or hinder the resolution of the challenges. Features that contribute to collaboration being supportive are:

- joint discussion,
- an intolerance of silence,
- specific resources put aside for discussion,
- building and maintaining relationships, and
- an acceptance of compromise.

At all stages of meeting challenges, joint discussion is essential. Becoming aware that there is a challenge to be met often requires people to talk together so that a normal practice can be problematised. This means that within the discussion, there must be a willingness to listen respectfully and to not use blame to take the focus away from finding alternative courses of action. The joint discussions at Te Koutu were more successful if the focus was on the strengths of the different participants.

For example, Te Koutu often employed teachers who had completed English-medium teacher education programmes. These teachers may have strong *te reo Māori* skills and so chose for a range of reasons to do their teacher education in English. In discussions with the teachers about their knowledge of the mathematics register, it became clear that the school had to take on responsibility for providing professional development about the mathematics register. It could not be assumed that the teachers would know the *te reo Māori* terms, as English-medium education would not have provided information on this. Further discussions at Te Koutu resulted in Tony Trinick attending a staff meeting to describe the background to some terms. Within these discussions, there was an opportunity for teachers to share some of their strategies for learning mathematical terms and expressions. Resistance to teaching compass directions because they were orientated to East–West, rather than to North–South as in English, was somewhat overcome by hearing Tony talk about the background to why an East–West orientation was chosen. As the main users of the language, it is important that teachers' concerns about some terms are heard by those who are making decisions about the mathematics register. Not providing teachers with space to discuss these concerns will only result in them resisting using terms whose meanings they do not agree with. The in-depth reflection needed for changing practices, both learning and teaching practices, is strongly supported by having regular, intensive discussions about what is occurring in classrooms.

In contrast, silence about topics is unhelpful as it contributes to challenges going unrecognised. In the wider community, the New Zealand Ministry of Education continues to provide data to show that students who attend Māori-immersion schools do significantly better than their peers at English-medium schools in relationship to

NCEA. This has been extremely well-received, especially by Māori, who wanted to prove to the many doubters that Māori-immersion education has resulted not only in the language being saved but also in Māori students achieving academically. There is much to lose if Māori-immersion education is presented as unsuccessful. In 2008, the Northern Territory government in Australia effectively scrapped the bilingual education programmes in remote Indigenous communities, because of the perception that bilingual programmes had not been successful in providing students with adequate English (Wilkins, 2008).

Yet the situation in relationship to the external examinations remains problematic, unless it can be discussed and a resolution worked towards. The challenge becomes one of how to open up the discussion so the successes achieved by Māori-immersion education are not lost in the discussion about what needs to be improved. For us a supportive collaboration does not tolerate silences about some topics because of their political sensitivity. As Hynds (2008) stated about her own research:

> Dominant but unexplored and unacknowledged discourses seemed to influence the acceptance and practice of teachers' collaborative partnership work within and across both school communities. There also appeared to be a lack of leadership that would have engaged participants in critical analysis and open communication on such issues. (p. 155)

If discussion is to be encouraged, resources must be made available for this to occur. Every *kura kaupapa Māori* is different because of the circumstances in which they arose, have matured, and see themselves in the future. In order for individual *kura* and teachers to deal with the challenges that they face, there is a need for resources. The biggest need is to have time made available for teachers to talk together, but also to reflect on their own teaching. We would argue that having outside researchers as discussion partners supported the reflections, both individual and joint. Time to talk means teachers being granted relief days, and the availability of discussion partners also requires money. We have been very fortunate over the years in gaining grants to assist. The Teaching and Learning Initiative grant from the Ministry of Education provided initial funding in 2005 and again in 2007. This funding was provided specifically for schools and teachers to work on issues that were important for them, in collaboration with researchers. As well, the school itself contributed money, with some extra funding coming from the researchers' universities at different times. Nevertheless, seeking and gaining funding may not be possible for all *kura*. Yet the isolation of many teachers in Māori-immersion education and the unique challenges that they face (Te Maro, Averill, Higgins, & Tweed, 2009) suggest that professional development within individual *kura* has greater potential for success than pre-packaged programmes.

One of the most important features that contributed to successful collaboration in meeting challenges was *whakawhānaungatanga,* which means building and maintaining relationships (Bishop, 2008). If collaborating to meet challenges is assumed to involve learning, then the need to have others to share ideas with is important.

> In plain terms—people learn from and with others in particular ways. They learn through practice (learning as doing), through meaning (learning as intentional), through community (learning as participating and being with others), and through identity (learning as changing who we are). Professional learning so constructed is rooted in the human need to feel a sense of belonging and of making a contribution to a community where experience and knowledge function as part of community property. Teachers' professional development should be refocused on the building of learning communities. (Lieberman & Mace, 2008, p. 227)

In responding to the challenges, there have usually been several groups of people working together. For example, to compile the first mathematics word lists, different individuals and groups came together. These included teachers, *kaumatua* (elders), mathematicians, mathematics educators, linguists, and education policy people. Without the combination of people, it would have been very difficult for the word lists to gain enough status needed for them to be accepted by those who needed to use them.

Building and maintaining relationship takes time and energy. Even teachers who work together carrying out their normal duties may not know each other very well. When many of the Te Koutu teachers attended a mathematics teacher conference in 2007, there was much discussion about how valuable it was to spend time together so that they could get to know each other (Meaney et al., 2009a).

A long-running project such as ours could not have continued without *whakawhānaungatanga*. Since 1998 when we began working together, the staff at Te Koutu, as well as the students, have changed constantly. For researchers, who visit intermittently, building relationships is an important part of how we spend our time. It is essential that we build trust with the teachers, so that they are willing to share their practices with us. Without this level of trust, any suggestions that we made about alternative mathematics education practices would have been less well received (see Meaney, 2005a, for a discussion of an outsider's role in Indigenous research). Good relationships also enable the understandings we gained from earlier work, such as the use of the MRA model, to be shared with new staff. In looking for his family, Māui too recognised that relationships are at the centre of any understanding of what is to be done.

The final feature that we see as essential in collaboration is an acceptance that compromises are necessary at times. If the primary aims of revitalising and maintaining the language, as well as supporting students to achieve academically, remain the focus of Te Koutu, then compromises do not have to become limitations of the possibilities for action. In becoming a state-funded school, Te Koutu not only received money for salaries and buildings but also had to accept the requirement to teach the state-mandated curriculum. This is perhaps not such a big compromise as it is unlikely that parents, even the most staunch of activists, would have sent their children to a school which did not provide them with qualifications for careers in the mainstream society of New Zealand.

Nonetheless, compromise emerged between needing funding to continue providing an education to Te Koutu students and wanting to maintain the integrity of a Māori-focused curriculum. This tension, which the compromise resolved, remains part of an ongoing discussion at Te Koutu. Although there is no talk about rejecting the funding, there is much debate about how the Māori focus can be implemented within a state-mandated curriculum, including the teaching of mathematics. Over the years, teachers have incorporated activities which reinforced cultural values, for example, the head being *tapu*, and showing the skills of the ancestors through highlighting the use of transformation in traditional designs. The integration of these activities into mathematics lessons is a reactive response to the reality, rather than a proactive response, which organically may have developed a new curriculum from traditional understandings about Māori culture. However, by keeping the focus on the revitalisation and maintenance of *te reo Māori*, Te Koutu has ensured that the mathematics teaching has not been a simple vehicle for the transmission of Western knowledge and values.

Compromises require concessions from all parties, and this is generally necessary when challenges are to be worked through. What is important is how much the concession forces participants away from what they view as important in the education of students. In many ways, the compromises from collaboration can be viewed as the spaces in which the external challenge from having to interact with the European society of New Zealand intersects with the internal challenge that the Te Koutu community set for itself in teaching mathematics in *te reo Māori*.

Collaboration requires many features to be successful. These features need to be present individually, but they also must interact effectively. The case studies showed that collaboration between different participants contributed to many challenges being recognised and worked through, some to a successful conclusion. Collaboration worked differently in the various stages of solving problems. Joint discussions were most prominent in having challenges recognised. Although joint discussions were also necessary in the other stages, building and maintaining relationships supported the overcoming of resistance. In the exploring of alternative practices, the acknowledgement of the importance of compromise was essential.

Conclusion

In using the story of Māui, throughout this book, we wanted to illustrate how the story of teaching mathematics in an Indigenous language cannot be considered as being similar to a Western fairy story. There is no happy ending with everyone living harmoniously together. As the situation within and without Te Koutu changes, the challenges for this learning community will also change. As one set of challenges are resolved, others will arise. Like the excerpts from the story of Māui, we have provided a snapshot over a particular period of time to illustrate the kinds of challenges faced when teaching mathematics in an Indigenous language.

The case studies presented within this book indicate how the collaboration of a range of different people can result in huge changes to what is considered normal. These changes do not happen overnight, and there is a need for significant collaboration in order for alternative practices to be seen as possibilities. Collaboration that builds on participants' strengths can achieve great things.

Setting up an alternative education system and individual schools like Te Koutu can only be effective if there are clear goals about what students are expected to achieve. Although learning about how mathematics is taught in an Indigenous language can be valued for the intellectual and theoretical input that it provides about what is possible, for the Te Koutu community it also *had* to be successful. Their children are not laboratory guinea pigs but humans with adult lives to be navigated. Te Koutu needed to provide them with the best start possible. In this book, we have provided information not just about the challenges faced, but also about how the caregivers, students, teachers, and researchers worked with outside organisations such as the Ministry of Education in order to ensure the school's success. Consequently, we hope that readers will read this first and foremost as a story of real people who came together to do extraordinary things. Therefore, it is fitting to conclude with a final Māori proverb that emphasises the role of people within this story.

Hutia te rito o te harakeke
Kei whea te kōmako e kō
Kī mai ki ahau
He aha te mea nui o te ao?
Māku e kī atu
He tāngata, he tāngata, he tāngata
Pluck the shoots of the flax and it will die
Then where will the kōmako be?
Say to me
What is the greatest thing in the world?
I will respond
It is the people, the people, the people.

Glossary

Māori Words

aho	mathematical chord
ahutoru	pyramid
Aotearoa	New Zealand
arā	linguistic marker that highlights that what follows will be surprising to the listener
aroha	love
eka	acre (transliteration)
hanga	shape
hangarite	symmetrical
herengi	shillings (transliteration)
hīnaki	eel trap
huihui	addition
ine	measurement
īnihi	inch (transliteration)
iwi	tribe
kaumatua	elders
kaute	count (transliteration)
kauwhata	graph
kē	linguistic marker that informs the listener that a definition is to follow
kura	school
kura kaupapa Māori	Māori-immersion education system based around the principles of *Te Aho Matua*
mahi whiki	figure-work
mahi nama	number work
mano	traditionally an indescribably large number, but now more commonly means a thousand
matipikeihana	multiplication
Māui	a great Polynesian hero
māhunga	head (on a coin)

T. Meaney et al., *Collaborating to Meet Language Challenges in Indigenous Mathematics Classrooms*, Mathematics Education Library 52,
DOI 10.1007/978-94-007-1994-1, © Springer Science+Business Media B.V. 2012

mārō	fathom
matua	father or male teacher
mātauranga	knowledge
meha	measure (transliteration)
miriona	million (transliteration)
mutunga	finite
nama	number (transliteration)
panoni	geometric transformation
pāuna	pounds sterling, English money (transliteration)
parahau	justification genre
Pākehā	non-Māori and predominantly European origin, generally referring to people
peni	pennies (transliteration)
Poutama Tau	professional development project that focussed on students' numeracy strategies. Te Koutu joined this project in 2005 but has had an intermittent exposure to it
pūhara	bushel
rite	corresponding
rau	traditionally a great many, but more commonly now means 100
rūnā	pare down
tamariki	children
tango	subtraction
taonga	treasure
tapa whā rite	square
tatau ake	count on
tatau māwhitiwhiti	skip count
tatau pāngatahi	one to one counting
tau	number
tau e whakareatia ana	multiplicand
tau matapōkere	random number
tau whakawehe	divisor
tauanga	statistics
taurau	a hundred number
taurea	multiple
taurua	even numbers
Te Aho Matua	document setting out the guiding principles for kura kaupapa Māori
te Taura Whiri	the Māori Language Commission
te reo Māori	the Māori language
te reo Tātaitati	the mathematics register in te reo Māori
tikanga	culture
tuaka	axis of a graph
tūpuna	ancestors
unahi	parabola

wehe	division
wero	ritual challenge presented to visitors as part of the pōwhiri
whaea	mother or female teacher
whakaahua	description genre
whakaaro	notion
whakamārama	explanation genre
whakarea	multiplication
whakarūnā	simplify
whakatauwehe	factorisation
whānau	family or wider school community
wharekura	Māori-immersion high school
whenu	cosine
whiore	tail (of a coin)
whika	figure (transliteration)

Acronyms

ESL	English as a Second Language
IRF	initiation–response–feedback exchange
NCEA	National Certificate in Educational Achievement
NCTM	National Council of Teacher of Mathematics
NZ	New Zealand
NZQA	New Zealand Qualifications Authority
MRA	mathematics register acquisition
OECD	Organisation for Economic Co-operation and Development
RAVE	Read (the question), Answer (the question), use mathematics Vocabulary, and use Examples. It is a structure for producing a written response to a mathematics question
TAM	Te Aho Matua

References

Addison, J., & Te Whare, W. (n.d.). *Arhythmetic-kanikani pāngarau.*

Adu-Ampona, S. M. (1975). African languages in the teaching of science, mathematics and technology. *Science Teacher, 19*(5–6), 21–26.

Albert, L. (2000). Outside-in – inside-out: Seventh grade students' mathematical thought processes. *Educational Studies in Mathematics, 41*, 109–141.

Alrø, H., & Skovsmose, O. (2002). *Dialogue and learning in mathematics education: Intention, reflection, critique.* Boston: Kluwer Academic Publishers.

Alrø, H., Skovsmose, O., & Valero, P. (2009). Inter-viewing foregrounds: Students' motives for learning in a multicultural setting. In M. César & K. Kumpulainen (Eds.), *Social interactions in multicultural settings* (pp. 13–37). Rotterdam: Sense Publishers.

Amir, G. S., & Williams, J. S. (1999). Cultural influences on children's probabilistic thinking. *Journal of Mathematical Behaviour, 18*(1), 85–107.

Amit, M., & Jan, I. (2006). Autodidactic learning of probabilistic concepts through games. In J. Novotná, H. Moraová, M. Krátká, & N. Stehlíková (Eds.), *Proceedings of the 30th conference of the international group for the psychology of mathematics education* (Vol. 2, pp. 49–56). Prague: PME.

Anghileri, J. (2002). Scaffolding practices that enhance mathematics learning. In A. Cockburn & E. Nardi (Eds.), *Proceedings of the 26th annual conference of the international group for the psychology of mathematics education* (Vol. 2, pp. 49–56). Norwich, UK: UEA.

Annamalai, E. (1979). Movement for linguistic purism. The case of Tamil. In E. Annamalai (Ed.), *Language movements in India* (pp. 35–39). Mysore: Centre for Indian Languages.

Anthony, G., & Walshaw, M. (2007). *Effective pedagogy in mathematics/Pāngarau: Best evidence synthesis iteration.* Wellington: New Zealand Ministry of Education.

Antonacopoulou, E. P. (2008). On the practise of practice: In-tensions and ex-tensions in the ongoing reconfiguration of practices. In D. Barry & H. Hansen (Eds.), *The Sage handbook of new approaches in management and organization* (pp. 112–131). Los Angeles: Sage.

Apple, M. W. (1992). Do the standards go far enough? Power, policy, and practice in mathematics education. *Journal for Research in Mathematics Education, 23*(5), 412–431.

Appleby, P. (2002). Kura kaupapa Māori: Tomorrow's schools and beyond. *New Zealand Annual Review of Education, 11*, 105–121.

Apthorp, H. S., D'Amato, E. D., & Richardson, A. (2003). *Effective standards-based practices for native American Indians: A review of the literature.* Aurora, CO: Mid-Continent Research for Education and Learning.

Au, K. H. (1980). Participation structures in a reading lesson with Hawaiian children: Analysis of a culturally appropriate instructional event. *Anthropology and Education Quarterly, 11*(2), 91–115.

Bakalevu, S. L. (1999). The language factor in mathematics education. *Journal of Educational Studies, 21*(2), 59–68.

Ball, D. L. (2000). Bridging practices: Intertwining content and pedagogy in teaching and learning to teach. *Journal of Teacher Education, 51*(3), 241–247.

Barbour, R., & Keegan, T. T. (1996). Education in technologies for LCTLs: Ngā tautono roro-hiko. *NFLRC 1996 summer institute symposium: New technologies for less commonly taught languages conference,* June–July, 1996. Retrieved from http://www.cs.waikato.ac.nz/~tetaka/PDF/EducationInTechnology.pdf.

Barton, B., & Cleave, P. (1989). He kupu tikanga tau ahuatanga. *Special edition of Nga Mauranga Auckland*: MECA: Auckland College of Education.

Barton, B., & Fairhall, U. (1995a). *Mathematics in Māori education.* Mathematics Education Unit, Department of Mathematics, University of Auckland, New Zealand.

Barton, B., & Fairhall, U. (1995b). Is mathematics a Trojan horse? In B. Barton & U. Fairhall (Eds.), *Mathematics in Māori mathematics.* Mathematics Education Unit, Auckland: University of Auckland.

Barton, B., Fairhall, U., & Trinick, T. (1995). *Whakatupu reo tatai: History of the development of Māori mathematics vocabulary SAMEpapers.* Hamilton: Centre for Science, Mathematics and Technology Education Research, University of Waikato.

Barton, B., Fairhall, U., & Trinick, T. (1998). Tikanga reo tatai: Issues in the development of a Māori mathematics register. *For the Learning of Mathematics, 18*(1), 3–9.

Begg, A. (1991, November). *Mathematika Pasefika UNESCO conference* Hamilton: New Zealand.

Begg, A. (1999, July). *Learning theories and mathematics: A, B, C, D and E.* Paper presented at the sixth biennial conference of the New Zealand Association of Mathematics Teachers, Dunedin, New Zealand.

Ben-Chaim, B., Lappan, G., & Houang, R. T. (1989). Adolescents' ability to communicate spatial information: Analyzing and effecting students' performance. *Educational Studies in Mathematics, 20,* 121–146.

Benko, P., & Maher, C. (2006). Students constructing representations for outcomes of experiments. In J. Novotná, H. Moraová, M. Krátká, & N. Stehlíková (Eds.), *Proceedings of the 30th conference of the international group for the psychology of mathematics education* (Vol. 2, pp. 137–143). Prague: PME.

Benton, R. (1996). The Māori language in New Zealand. In S. Wurm, P. Mühlhäusler, & T. Tryon (Eds.), *Atlas of languages of intercultural communication in the Pacific, Asia and the Americas* (pp. 167–172). Berlin: Walter de Gruyter.

Benton, R. A. (1991). *The Māori language: Dying or reviving*? Working paper prepared for the East–West Centre Alumni – in-Residence Working Paper. Hawaii: East-West Centre.

Bernstein, B. (1990). *Class, codes and control: The structuring of pedagogic discourse* (Vol. IV). London: Routledge.

Bernstein, B. (2000). *Pedagogy, symbolic control and identity: Theory, research, critique.* Lanham, MD: Rowman & Littlefield.

Berry, J. (1985). Learning mathematics in a second language: Some cross cultural issues. *For the Learning of Mathematics, 5*(2), 18–21.

Best, E., (1907). Māori numeration. The vigesimal system. *Journal of the Polynesian Society, 16,* 94–98.

Bickmore-Brand, J., & Gawned, S. (1990). Scaffolding for improved mathematical understanding. In J. Bickmore-Brand (Ed.), *Language in mathematics.* Melbourne: Australian Reading Association.

Bicknell, B. (1999). The writing of explanations and justifications in mathematics: Differences and dilemmas. In J. M. Truran & K. M. Truran (Eds.), *Making the difference: Proceedings of the 22nd annual conference of the mathematics education research group of Australasia* (pp. 75–83), 4–7 July 1999. Adelaide: MERGA.

Bill, V. L., Leer, M. N., Reams, L. E., & Resnick, L. B. (1992). From cupcakes to equations: The structure of discourse in a primary mathematics classroom. *Verbum, 1–2,* 63–85.

Bills, C. (2002). Linguistic pointers in young children's descriptions of mental calculations. In A. Cockburn & E. Nardi (Eds.), *Proceedings of the 26th annual conference of the international group for the psychology of mathematics education* (Vol. 2, pp. 97–104). Norwich: PME.

Bishop, A. (1988). *Mathematical enculturation: A cultural perspective on mathematics education*. Dordrecht, Netherlands: Kluwer Academic.

Bishop, R. (2007). Māori education. In A. W. H. Timperley, H. Barrar, & I. Fung (Eds.), *Best evidence synthesis: Teacher professional development* (pp. xvi–xix). Wellington: Ministry of Education.

Bishop, R. (2008). A culturally responsive pedagogy of relations. In C. McGee & D. Fraser (Eds.), *The professional practice of teaching* (3rd ed., pp. 154–171). Melbourne: CENGAGE Learning.

Bishop, R., Berryman, M., Tiakiwai, S., & Richardson, C. (2003). *Te Kotahitanga: The experiences of year 9 and 10 Māori students in mainstream classrooms*. Report to the Ministry of Education. Wellington: Ministry of Education.

Black, P., & Wiliam, D. (2001, September). *Inside the blackbox: Raising the standards through classroom assessment*. British Education Research Association Conference, 13–15 September 2001, Leeds University. Retrieved from http://www.collegenet.co.uk/admin/download/inside%20the%20black%20box_23_doc.pdf.

Bonner, P. J. (2006). Transformation of teacher attitude and approach to math instruction through collaborative action research. *Teacher Education Quarterly, 33*(3), 27–44.

Boston, J., Martin, J., Pallot, J., & Walsh, P. (1996). *Public management: The New Zealand model*. Auckland: Oxford University Press.

Boyce, M. (2005). Attitudes to Māori. In A. Bell, R. Harlow, & D. Starks (Eds.), *Languages of New Zealand* (pp. 86–110). Wellington: Victoria University Press.

Brenner, M. E. (1998). Adding cognition to the formula for culturally relevant instruction in mathematics. *Anthropology and Education Quarterly, 29*(2), 214–244.

Brett, A., Rothlein, L., & Hurley, M. (1996). Vocabulary acquisition from listening to stories and explanations of target words. *The Elementary School Journal, 96*(4), 415–422.

Brewer, J. (2000). *Ethnography*. Buckingham: Open University Press.

Brislin, R. (1970). Back-translation for cross-cultural research. *Journal of Cross-Cultural Research, 1*(3), 185–216.

Brown, M., Jones, K., Taylor, R., & Hirst, A. (2004). Developing geometrical reasoning. In I. Putt, R. Faragher, & M. McLean (Eds.), *Mathematics education for the third millennium: Towards 2010: Proceedings of the 27th annual conference of the mathematics education research group of Australasia* (Vol. 1, pp. 127–134). Townsville: MERGA.

Brown, T. (1997). *Mathematics education and language: Interpreting Hermeneutics and Post-structuralism*. Dordrecht: Kluwer Academic Press.

Bull, A., Brooking, K., & Campbell, R. (2008). *Successful home-school partnerships: A report to the ministry*. Wellington: New Zealand Council for Educational Research.

Burkhill, R., & Bye, K. (2005). *Action research: Using formative assessment strategies to raise student achievement*. Retrieved from http://www.nzqa.govt.nz/ncea/for.../actionresusingformassessment.pdf.

Burns, M. (2005). The building blocks of math. *Instructor, 115*(3), 42–43.

Burton, L., & Morgan, C. (2000). Mathematicians writing. *Journal for Research in Mathematics Education, 31*(4), 429–453.

Butterworth, B., & Reeve, R. (2008). Verbal counting and spatial strategies in numerical tasks: Evidence from Indigenous Australia. *Philosophical Psychology, 21*(4), 443–457.

Campbell, A. E., Adams, V. M., & Davis, G. (2007). Cognitive demands and second language learners: A framework for analysing mathematics instructional contexts. *Mathematical Thinking and Learning, 9*(1), 3–30.

Campbell, P., & Rowan, T. (1997). Teacher questions + student language + diversity = mathematical power. In J. Trentacosta & M. Kenney (Eds.), *Multicultural and gender equity in the mathematical classroom the gift of diversity* (pp. 60–70). Reston, VA: National Council of Teachers of Mathematics.

Cantoni, G. (1991, Fall). Applying a cultural compatibility model to the teaching of mathematics to indigenous populations. *Journal of Navojo Education, IX*(1), 33–42.

Carr, M., McGee, C., Jones, A., Bell, B., Barr, H., & Simpson, T. (2005). *The effects of curricula and assessment on pedagogical approaches and on educational outcomes*. Wellington: Ministry of Education.

Carrasquillo, A. L., & Rodriguez, V. (1996). *Language minority students in the mainstream classroom*. Clevedon, UK: Multilingual Matters.

Castle, K., & Aichele, D. B. (1994). Professional development and teacher autonomy. In D. B. Aichele & A. F. Coxford (Eds.), *Professional development for teachers of mathematics 1994 yearbook* (pp. 1–8). Reston, VA: National Council of Teachers of Mathematics Inc.

Cavalcanti, M. C. (2004). 'It's not writing by itself that is going to solve our problems': Questioning a mainstream ethnocentric myth as part of a search for self-sustained development. *Language and Education, 18*(4), 317–325.

Chapman, A. (1997). Towards a model of language shifts in mathematics learning. *Mathematics Education Research Journal, 9*(2), 152–172.

Chapple, S., Jeffries, R., & Walker, R. (1997). *Māori participation and performance in education. A literature review and research programme*. Wellington: Ministry of Education.

Chick, H., & Baker, M. (2005). Teaching elementary probability: Not leaving it to chance. In P. Clarkson, A. Downton, D. Gronn, M. Horne, A. McDonough, R. Pierce, & A. Roche (Eds.), *Building connections: Theory, research and practice: Proceedings of the 28th annual conference of the mathematics education research group of Australasia* (pp. 233–231). Melbourne: MERGA.

Christensen, I. (2003). *Exploring issues in mathematics education*. Wellington: Ministry of Education.

Christensen, I., Trinick, T., & Keegan, P. J. (2003, June). *Pāngarau curriculum framework and map: Levels 2–6: asTTle technical report 38*. Auckland: University of Auckland/Ministry of Education.

Clarke, D. J. (1996). Assessment. In A. J. Bishop, M. A. Clements, C. Keitel, J. Kilpatrick, & C. Laborde (Eds.), *International handbook of mathematics education* (pp. 327–370). Dordrecht: Kluwer Academic Publisher.

Clarke, D. J., Waywood, A., & Stephens, M. (1993). Probing the structure of mathematical writing. *Educational Studies in Mathematics, 25*, 235–250.

Clarkson, P. C. (1991). *Bilingualism and mathematics learning*. Geelong: Deakin University.

Cleghorn, A. (1992). Primary level science in Kenya: Constructing meaning through English and Indigenous languages. *International Journal of Qualitative Studies in Education, 5*(4), 311–323.

Closs, M. P. (1977). *A survey of mathematics development in the new world: Report 410-77-0222*. Ottawa, Canada: University of Ottawa.

Clyne, M. (1997). Introduction. In M. Clyne (Ed.), *Undoing and redoing corpus planning* (pp. 1–10). Berlin: Mouton de Gruyter.

Cobarrubias, J., & Fisherman, J. A. (Eds.). (1983). *Progress in language planning: International perspectives*. Berlin: Mouton.

Cocking, R. R., & Mestre, J. P. (Eds.). (1988). *Linguistic and cultural influences on mathematics learning*. Hillsdale, NJ: Lawrence Erlbaum Associates.

Cooper, R. L. (1989). *Language planning and social change*. Cambridge: Cambridge University Press.

Creswell, J. W. (1998). *Qualitative inquiry and research design: Choosing among five traditions*. Thousand Oaks: Sage.

Crowley, T. (1998). How many languages will survive in the Pacific? *Te Reo, 41*, 116–125.

Dale, D. C., & Cueras, G. J. (1987). Integrating language and mathematics learning. In J. Crandell (Ed.), *ESL through content-area instruction* (pp. 9–23). Regents, NJ: Prentice Hall.

Darder, A. (1991). *Culture and power in the classroom: A critical foundation for bicultural education*. London: Bergin & Garvey.

Dawe, L. (1983). Bilingualism and mathematical reasoning in English as a second language. *Educational Studies in Mathematics, 14*, 325–353.

Denny, J. P. (1986). Cultural ecology of mathematics: Ojibway and Inuit hunters. In M. Closs (Ed.), *Native American mathematics* (pp. 129–180). Austin: University of Texas Press.

Dewaele, J.-M. (2007). Multilinguals' language choice for mental calculation. *Intercultural Pragmatics, 3–4*, 343–376.

Dewalt, M. W., & Troxell, B. K. (1989). Old order Mennonite one-room school: A case study. *Anthropology and Education Quarterly, 20*(4), 308–325.

Dixon, R. (1991). The endangered language of Australia, Indonesia and Oceania. In R. Robins & E. Uhlenbeck (Eds.), *Endangered languages*. Oxford, NY: Berg.

Doerr, H. M., & Chandler-Olcott, K. (2009). Negotiating the literacy demands of standards based curriculum materials: A site for teachers' learning. In A. Schoenfeld (Series Ed.) & J. Remillard, B. Herbel-Eisenmann, & G. Lloyd (Vol. Ed.), *Mathematics teachers at work: Connecting curriculum materials and classroom instruction* (pp. 282–301). New York: Routledge.

Drake, B. M., & Amspaugh, L. B. (1994). What writing reveals in mathematics. *Focus on learning problems in mathematics, 16*(3), 43–50.

Draper, R. J. (2002). School mathematics reform, constructivism, and literacy: A case for literacy instruction in the reform-orientated math classroom. *Journal of Adolescent and Adult Literacy, 45*(6), 520–529.

Duncombe, R., & Kathleen, A. (2004). Collaborative professional learning: From theory to practice. *Journal of In-Service Education, 30*(1), 141–166.

Durie, M. (2003). *Ngā kāhui pou: Launching Māori futures*. Wellington: Huia Publishers.

Durkin, K., & Shire, B. (1991). *Language in mathematics education: Research and practice*. Milton Keynes: Open University Press.

Earp, R. (2004, September). *The treaty of Waitangi and Māori health*. Paper presented at the National Forum on Indigenous Health and the Treaty Debate, University of New South Wales, Sydney.

Education Review Office. (2002). *The performance of Kura Kaupapa Māori*. Wellington: Author.

Ellerton, N., & Clements, M. A. (1990). Language factors in mathematics learning: A review. In K. Milton & H. McCann (Eds.), *Mathematical turning points: Strategies for the 1990s: Proceedings of the13th biennial conference of the Australian Association of mathematics teachers* (Vol. 1, pp. 230–258). Hobart: The Australian Association of Mathematics Teachers.

Esty, W. (1992). Language concepts of mathematics. *Focus on Learning Problems in Mathematics, 14*(4), 31–54.

Evans, S. W. (2006). Differential performance of items in mathematics assessment materials for 7-year-old pupils in English-medium and Welsh-medium versions. *Educational Studies in Mathematics, 64*, 145–168.

Fairhall, U., & Keegan, P. (2002, June). *Mapping the pāngarau curriculum: Levels 2–4: asTTle technical report 13*. Auckland: University of Auckland/Ministry of Education.

Faure, E., Herrera, F., Kaddoura, A., Lopes, H., Petrovsky, A. V., Rahnema, M., et al. (1972). *Learning to be: The world of education today and tomorrow*. Paris: UNESCO.

Finlayson, R., & Madiba, M. (2002). The intellectualisation of the Indigenous languages of South Africa: Challenges and prospects. *Current Issues in Language Planning, 3*(1), 40–61.

Fishman, J. (2006). *Do not leave your language alone. The hidden status agendas within corpus planning in language policy*. London: Lawrence Erlbaum Associates.

Flockton, L., Crooks, T., Smith, J., & Smith, L. F. (2006). *National education monitoring report 37. Mathematics: Assessment results 2005*. Dunedin, New Zealand: University of Otago, Education Assessment Research Unit.

Forbes, S. (2005). *Potential uses of longitudinal analyses to investigate education outcomes*. Paper presented at the IASE sessions at the 55th session of the International Statistical Institute, 5–12 April 2005, Sydney. Retrieved from http://www.stat.auckland.ac.nz/~iase/publications.php?show=13.

Franke, M. L., Carpenter, T. P., Levi, L., & Fennema, E. (2001). Capturing teachers' generative change: A follow-up study of professional development in mathematics. *American Education Research Journal, 38*(3), 653–689.

Freire, P. (1996). *Pedagogy of the oppressed*, M. Bergman Ramos (trans., reprint ed.). London: Penguin Books.

Fryer, K., Kalafatelis, E., & Palmer, S. (2009). *New Zealanders' use of broadcasting and related media: Final report*. Wellington: Te Puni Kōkiri.

Garet, M. S., Porter, A. C., Desimone, L., Birman, B. F., & Yoon, K. S. (2001). What makes professional development effective? Results from a national samples of teachers. *American Educational Research Journal, 38*(4), 915–945.

Garrett, M. (1996). Two people: An American Indian narrative of bicultural identity. *Journal of American Indian Education, 36*(1), 1–21.

Garvin, P. (1973). Some comments on language planning. In J. Rubin & R. Shy (Eds.), *Language planning: Current issues and research* (pp. 24–73). Washington: George Town University Press.

Gass, S. M. (1997). *Input, interaction and the second language learner*. Mahwah, NJ: Lawrence Erlbaum Associates.

Gee, J. P. (1989). Orality and literacy: From the savage mind to ways with words. *Journal of Education, 171*(1), 39–60.

Gergen, K. (1985). The social constructionist movement in modern psychology. *American Psychologist, 40*(3), 266–275.

Gibbons, P. (1998). Classroom talk and the learning of new registers in a second language. *Language and Education, 12*(2), 99–118.

Gibbs, W., & Orton, J. (1994). *Language and mathematics: Issues in mathematics*. London: Cassells.

Good, T. L., & Brophy, J. E. (1990). *Educational psychology: A realistic approach*. White Plains, NY: Longman.

Goodson, I. F. (1999). Representing teachers. In M. Hammersley (Ed.), *Researching school experience*. London: Falmer Press.

Gordon, C. (Ed.). (1980). *Power/knowledge: Selected interviews and other writings, 1972–1977, Michel Foucault*, C. Gordon, L. Marshall, J. Mepham, & K. Soper (trans.). New York: Harvester Wheatsheaf.

Graham, B. (1988). Mathematical education and Aboriginal children. *Educational Studies in Mathematics, 19*, 119–135.

Green, D. R. (1983). A survey of probability concepts in 3000 pupils aged 11–16 years. In D. R. Grey, P. Holmes, V. Barnett, & G. M. Constable (Eds.), *Proceedings of the first international conference on teaching statistics* (pp. 766–783). Sheffield, UK: Teaching Statistics Trust.

Grin, F., & Vaillancourt, F. (1998). *Language revitalisation policy: An analytical survey. Theoretical framework, policy experience and revitalisation of Te Reo Māori (Report to treasury)*. Wellington: New Zealand Treasury.

Gutstein, E. (2003). Teaching and learning mathematics for social justice in an urban Latino school. *Journal for Research in Mathematics Education, 34*(1), 37–73.

Halliday, M., & Hasan, R. (1985). *Language, context and text: Aspects of language in a social-semiotic perspective*. Melbourne: Deakin University Press.

Halliday, M. A. K. (1978). *Language as social semiotic*. London: Edward Arnold.

Halliday, M. A. K. (1988). On the language of physical science. In M. Ghadessy (Ed.), *Registers of written English: Situational factors and linguistic features* (pp. 162–178). London: Pinter Publishers.

Halliday, M. A. K., & Martin, J. R. (1993). *Writing science: Literacy and discursive power*. London: Falmer.

Halliday, M. K. (1993). Towards a language-based theory of learning. *Linguistics and Education 5*(2), 93–116.

Hamilton, D., & McWilliam, E. (2001). *Ex-centric voices that frame research on teaching* (4th ed.). Washington: AERA.

Harlow, R. (1993). A science and mathematics terminology for Māori *SAMEpapers*. Hamilton: Centre for Science, Mathematics and Technology Education Research, University of Waikato.

Harlow, R. (2001). *A Māori reference grammar*. Auckland: Pearson Education Ltd.

Harlow, R. (2003). Issues in Māori language planning and revitalisation. He puna kōrero. *Journal of Māori & Pacific Development, 4*(1), 32–43.

Harris, S. (1980). *Culture and learning: Tradition and education in northern Arnhem Land*. Darwin: Professional Services Branch, Northern Territory Education Department.

Hemara, W. (2000). *Māori Pedagogies. A view from the literature*. Wellington: New Zealand Council for Educational Research.

Herbel-Eisenmann, B. (2002). Using student contributions and multiple representations to develop mathematical language. *Mathematics Teaching in the Middle School, 8*(2), 100–105.

Hindley, R. (1990). *The death of the Irish language: A qualified obituary*. London: Routledge.

Hinton, L., & Hale, K. (2001). *The green book of language revitalization in practice*. San Diego: Academic Press.

Hipkins, R., & Neill, A. (2006). *Shifting balances – The impact of level 1 NCEA on the teaching of mathematics and science*. Wellington: Ministry of Education.

Hohepa, M. K. (2006). Biliterate practices in the home: Supporting language regeneration. *Journal of Language, Identity and Education, 5*(4), 293–315.

Hollings, M. (2005). Māori language broadcasting: Panacea or pipedream. In A. Bell, R. Harlow, & D. Starks (Eds.), *Languages of New Zealand* (pp. 111–130). Wellington: Victoria University Press.

Hornberger, N. H. (Ed.). (1996). *Indigenous literacies in the Americas: Language planning from the bottom up*. Berlin: Mouton.

Hornberger, N. H. (2002). Multilingual language policies and the continua of biliteracy: An ecological approach. *Language Policy, 1*, 27–51.

Hurrell, A. (1981). *Language in mathematics lessons*. Nairobi: National University of Lesotho.

Hynds, A. (2008). Developing and sustaining open communication in action research initiatives: A response to Kemmis (2006). *Educational Action Research, 16*(2), 149–162.

Irwin, K. (1992, November). *Māori research methods and processes: An exploration and discussion*. NZARE/AARE Annual Conference, November, 1992, Geelong. Retrieved from http://www.aare.edu.au/92pap/irwik92373.txt.

Irwin, K., & Davis, L. (1994). Māori education in 1994: A review and discussion. *New Zealand Annual Review of Education, 4*, 77–104.

Jaberg, P., Lubinski, C., & Yazujian, T. (2002). One teacher's journey to change her mathematics teaching. *Mathematics Teacher Education and Development, 4*, 3–14.

Jackson, W. (2009). *Māori should be compulsory*. Retrieved from http://www.stuff.co.nz/auckland/opinion/2703436/Maoori-should-be-compulsory.

Jernudd, B. H., & Neustupny, J. V. (1987). *Language planning: for whom?* Paper presented at the Actes du Colloque international sur l'aménagement linguistique/Proceedings of the International Colloquium on Language Planning, 25–29 May 1986, Quebec.

Johanning, D. I. (2000). An analysis of writing and postwriting group collaboration in middle school pre-algebra. *School Science and Mathematics, 100*(3), 151–160.

Johnson, T. M., Jones, G. A., Thornton, C. W., Langrall, C. W., & Rous, A. (1998). Students' thinking and writing in the context of probability. *Written Communication, 15*(2), 203–229.

Jones, A., McCulloch, G., Marshall, J. D., Smith, G. H., & Smith, L. T. (1990). *Myths and realities: Schooling in New Zealand*. Palmerston North: Dunmore Press.

Jones, G. A. (1974). *The performance of first, second, and third grade children on five concepts of probability and the effects of grade, IQ, and embodiments on their performances*. Unpublished Doctoral dissertation, Indiana University, Bloomington, IN.

Jones, G. A., Langrall, C. W., Thornton, C. A., & Mogill, A. T. (1997). A framework for assessing young children's thinking in probability. *Educational Studies in Mathematics, 32*, 101–125.

Jones, M. C. (1998). *Language obsolescence and revitalisation: Linguistic change in two sociolinguistically contrasting Welsh communities*. Oxford: Oxford University Press.

Kamira, R. (2003). Te mata o te tai – The edge of the tide: Rising capacity in information technology of Māori in Aotearoa/New Zealand. *The Electronic Library, 21*(5), 465–475. Retrieved from www.emeraldinsight.com/journals.htm?articleid=862026&show=pdf.

Kaplan, R., & Baldauf, R. (1997). *Language planning: From practice to theory*. Clevedon: Multilingual Matters.

Kaplan, R. B. (1989). Language planning vs planning language. In C. N. Candlin & T. F. McNamara (Eds.), *Language, learning and community* (pp. 193–203). Sydney: NCELTR, Macquarie University.

Karetu, T. (1995). Māori language rights in New Zealand. In T. Skutnabb-Kangas & R. Pillipson (Eds.), *Linguistic human rights: Overcoming linguistic discrimination* (pp. 209–218). Berlin: Mouton de Gruyter.

Kazima, M. (2006). Malawian students' meanings for probability vocabulary. *Educational Studies in Mathematics, 64*, 169–189.

Keegan, P. (2005). The development of Māori vocabulary. In A. Bell, R. Harlow, & D. Starks (Eds.), *Languages of New Zealand* (pp. 131–150). Wellington: Victoria University Press.

Keegan, T. T. (1998). Reflections on 5 years of teaching computing through the medium of the Māori language. In M. S. Brown (Ed.), *Conference proceedings for beyond the Fringe: Learning development conference*, 2–3 July, 1998 (pp. 51–55). Hamilton: Waikato University.

Kemmis, S. (2009, October). *What is to be done? The place of action research*. Paper presented at the Collaborative Action Research Network Annual Conference Athens, 30 October-1 November 2009, Greece.

Kemmis, S., & Grootenboer, P. (2008). Situating praxis in practice. In S. Kemmis & T. Smith (Eds.), *Enabling Praxis: Challenges for education* (pp. 37–64). Rotterdam: Sense Publications.

Kennedy, M. (1997). *Defining optimal knowledge for teaching science and mathematics: Research monograph no. 10*. Wisconsin-Madison: University of Wisconsin-Madison, National Institute for Science Education.

Kermani, H., & Brenner, M. E. (1996, April). *Maternal scaffolding in the child's zone of proximal development: Cultural perspectives*. Paper presented at the Annual Meeting of the American Educational Research Association, New York City, NY.

Kerswill, P. (1996). Children, adolescents and language change. *Language Variation and Change, 8*, 177–202.

Khisty, L. L., & Chevl, K. B. (2002). Pedagogic discourse and equity in mathematics: When teachers' talk matters. *Mathematics Education Research Journal, 14*(3), 154–168.

Kimura, L., & April, I. (2009). Indigenous new words creation: Perspectives from Alaska and Hawai'i. In J. Reyhner & L. Lockard (Eds.), *Indigenous language revilitzation: Encouragement, guidance and lessons learned* (pp. 121–139). Flagstaff, AZ: Northern Arizona University.

Knight, L., & Hargis, C. (1977). Math language ability: Its relationship to reading in math. *Language Arts, 54*, 423–428.

Knijnik, G. (2000, March). *Re-searching mathematics education from a critical perspective*. Paper presented at the Second Mathematics Education and Society Conference (MES2), Montechoro, Algarve, Portugal.

Kokiri, T. P. (2007). *Survey of the health of the Māori language in 2006*. Wellington: Te Puni Kokiri.

Kura Kaupapa Māori Working Group, & Katarina Mataira (Chair). (1989). *Te Aho Matua*. Wellington: Ministry of Education.

Lange, D. (1988). *Tomorrow's schools: The reform of educational administration in New Zealand*. Wellington: Government Printer.

Lange, T. (2008). Homework and minority students in difficulties with learning mathematics: The influence of public discourse. *Nordic Studies in Mathematics Education, 13*(4), 51–68.

Leap, W. (1982). Semilingualism as a form of linguistic proficiency. In R. S. Clair & W. Leap (Eds.), *Language renewal among American Indian tribes: Issues, problems and prospects* (pp. 149–159). Rosslyn, VA: National Clearinghouse for Bilingual Education.

Learning Media. (1991). *Nga Kupu Tikanga Pangarau: Mathematics vocabulary*. Wellington: Ministry of Education.

Leont'ev, A. N. (1978). *Activity, consciousness and personality*, M. J. Hall (trans.). Englewood Cliffs, NJ: Prentice Hall.

Lerman, S. (2007). Directions for literacy research in science and mathematics education. *International Journal for Science and Mathematics Education, 5*, 755–759.

Leung, C. (2005). Mathematical vocabulary: Fixers of knowledge or points of exploration. *Language and Education, 19*(2), 1227–1135.

Liddicoat, A. J., & Bryant, P. (2002). Intellectualisation: A current issue in language planning. *Current Issues in Language Planning, 3*(1), 1–4.

Lieberman, A., & Mace, D. H. P. (2008). Teacher learning: The key to educational reform. *Journal of Teacher Education, 59*, 226–234.

Lim, C. S., & Ellerton, N. (2009). Malaysian experiences of teaching mathematics in English: Political dilemma versus reality. In M. Tzekaki, M. Kaldrimidou, & H. Sakonidis (Eds.), *Proceedings of the 33rd conference of the international group for the psychology of mathematics education* (Vol. 1, pp. 9–16). Thessaloniki, Greece: PME.

Lipka, J. (1994, May). Culturally negotiated schooling: Toward a Yup'ik mathematics classroom. *Journal of American Indian Education, 33*(3), 14–30.

Lipka, J., Hogan, M. P., Webster, J. P., Yanez, E., Adams, B., Clark, S., et al. (2005). Math in a cultural context: Two case studies of a successful culturally based math project. *Anthropology and Education Quarterly, 36*(4), 367–385.

Lipka, J., Mohatt, G. V., & Cisulistet Group. (1998). *Transforming the culture of schools: Yup'ik Eskimo examples*. Mahwah, NJ: Lawrence Erlbaum Associates.

Love, E., & Tahta, D. (1991). Reflections on some words used in mathematics education. In D. Pimm & E. Love (Eds.), *Teaching and learning school mathematics*. London: Hodder & Stoughton.

Lozare, D. K. (1993). *The Mohawk standardization project: Conference report*, A. K. Jacobs, N. K. Thompson, & M. K. Leaf (trans. into English). Retrieved from http://www.kanienkehaka.com/msp/msp.htm.

MacFarlane, A. (2004). *Kia Hiwa Ra! Listen to culture*. Wellington: New Zealand Council for Educational Research.

Marks, G., & Mousley, J. (1990). Mathematics education and genre: Dare we make the process writing mistake again? In F. Christie (Ed.), *Literacy in social processes: Papers from the inaugural Australian systemic linguistic conference* (pp. 142–158). Geelong: Deakin University.

Matthews, N. (2004). The physicality of the Māori message transmission: Ko te tinana, he waka tuku kōrero. *Junctures, 3*, 9–18.

May, S. (2000). Accommodating and resisting minority language policy: The case of Wales. *International Journal of Bilingual Education and Bilingualism, 3*, 101–128.

May, S. (2004). Māori-medium education in Aotearoa/New Zealand. In J. W. Tollefson & A. B. M. Tsu (Eds.), *Medium of instruction policies* (pp. 21–42). Mahwah, NJ: Lawrence Erlbaum Associates.

May, S., & Hill, R. (2005). Bilingual education in Aotearoa/New Zealand: At the cross-roads. In J. Cohen, K. T. McAlister, K. Rolstad, & J. MacSwan (Eds.), *ISB4: Proceedings of the 4th international symposium on Bilingualism*, 30 April–May 3 2003, Arizona State University (pp. 1567–1573). Somerville, MA: Cascadilla Press.

McCarty, T. L. (1989). School as community: The Rough Rock demonstration. *Harvard Educational Review*, 59(4), 484–503.

McDonald, M. (1989). *We are not French! Language, culture and identity in Brittany*. London: Routledge.

McKinley, E. (1995). *A power/knowledge nexus: Writing a science curriculum in Māori, science education*. Hamilton, New Zealand: University of Waikato.

McKinley, E., & Keegan, P. (2008). Curriculum and language in Aotearoa New Zealand: From science to pūtaiao. *L1 – Educational Studies in Language and Literature, 8*(1), 135–147.

McMurchy-Pilkington, C. (2004). *Pangarau Māori medium mathematics curriculum: Empowerment or New Hegemonic Accord*. Unpublished Ed.D., University of Auckland, Auckland.

McMurchy-Pilkington, C. (2008). Indigenous people: Emancipatory possibilities in curriculum development. *Canadian Journal of Education, 31*(3), 614–638.

McMurchy-Pilkington, C., & Trinick, T. (2002). Horse power or empowerment? Mathematics curriculum for Māori – Trojan horse revisited. In B. Barton, K. C. Irwin, M. Pfannkuch, & M. O. J. Thomas (Eds.), *Mathematics education in the South Pacific: Twenty-fifth annual conference of the mathematics education research group of Australasia* (pp. 465–472). Auckland: MERGA.

McMurchy-Pilkington, C., Trinick, T., Dale, H., & Tuwhangai, R. (2009). Mahi tahi: Collaborating partnerships. In T. Trinick (Ed.), *Quality teaching research and development project: Māori medium and Kura Kaupapa Māori research projects* (pp. 63–75). Auckland: The University of Auckland.

McNaughton, S., MacDonald, S., Barber, J., Farry, S., & Woodard, H. (2006). *Ngā Taumatua: Research on literacy practices and language development (Te Reo) in years 0–1 in Māori medium classrooms*. Retrieved from http://www.educationcounts.govt.nz/publications/maori_education/5251.

Meaney, T. (2001). *An ethnographic case study of a community-negotiated mathematics curriculum development project*. Unpublished PhD thesis dissertation, University of Auckland, Auckland.

Meaney, T. (2002). Aspects of written performance in mathematics learning. In K. Irwin, B. Barton, M. Pfannkuch, & M. Thomas (Eds.), *Mathematics in the South Pacific: Proceedings of the 25th mathematics education research group of Australasia conference*, 7–10 July 2002, Auckland (Vol. 2, pp. 481–488). Auckland: University of Auckland.

Meaney, T. (2005a). Mathematics as text. In A. Chronaki & I. M. Christiansen (Eds.), *Challenging perspectives in mathematics classroom communication* (pp. 109–141). Westport, CT: Information Age.

Meaney, T. (2005b). The use of algebra in senior high school students' justifications. In P. Clarkson, A. Downton, D. Gronn, M. Horne, A. McDonough, R. Pierce, & A. Roche (Eds.), *Building connections: Theory, research and practice: Proceedings of the 28th mathematics education research group of Australasia,* Melbourne (pp. 537–544). Sydney: MERGA.

Meaney, T. (2006a). Really that's probably about roughly what goes down: Hesitancies and uncertainties in mathematics assessment interactions. *Language and Education, 20*(5), 374–390.

Meaney, T. (2006b). Mathematics register acquisition. *Set, 3*, 39–43.

Meaney, T. (2007). Weighing up the influence of context on judgements of mathematical literacy. *International Journal of Mathematics and Science Education, 5*(4), 681–704.

Meaney, T., & Fairhall, U. (2003). Tensions and possibilities: Indigenous parents doing mathematics education curriculum development. In L. Bragg, C. Campbell, G. Herbert, & J. Mousley (Eds.), *Mathematics education research: Innovation, networking, opportunities: Proceedings of the 26th annual conference of the mathematics education research group of Australasia* (pp. 507–514). Sydney: MERGA.

Meaney, T., Fairhall, U., & Trinick, T. (2007). *Te Reo Tātaitai: Developing rich mathematical language in Māori immersion classrooms*. Wellington: NZCER.

Meaney, T., Fairhall, U., & Trinick, T. (2008). The role of language in ethnomathematics. *Journal of Mathematics and Culture, 3*(1). Available from http://nasgem.rpi.edu/pl/journal-mathematics-culture-volume-3-number-1.

Meaney, T., Lange, T., & Valero, P. (2009). *Dispositions and changed teacher practice in mathematics*. AARE 2009 Conference Proceedings, 29 November-3 December 2009 Canberra. Retrieved from http://www.aare.edu.au/09pap/mea091367.pdf.

Meaney, T., McMurchy-Pilkington, C., & Trinick, T. (2008). Mathematics education and Indigenous students. In H. Forgasz, A. Barkatsas, A. Bishop, B. Clarke, S. Keast, W.-T. Seah, P. Sullivan, & S. Willis (Eds.), *Research in mathematics education in Australasia 2004–2007* (pp. 119–139). Rotterdam: Sense Publications.

Meaney, T., Trinick, T., & Fairhall, U. (2009a). 'The conference was awesome': social justice and a mathematics teachers conference. *Journal of Mathematics Teacher Education, 12*(6), 445–462.

Meaney, T., Trinick, T., & Fairhall, U. (2009b). *Mathematics: She'll be write!* Wellington: NZCER.

Mehan, H. (1979). *Learning lessons: Social organization in the classroom.* Cambridge, MA: Harvard University Press.

Merriam, S. B. (1988). *Case study research in education: A qualitative approach.* San Francisco: Josey-Bass Publishers.

Middleton, J. (2010). Ka rangona te reo: The development of Māori-language television boadcasting in Aotearoa New Zealand. *Te Kaharoa, 3*, 146–176. Retrieved from http://www.tekaharoa.com/index.php/tekaharoa/article/viewPDFInterstitial/73/43.

Ministry of Education. (1992). *Mathematics in the New Zealand curriculum.* Wellington: Learning Media.

Ministry of Education. (1996). *Pāngarau i roto i te marautanga o Aotearoa.* Te Whanganui ā Tara: Te Pou Taki Kōrero.

Ministry of Education. (2004). *Te Reo Pāngarau.* Wellington: Learning Media.

Ministry of Education. (2006). *The number framework.* Wellington: Learning Media.

Ministry of Education. (2008a). *Te Marautanga o Aotearoa.* Wellington: Learning Media.

Ministry of Education. (2008b). *Book 1: The number framework.* Wellington: Author.

Ministry of Education. (2009a). *Te reo Pūtaiao. A Māori language dictionary of science.* Palmerston North: He Kupenga Hao i te Reo.

Ministry of Education. (2009b). *Poutama Tau Evaluations 2008: Research findings in Pāngarau for years 1–10.* Wellington: Learning Media.

Ministry of Education. (2010). *Te Reo Pāngarau. Putanga Tuarua. A Māori language dictionary of mathematics.* Wellington: Learning Media.

Morgan, C. (1998). *Writing mathematically: The discourse of investigation.* London: Falmer Press.

Morris, R. W. (1975). *Linguistic problems encountered by contemporary curriculum projects in mathematics.* Paper presented at the Interactions between linguistics and mathematical education – September 1–11, 1974, Nairobi, Kenya.

Moskal, B. M., & Magone, M. E. (2000). Making sense of what students know: Examining the referents, relationships and modes students displayed in response to a decimal task. *Educational Studies in Mathematics, 43*, 313–335.

Mousley, J. (1999). Perceptions of mathematical understanding. In J. M. Truran & K. M. Truran (Eds.), *Making the difference: Proceedings of the 22nd annual conference of the mathematics education research group of Australasia* (pp. 388–395). Adelaide: MERGA.

Murphy, H., McKinley, S., & Bright, N. (2008). *Whakamanahia Te Reo Māori.* Wellington: Te Pouherenga Kaiako o Aotearoa (New Zealand Teachers Council).

Murray, D. (2000). Haka fracas: The dialectics of identity in discussions of a contempory Māori dance. *The Australian Journal of Anthropology, 11*(3), 345–357.

Murray, S. (2007). *Achievement at Māori immersion and bilingual schools: Update for 2005 results.* Wellington: Ministry of Education.

Nathan, M. J., & Knuth, E. J. (2003). A study of whole classroom mathematical discourse and teacher change. *Cognition and Instruction, 21*(2), 175–207.

National Council of Teachers of Mathematics. (2000). *Principles and standards for school mathematics.* Reston, VA: National Council of Teachers of Mathematics.

Nelson-Barber, S., & Estrin, E. (1995). *Culturally responsive mathematics and science education for native students.* San Francisco: Regional Educational Laboratory Network.

New Zealand Qualifications Authority. (2001). *Ao kawe kupu.* Retrieved 17/10/09, from http://www.nzqa.govt.nz/for-maori/publications/akk/akk-0102/count.html#te.

Ní Rîordaín, M. (2010, March). When did it all go right? The socio-political development of Gaeilge as a medium for learning mathematics in Ireland. In U. Gellert, E. Jablonka, & C. Morgan (Eds.) *Proceedings of 6th international mathematics education and society conference*, 20–25 March 2010, Berlin (pp. 340–349). Berlin: Freie Universität. Available from http://www.ewi-psy.fu-berlin.de/en/v/mes6/research_papers/index.html.

Nickson, M. (2000). *Teaching and learning mathematics: A teacher's guide to recent research and its application*. London: Cassel.

Nunan, D. (1992). *Research methods in language learning*. New York: Cambridge University Press.

O'Halloran, K. L. (2000). Classroom discourse in mathematics: A multisemiotic analysis. *Linguistics and Education, 10*(3), 359–388.

Olssen, M., & Mathews, K. M. (Eds.). (1997). *Education policy in New Zealand: The 1990s and beyond*. Palmerston North: Dunmore Press.

Onyango, J. (2005). Issues in national language terminology development in Kenya. *Swahili Forum, 12*, 219–234.

Peddie, R. (2005). Planning for the future? Language policy in New Zealand. In A. Bell, R. Harlow, & D. Starks (Eds.), *Languages in New Zealand* (pp. 30–58). Wellington: Victoria University Press.

Pere, R. R. (1982). *Ako: Concepts and learning in the Māori tradition. Working paper no. 17.* Hamilton: University of Waikato Department of Sociology.

Pere, R. T. (1997). *Te Wheke: A celebration of infinite wisdom*. Gisborne: Ao Ako Global Learning New Zealand.

Pihama, L., Smith, K., Taki, M., & Lee, J. (2004). *A literature review on Kaupapa Māori and Māori Education Pedagogy*. Retrieved from http://elearning.itpnz.ac.nz/.

Pimm, D. (1987). *Speaking mathematically*. London: Routledge & Kegan Paul.

Pimm, D., & Wagner, D. (2003). Investigation, mathematics, mathematics education and genre: An essay review of Candia Morgan's writing mathematically: The discourse of investigation. *Educational Studies in Mathematics, 53*, 159–117.

Pohatu, T. (1996). *Tipu ai Tatau i Ngā Turi ō Tatau Mātaua-Tīpuna. Transmission and Acquisition processes within Kāwai Whakapapa*. Unpublished thesis for Education Masters University of Auckland, Auckland.

Pollitt, A., & Ahmed, A. (1999). *A new model of the question answering process*. Paper presented at the International Association for Educational Assessment, May 1999, Bled, Slovenia. Retrieved from http://www.cambridgeassessment.org.uk/ca/digitalAssets/113788_A_New_Model_of_the_Question_Answering_Process.pdf.

Pollitt, A., Marriott, C., & Ahmed, A. (2000). *Language, contextual and cultural constraints on examination performance*. Paper presented at the International Association for Educational Assessment, May, 2000, Jerusalem, Israel. Retrieved from http://www.cambridgeassessment.org.uk/ca/digitalAssets/113940_Language__Contextual___Cultural_Constraints_on_Exam_Performa.pdf.

Pugalee, D. (2004). A comparison of verbal and written descriptions of students' problem solving processes. *Educational Studies in Mathematics, 55*, 27–47.

Radford, L. (2003a). On the epistemological limits of language: Mathematical knowledge and social practice during the Renaissance. *Educational Studies in Mathematics, 52*, 123–150.

Radford, L. (2003b). Gestures, speech, and the sprouting of signs: A semiotic-cultural approach to students' types of generalisations. *Mathematical Thinking and Learning, 5*(1), 37–70.

Radford, L. (2008). The ethics of being and knowing: Towards a cultural theory of learning. In L. Radford, G. Schubring & F. Seeger (Eds.), *Semiotics in mathematics education: Epistemology, history, classroom and culture* (pp. 218–234). Rotterdam: Sense Publishers.

Radford, L., & Empey, H. (2007). Culture, knowledge and the self: Mathematics and the formation of new social sensibilities in the Renaissance and Medieval Islam. *Revista Brasileira de História da Matemática, Especial 1*, 231–254.

Rau, C. (2001). *He Ara Angitu, a pathway to success: A framework for capturing the literacy achievement of year one students in Māori medium*. Paper presented at the Assessment Hui for the Ministry of Education, Wellington, September 2001. Retrieved from http://www.kiaatamai.org.nz/angitu/index.htm.

Reckwitz, A. (2002). Toward a theory of social practices. A development in culturalist theorizing. *European Journal of Social Theory, 5*, 243–263.

Reedy, S. K. (1999). *Word problems*. Mahwah, NJ: Lawrence Erlbaum Associates.

Reedy, T. (2000). Te reo Māori: The past 20 years and looking forward. *Oceanic Linguistics, 39*(1), 157–169.

Roberts, T. (1998). Mathematical registers in Aboriginal languages. *For the Learning of Mathematics, 18*(1), 10–16.

Rogoff, B. (1988). The joint socialisation of development by young children and adults. In A. Gellatly, D. Rogers, & J. A. Sloboda (Eds.), *Cognition and social worlds* (pp. 57–82). Oxford: Oxford University Press.

Roth, W.-M. (2001). Gestures: Their role in teaching and learning. *Review of Educational Research, 73*(3), 365–392.

Rutene, J., Candler, G., & Watson, S. (2003). Māori-medium education. *Education Now, 2*, 1–11.

Salmond, A. (1975). *Hui: A study of Māori ceremonial gatherings*. Wellington: A. H. & A. W. Reed Ltd.

Salmond, A. (1997). *Between worlds: Early exchanges between Māori and Europeans*. Auckland: Penguin Books (NZ) Ltd.

Schäfer, M. (2010). Mathematics registers in Indigenous languages: Experiences from South Africa. In L. Sparrow, B. Kissane, & C. Hurst (Eds.), *Shaping the future of mathematics education: Proceedings of the 33rd annual conference of the mathematics education research group of Australasia* (pp. 509–514). Fremantle, WA: MERGA.

Schatzchi, T. R. (2005). Peripheral vision: The sites of organizations. *Organization Studies, 26*(3), 465–484. doi: 10.1177/0170840605050876.

Schindler, D. E., & Davison, D. M. (1985). Language, culture and the mathematics concepts of American Indian learners. *Journal of American Indian Education, 24*(3), 27–34.

Schleppegrell, M. J. (2007). The linguistic challenges of mathematics teaching and learning: A research review. *Reading and Writing Quarterly, 23*, 139–159.

Schon, D. (1983). *The reflective practitioner*. USA: Basic Books.

Sellarés, J. A., & Toussaint, G. (2003). On the role of kinesthetic thinking in computational geometry. *International Journal of Mathematics in Science and Technology, 34*(2), 219–237.

Setati, M. (1998). Code-switching in a senior primary class of second language mathematics learners. *For the Learning of Mathematics, 18*(1), 34–40.

Setati, M. (2002). Researching mathematics education and language in multilingual South Africa. *The Mathematics Educator, 12*(2), 6–20.

Sfard, A., Nesher, P., Streefland, L., Cobb, P., & Mason, J. (1998). Learning mathematics through conversation: Is it as good as they say? *For the Learning of Mathematics, 18*(1), 41–51.

Sfard, A., & Prusak, A. (2005). Telling identities: In a search of an analytical tool for investigating learning as a culturally shaped activity. *Educational Researcher, 34*(4), 2–24.

Shulman, L. S. (1986). Those who understand: Knowledge growth in teaching. *Educational Researcher, 15*, 4–14.

Shulman, L. S. (1987). Knowledge and teaching: Foundations of the new reform. *Harvard Educational Review, 57*, 1–22.

Simon, J. (1998). *Ngā Kura Māori: The native schools system 1867–1969*. Auckland: Auckland University Press.

Skovsmose, O. (2005a). Foreground and politics of learning obstacles. *For the Learning of Mathematics, 25*(1), 4–10.

Skovsmose, O. (2005b). Meaning in mathematics education. In J. Kilpatrick, C. Hoyles, & O. Skovsmose (Eds.), *Meaning in mathematics education*. New York: Springer Science.

Smith, G. H. (1991). *Tomorrow's schools and the development of Māori education*. Monograph no. 5. Auckland: University of Auckland, International Research Institute for Māori and Indigenous Education.

Smith, G. H. (1997). *The development of Kaupapa Māori: Theory and Praxis*. Auckland: Department of Education, The University of Auckland.

Smith, G. H. (2003a). *Kaupapa Māori theory: Theorizing indigenous transformation of education and schooling*. Paper presented at the Annual Conference of NZARE/AARE, December, 2003 (Kaupapa Māori Symposium), Auckland.

Smith, G. H. (2003b, October). *Indigenous struggle for the transformation of education and schooling*. Paper presented at the Alaskan Federation and Natives (AFN) Convention, 23 October, 2003, Anchorage, Alaska, US. Retrieved from http://www.kaupapamaori.com/assets//indigenous_struggle.pdf.

Smith, L. T., & Cram, F. (1997). *An evaluation of the community panel diversion pilot programme*. Commissioned report for the Crime Prevention Unit, Wellington.

Solomon, Y., & O'Neill, J. (1998). Mathematics and narrative. *Language and Education, 12*(3), 210–221.

Spanos, G., Rhodes, N. C., Dale, T. C., & Crandall, J. (1988). Linguistic features of mathematical problem solving. In R. Cocking & J. P. Mestre (Eds.), *Linguistic and cultural influences on learning mathematics*. Hillsdale, NJ: Lawrence Erlbaum.

Spolsky, B. (2003). Reassessing Māori regeneration. *Language in Society, 32*, 553–578.

Spolsky, B. (2004). *Language policy: Key topics in sociolinguistics*. Cambridge: Cambridge University Press.

Starks, D., & Barkhuizen, G. (2003). Students as fact gatherers in language-in-education planning. In R. Barnard & T. Glynn (Eds.), *Bilingual children's language and literacy development* (pp. 247–273). Clevedon: Multilingual Matters.

Statistics New Zealand. (2002). *2001 Survey on the health of the Māori language*. Retrieved from http://www.stats.govt.nz/browse_for_stats/people_and_communities/maori/2001-survey-on-the-health-of-the-maori-language.aspx.

Stewart, G. (2005). Māori in the science curriculum: Developments and possibilities. *Educational Philosophy and Theory, 37*(6), 851–870.

Stewart, G. (2007). Science in the Māori-medium curriculum: Assessment of policy outcomes in pūtaiao education. *Philosophy of Education Society Conference,* 6–9 December 2007, Wellington. Retrieved from www.pesa.org.au/papers/2010-papers/PESA%202010%20Paper%2021.pdf.

Street, B. (1995). *Social literacies: Critical approaches to literacy in development, ethnography and education*. London: Longman Group Limited.

Sweiry, E., Crisp, V., Ahmed, A., & Pollitt, A. (2002, September). *Tales of the unexpected: The influence of students' expectations on exam validity*. British Educational Research Association Conference, 11–14 September 2002, Exeter University. Retrieved from http://www.cambridgeassessment.org.uk/ca/digitalAssets/113882_Tales_of_the_Expected._The_Influence_of_Students__Expectatio.pdf.

Tahana, N., Grant, K. T. O. K., Simmons, D. G., & Fairweather, J. R. (2000). *Tourism and Māori development in Rotorua. Tourism and research education centre (TREC) technical report no.15*. Christchurch: Lincoln University.

Tanner, H., & Jones, S. (2000). Scaffolding for success: Reflective discourse and the effective teaching of mathematical thinking skills. In T. Rowland & C. Morgan (Eds.), *Research in mathematics education: Papers of the British society for research into learning mathematics* (Vol. 2, pp. 19–32). London: British Society for Research into Learning Mathematics.

Taratoa, W. (1858). *He pukapuka whiha tēnei hei ako ma nga tangata (Māori arithmetic book for schools)*. Wellington: George Watene.

Te Maro, P., Averill, R., Higgins, J., & Tweed, B. (2009). Fostering the growth of teacher networks within professional development: Kaiako wharekura working in pāngarau. In Ministry of Education (Ed.), *Poutama Tau evaluations 2008: Research Findings in Pāngarau for years 1–10* (pp. 34–46). Wellington: Learning Media.

Te Puni Kōkiri. (1993). *The benefits of kura kaupapa Māori*. Wellington: Te Puni Kōkiri.

Te Puni Kōkiri. (2001). *Māori students learning Te Reo in mainstream secondary schools*. Wellington: Te Puni Kōkiri.

Te Puni Kōkiri. (2004). *Māori language in the community*. Wellington: Te Puni Kōkiri.

Te Puni Kōkiri. (2006). *Attitudes to the Māori language: Fact sheet 27*. Wellington: Te Puni Kōkiri. Retrieved from http://www.tpk.govt.nz/en/in-print/our-publications/fact-sheets/attitudesmaorilang/.

Te Puni Kōkiri. (2008a). *Te oranga o te reo Māori: The health of the Māori language in 2006*. Wellington: Te Puni Kōkiri.

Te Puni Kōkiri. (2008b). *Te oranga o te reo Māori i te rāngai pāpāho 2006: The health of the Māori language in the broadcasting sector 2006*. Wellington: Te Puni Kōkiri.

Te Puni Kōkiri. (2007a). *The Māori language survey fact sheet*. Wellington: Te Puni Kōkiri. Retrieved from http://www.tpk.govt.nz/en/in-print/our-publications/fact-sheets/the-maori-language-survey-factsheet/.

Te Puni Kōkiri. (2007b). *Survey of the health of the Māori language in 2006*. Wellington: Te Puni Kōkiri.

Te Puni Kōkiri. (2010). *Te Reo Pāhu: Use of broadcasting and e-media, Māori language and culture*. Wellington: Te Puni Kōkiri.

Te Ua, H. R. (n.d.). *A brief history of Māori radio broadcasting*. Retrieved from http://www.irirangi.net/roopu-tautoko/history/whakaaro-ake.aspx.

Thornbury, S. (2002). *How to teach vocabulary*. Harlow: Longman.

Timperley, H., Wilson, A., Barrar, H., & Fung, I. (2007). *Best evidence synthesis: Teacher professional development*. Wellington: New Zealand Ministry of Education.

Touval, A., & Westreich, G. (2003). Teaching sums of angle measures: A kinaesthetic approach. *The Mathematics Teacher, 96*(4), 230–233.

Trinick, T. (1994). *Mathematical Māori discourse and the language of quantification*. Unpublished Dip. Maths Ed. Investigation, The University of Auckland, Auckland.

Trinick, T. (1999). *The relationships between Māori culture and Māori mathematical language*. Unpublished Masters Thesis, University of Auckland, Auckland.

Trinick, T. (2005). Te Poutama Tau: A case study of two schools. In J. Higgins, K. C. Irwin, G. Thomas, T. Trinick, & J. Young-Loveridge (Eds.), *Findings from the New Zealand numeracy development project 2004* (pp. 80–88). Wellington: Ministry of Education.

Trinick, T. (2006). Te Poutama Tau: A case study of two schools. In F. Ell, J. Higgins, K. C. Irwin, G. Thomas, T. Trinick, & J. Young-Loveridge (Eds.), *Findings from the New Zealand numeracy development project 2004* (pp. 103–114). Wellington: Ministry of Education.

Trinick, T. (forthcoming). *Te reo tātai (The Māori medium mathematics register) – challenges for students and teachers*. Unpublished Doctoral dissertation, University of Waikato, Hamilton.

Trinick, T., & Stevenson, B. (2007). Te Poutama Tau: Trends and patterns. In Ministry of Education (Eds.), *Findings from the New Zealand numeracy development projects 2006* (pp. 44–53). Wellington: Learning Media.

Trinick, T., & Stevenson, B. (2008). Te Ara Poutama Tau: An evaluation of Te Poutama Tau 2007. In Ministry of Education (Eds.), *Te Poutama Tau evaluation report 2007. Research findings in Pāngarau for years 1–10* (pp. 2–11). Wellington: Ministry of Education.

Unsworth, L. (2001). *Teaching multiliteracies across the curriculum: Changing contexts of text and image in classroom practice*. Buckingham: Open University Press.

Valero, P. (2002). *Reform, democracy, and mathematics education*. Ph.D. dissertation, The Danish University of Education, Copenhagen.

Valero, P. (2007). In between the global and the local: The politics of mathematics education reform in a globalized society. In B. Atweh, A. C. Barton, M. Borba, N. Gough, C. Keitel, & C. Vistro-Yu (Eds.), *Internationalisation and globalisation in mathematics and science education*. Dordrecht, The Netherlands: Springer Verlag.

Valero, P. (2009). Mathematics education as a network of social practices. In V. Durand-Guerrier, S. Soury-Lavergn, & F. Arzarello (Eds.) *Proceedings of CERME 6, Lyon 28th January to 1st February, 2009* (pp. LV–LXXX). Available from http://www.inrp.fr/editions/editions-electroniques/cerme6/working-group-14.

Waitangi Tribunal. (1986). *Report of the Waitangi Tribunal on the Te Reo Māori Claim (Wai 11)*. Wellington: Waitangi Tribunal.

Waitangi Tribunal. (2010). *Pre-publication Waitangi Tribunal report 262: Te Reo Māori*. Wellington: Author.

Walkerdine, V. (1988). *The mastery of reason*. London: Routledge.

Wallace, M. L., & Ellerton, N. F. (2004). *Language genre and school mathematics*. Paper presented at the Topic Study Group 25 at International Congress of Mathematics Education-10, 4–11 July 2004, Copenhagen, Denmark. Available from http://www.icme-organisers.dk/tsg25/distribution/wallace.pdf.

Walsh, M. (2005). Will indigenous languages survive? *The Annual Review of Anthropology, 34*, 293–315.

Walshaw, M., & Anthony, G. (2008). Creating productive learning communities in the mathematics classroom: An international literature review. *Pedagogies. An International Journal, 3*, 133–149.

Wang, H., & Harkess, C. (2007). *Senior secondary students' achievement at Māori medium schools – 2004–2006 fact sheet*. Wellington: Ministry of Education.

Watson, J. (2006). *Statistical literacy in schools: Growth and goals*. Lawrence Erlbaum Associates Inc.

Weiss, H. M., & Edwards, M. E. (1992). The family-school collaboration project: Systemic interventions for school improvement. In S. L. Christenson & J. C. Conoley (Eds.), *Home school collaboration: Enhancing children's academic and social competence* (pp. 215–243). Maryland: National Association of School Psychologists.

Wells, G. (1999). *Dialogic inquiry: Towards a sociocultural practice and theory of education*. Cambridge: Cambridge University Press.

Wheelahan, L. (2007). Blending activity theory and critical realism to theorise the relationship between the individual and society and the implications for pedagogy. *Studies in the Education of Adults, 39*(2), 183–196.

White, D. Y. (2003). Promoting productive mathematical discourse with diverse students. *Mathematical Behaviour, 22*, 37–53.

White, P., Mitchelmore, M., Branca, N., & Maxon, M. (2004). Professional development: Mathematical content versus pedagogy. *Mathematics Teacher Education and Development, 6*, 49–60.

Wilkins, D. P. (2008). W(h)ither language, culture and education in remote Indigenous communities of the Northern Territory? *Australian Review of Public Affairs*. Retrieved from http://www.australianreview.net/digest/2008/10/wilkins.html.

Wilkinson, A., Barnsley, G. P., Hanna, P., & Swan, M. (1980). *Assessing language development*. Oxford: Oxford University Press.

Williams, S. R., & Baxter, J. A. (1996). Dilemmas of discourse-orientated teaching in one middle school mathematics classroom. *The Elementary School Journal, 97*(1), 21–38.

Winch, G., Johnston, R. R., March, P., Ljungdahl, L., & Holliday, M. (2004). *Literacy: Reading, writing and children's literature*. Melbourne: Oxford University Press.

Wood, D., Bruner, J. S., & Ross, G. (1976). The role of tutoring in problem solving. *Journal of Child Psychology and Psychiatry, 25*, 89–100.

Wood, T. (1998). Alternative patterns of communication in mathematics classes: Funneling or focusing? In H. Steinbring, M. G. B. Bussi, & A. Sierpinska (Eds.), *Language and communication in the mathematics classroom*. Reston, VA: NCTM.

Woods, P. (1992). Symbolic interactionism: Theory and method. In M. LeCompte, W. Millroy, & J. Preissle (Eds.), *The handbook of qualitative research in education* (pp. 337–404). London: Academic Press Ltd.

Yackel, E. (2001). Explanation, justification and argumentation in mathematics classrooms. In A. Cockburn & E. Nardi (Eds.), *Proceedings of the 26th annual conference of the international group for the psychology of mathematics education* (Vol. 1, pp. 9–25). Norwich: PME.

Yin, R. K. (2003). *Applications of case study research*. Thousand Oaks, CA: Sage.

Young-Loveridge, J. (2005). *Seeing the mathematics of Maori boys in a new light: The benefits of diagnostic interviews*. Paper presented at the NZARE National Conference, December, 2005, Dunedin, New Zealand.

Index

MIX
Papier aus verantwortungsvollen Quellen
Paper from responsible sources
FSC® C105338

If you have any concerns about our products,
you can contact us on
ProductSafety@springernature.com

In case Publisher is established outside the EU,
the EU authorized representative is:
Springer Nature Customer Service Center GmbH
Europaplatz 3, 69115 Heidelberg, Germany

Printed by Libri Plureos GmbH
in Hamburg, Germany

In einer Flasche, und die hängt am untersten Ast! Und da, da drüben, an der Hecke, da hängen Gummibärchen, fein aufgespießt an Heckenästchen.
Frieder hat einen knallroten Kopf vor Aufregung.
„Oma“, schreit er, so laut er kann. „Oma, komm schnell und schau!“
Da ist die Oma auch schon da, und sie staunt. „Ja da schau her, ja so was!“ Und sie lacht übers ganze Gesicht und schwenkt ein Senfglas in der Hand und jammert: „Der Senf hat nicht wachsen wollen. Glas gedeiht so schlecht im Garten, schad, gell?“ Und dabei lacht sie wieder übers ganze Gesicht. Da muss der Frieder auch lachen! Er greift nach Omas Hand und lacht: „Mensch, Oma!“
„So schnell wächsts aber nicht jeden Tag, dass dus nur weißt“, grinst die Oma. Sie packt Frieders Hand fester und gemeinsam gehen sie ernten, alles, was so schnell gewachsen ist … nur die Gummibärchen, die darf der Frieder alleine ernten, und vorsichtig pickt er eins nach dem anderen von der Hecke und isst sie alle auf.
Sehnsüchtig denkt er: „Von mir aus könnten hier immer Gummibärchen wachsen, jeden Tag!“

1 Im **Garten**.

2
- [ ] Oma liegt auf dem Liegestuhl.
- [x] Frieder langweilt sich.
- [ ] Frieder und Oma spielen Fußball.
- [x] Oma pflanzt Salat.
- [x] Frieder will ernten.

3
- [ ] bis zum Sommer
- [x] drei bis vier Wochen
- [ ] zwei Monate

4 Er träumt von einem (himmelhohen) Spuckebaum.

5

6 Oma hat die Sachen aufgehängt und eingegraben, als Frieder geschlafen hat.

1099411L